Chemistry Didactics Compact

Hans-Dieter Barke · Günther Harsch ·
Simone Kröger · Annette Marohn

Chemistry Didactics Compact

Learning Processes in Theory and Practice

Hans-Dieter Barke
Westfälische Universität Münster
Münster, Germany

Günther Harsch
Westfälische Universität Münster
Münster, Germany

Simone Kröger
Westfälische Universität Münster
Münster, Germany

Annette Marohn
Westfälische Universität Münster
Münster, Germany

ISBN 978-3-662-70079-2   ISBN 978-3-662-70080-8   (eBook)
https://doi.org/10.1007/978-3-662-70080-8

Translation from the German language edition: "Chemiedidaktik kompakt" by Hans-Dieter Barke et al., © Springer-Verlag GmbH Deutschland, ein Teil von Springer Nature 2018. Published by Springer Berlin Heidelberg. All Rights Reserved.

This book is a translation of the original German edition "Chemiedidaktik kompakt," 3rd edition, by Hans-Dieter Barke, published by Springer-Verlag GmbH, DE in 2018. The translation was done with the help of an artificial intelligence machine translation tool. A subsequent human revision was done primarily in terms of content, so that the book will read stylistically differently from a conventional translation. Springer Nature works continuously to further the development of tools for the production of books and on the related technologies to support the authors.

© The Editor(s) (if applicable) and The Author(s), under exclusive license to Springer-Verlag GmbH, DE, part of Springer Nature 2025

This work is subject to copyright. All rights are solely and exclusively licensed by the Publisher, whether the whole or part of the material is concerned, specifically the rights of translation, reprinting, reuse of illustrations, recitation, broadcasting, reproduction on microfilms or in any other physical way, and transmission or information storage and retrieval, electronic adaptation, computer software, or by similar or dissimilar methodology now known or hereafter developed.
The use of general descriptive names, registered names, trademarks, service marks, etc. in this publication does not imply, even in the absence of a specific statement, that such names are exempt from the relevant protective laws and regulations and therefore free for general use.
The publisher, the authors and the editors are safe to assume that the advice and information in this book are believed to be true and accurate at the date of publication. Neither the publisher nor the authors or the editors give a warranty, expressed or implied, with respect to the material contained herein or for any errors or omissions that may have been made. The publisher remains neutral with regard to jurisdictional claims in published maps and institutional affiliations.

This Springer imprint is published by the registered company Springer-Verlag GmbH, DE, part of Springer Nature.
The registered company address is: Heidelberger Platz 3, 14197 Berlin, Germany

If disposing of this product, please recycle the paper.

Preface

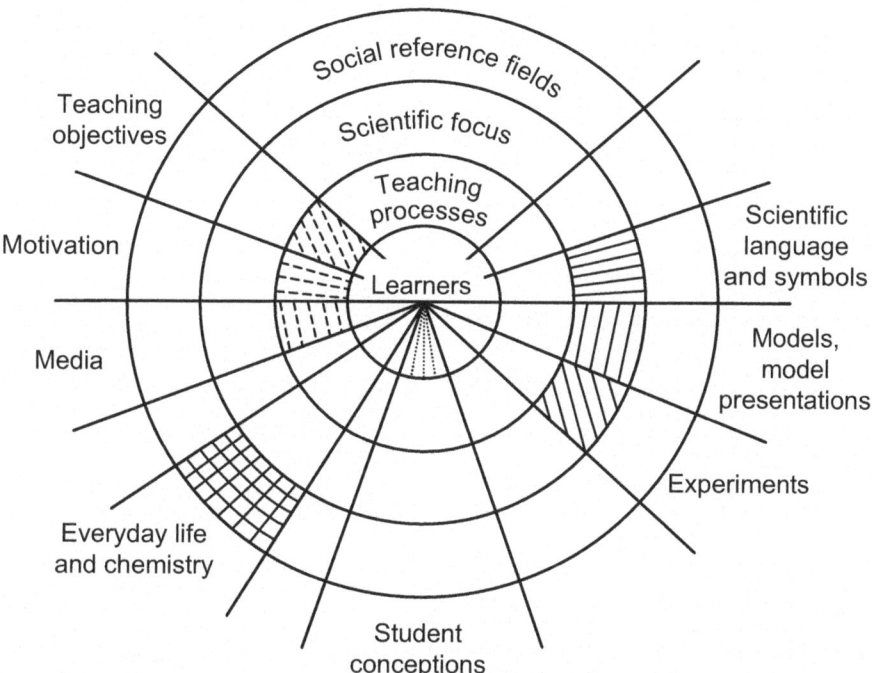

It is not easy to clearly arrange content in chemistry didactics: Topics such as *educational objectives, student conceptions, motivation, media, experiments, models, scientific language and symbols* or *everyday life and chemistry* are intricately intertwined and do not build upon each other linearly. A pie chart has proven to be the best organizational structure for the eight basic topics in our lectures and seminars (see picture). Questions and problems are discussed at certain levels of reflection for each topic: learners, scientific focus, mediation processes, social reference fields. The didactic analyses, reflections, and recommendations are carried out according to the pie chart for the individual topics always also include many examples from teaching practice.

This type of structure is intended to make it clear that there is no set order of topics and that additional topics can be incorporated into the pie chart. This

Übersicht Sicheres Experimentieren

Die Versuche auf den Praktikumsseiten sind mit einer **Symbolleiste** versehen. Zunächst erscheinen als rote Rauten die *Gefahrenpiktogramme*. Dann folgen Symbole, die unbedingt einzuhaltende Hinweise zur *Arbeitssicherheit* geben (Schutzbrille, Handschuhe, Abzug). Das letzte Symbol bezieht sich auf die korrekte Entsorgung der Stoffe. Die genaue Bedeutung dieser Symbole ist in der Übersicht *Richtig entsorgen* beschrieben.

Sicherheitshinweise. Wegen des möglichen Umgangs mit Gefahrstoffen im Chemieunterricht sind Sicherheits- und Schutzhinweise zu beachten:

- Versuchsvorschriften und Hinweise müssen genau befolgt werden. Bei Unklarheiten frage deine Lehrkraft.
- Im Chemieraum darf nicht gegessen oder getrunken werden.
- Ein Versuch darf erst dann durchgeführt werden, wenn ihr dazu aufgefordert werdet. Die Geräte und Substanzen dürfen nicht ohne ausdrückliche Genehmigung und nur sachgerecht verwendet werden.
- Beim Experimentieren sind Schutzbrillen zu tragen.
- Die Anlagen für Strom, Gas und Wasser dürfen erst nach Freigabe durch die Lehrkraft eingeschaltet werden.
- Die Geräte müssen in sicherem Abstand von der Tischkante standfest aufgebaut werden.
- Ausgehändigte Schutzhandschuhe müssen benutzt werden.
- Geschmacksproben sind verboten. Geruchsproben sind erst nach besonderer Anleitung erlaubt.
- Chemikalien dürfen nicht mit den Händen berührt werden.
- Die Haare sind so zu tragen, dass sie nicht in die Brennerflamme geraten können.
- Chemikalien dürfen nicht umgefüllt werden. Gefäße, die zur Aufbewahrung oder Aufnahme von Speisen und Getränken bestimmt sind, sind in Fachräumen verboten.
- Der Arbeitsplatz ist stets sauber, aufgeräumt und übersichtlich.
- Glasgeräte werden ausgespült.
- Chemikalienreste müssen vorschriftsmäßig entsorgt werden (siehe Seite 14).

Gefahrenpiktogramme. Beim Experimentieren arbeitet man auch mit Gefahrstoffen. Welche Gefahren von einem solchen Stoff ausgehen, wird unter anderem durch die Piktogramme angegeben.

 Explosive Stoffe: explosive oder selbstzersetzliche Stoffe, sowie organische Peroxide.

 Entzündbare Stoffe: Stoffe, die entzündbar sind oder bei Berührung mit Wasser entzündbare Gase bilden. Selbsterhitzungsfähige, selbstzersetzliche oder pyrophore Stoffe. Organische Peroxide.

 Oxidierende Stoffe: Stoffe, die einen Brand anderer Materialien verursachen und unterstützen können.

 Gase unter Druck: Gase, die unter erhöhtem Druck in entsprechenden Behältern aufbewahrt werden.

 Ätzende Stoffe: Stoffe, die nach kurzer Einwirkung Haut oder Augen schädigen. Stoffe, die korrosiv gegenüber Metallen wirken.

 Giftige Stoffe: Stoffe, die beim Verschlucken, bei Hautkontakt oder beim Einatmen akut giftig wirken oder zum Tode führen können.

 Gesundheitsschädliche / reizende Stoffe: Stoffe, die beim Verschlucken, bei Hautkontakt oder beim Einatmen gesundheitsschädlich sind, die Haut oder die Augen reizen oder auf die Haut allergen wirken. Stoffe, die die Atemwege reizen oder Schläfrigkeit und Benommenheit verursachen oder die Ozonschicht schädigen.

 Chronisch gefährliche Stoffe: Stoffe, die das Erbgut verändern, Krebs erzeugen, die Fruchtbarkeit beeinträchtigen, das Kind im Mutterleib schädigen oder in bestimmten Organen toxisch wirken. Stoff, die beim Verschlucken oder Eindringen in die Atemwege tödlich sein können (Aspirationsgefahr) oder beim Einatmen Allergien, asthmaartige Symptome oder Atembeschwerden verursachen.

 Umweltgefährdende Stoffe: Stoffe, die für Wasserorganismen sehr giftig oder giftig sind.

Safety concept and hazard symbols. (With kind permission from Schroedel Publishing [1])

Preface

Übersicht   **Sicher entsorgen**

Jeder weiß es: **Gefährliche Stoffe müssen ordnungsgemäß entsorgt werden.** Chemikalienreste und Reaktionsprodukte dürfen also nicht ohne weiteres in den Abfluss oder den Abfalleimer gegeben werden. Folgende Regeln sind genau zu beachten:

Gefährliche Abfälle vermeiden. Eine der wichtigsten Regeln für einen verantwortungsbewussten Umgang mit Stoffen ist ganz einfach: *Die Entstehung unnötiger Abfälle oder unnötig großer Mengen an Abfällen muss vermieden werden.* Die Anwendung dieser Regel setzt eine sorgfältige Versuchsplanung im Hinblick auf Art und Menge der verwendeten Stoffe voraus.

Gefährliche Abfälle umwandeln. Nicht vermeidbare gefährliche Abfallstoffe sollen in weniger gefährliche umgewandelt werden:

Saure und alkalische Lösungen werden neutralisiert. Es ist zweckmäßig, saure und alkalische Lösungen in einem gemeinsamen Behälter zu sammeln. Sie brauchen dann nicht einzeln neutralisiert zu werden.

Viele *lösliche Stoffe* können zu schwer löslichen Stoffen umgesetzt werden.

Gefährliche Abfälle sammeln. Chemikalienreste von Praktikumsversuchen, die nicht sofort in ungefährliche Produkte umgewandelt werden können, werden in besonders beschrifteten Abfallbehältern gesammelt. Von Zeit zu Zeit werden die Abfallbehälter dann durch ein *Entsorgungsunternehmen* abgeholt.

Durch das Sammeln in getrennten Behältern verringern sich die Kosten für eine endgültige Beseitigung.

Behälter 1: Saure und alkalische Lösungen
In Behälter 1 gehören nur saure und alkalische Lösungen, in denen *keine Schwermetallverbindungen* enthalten sind. Saure und alkalische Lösungen neutralisieren sich dann gegenseitig. In der Regel reagiert die Lösung im Behälter 1 nicht neutral, sondern leicht sauer oder alkalisch. Bevor der Behälter ganz gefüllt ist, muss er deshalb neutralisiert werden. Der neutralisierte Inhalt kann dann in den Ausguss geschüttet werden.

Behälter 2: Giftige anorganische Stoffe
In Behälter 2 werden giftige und umweltschädliche Schwermetallsalze und ihre Lösungen gesammelt – auch dann, wenn es sich um saure oder alkalische Lösungen handelt. Die endgültige Entsorgung erfolgt durch ein Entsorgungsunternehmen.

Behälter 3: Halogenfreie organische Stoffe
Wasserlösliche und wasserunlösliche halogenfreie organische Stoffe werden gemeinsam in einem Abfallbehälter gesammelt. Damit sich kein zu großes Volumen an leicht entzündlichen Flüssigkeiten ansammelt, kann die Lehrkraft im Einzelfall anordnen, geringe Mengen nicht giftiger wasserlöslicher organischer Abfälle wie Ethanol oder Aceton in den Ausguss zu geben. Behälter 3 muss von einem Entsorgungsunternehmen ordnungsgemäß entsorgt werden.

Behälter 4: Halogenhaltige organische Stoffe
Organische Halogenverbindungen und Abfälle, die bei Halogenierungsreaktionen entstehen, werden im Behälter 4 gesammelt. Dieser Behälter muss von einem Entsorgungsunternehmen ordnungsgemäß entsorgt werden.

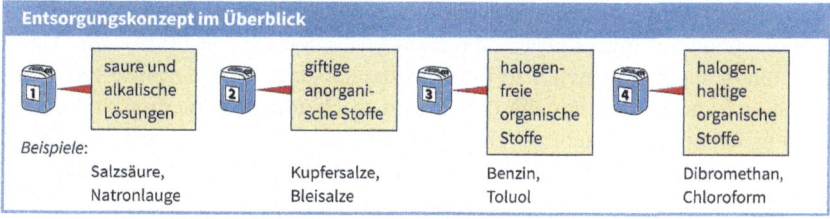

Disposal concept and waste symbols. (With kind permission from Schroedel Publishing [2])

ensures that colleagues who also want to train their students on this basis can link their own ideas with the content suggested here. This concept has been very well received with our 2001 published, but currently out of print, German textbook "Chemiedidaktik Heute".

In many conversations with chemistry education students, with chemistry teachers, with heads of study seminars, and with university lecturers, it was expressed that these chapters of chemistry didactics are often used as the basis for seminars. We hope that the present version, which has been changed in many chapters, "Chemistry Didactics Compact," is well received by the readers.

This new edition was necessitated by new results of empirical surveys, which are reflected in the eight chapters of chemistry didactics. Since the contents of almost every chapter are made clear with experiments, but so far the safety instructions demanded by the readers are missing, these instructions have been supplemented with the new GHS-hazard symbols. With the permission of Schroedel-Verlag, they have been taken from the German textbook "Chemie heute, NRW" (Braunschweig 2016), as well as elements of the disposal concept (see following illustrations). In reference to these graphics, the safety bars for the experiments were designed and inserted.

Since chemistry cannot be understood solely at the "Macro level", many illustrations of structural models are printed in the chapter on models and model conceptions for a good understanding of chemistry at the "Submicro-Level"—beyond molecule models, the sphere packs and space lattices for metal and salt structures. In particular, special beaker glass models are offered for the mediation of solutions of acids, bases and salts, which visualize the involved ions—otherwise, there will inevitably be related misconceptions about molecules. The transition to the "Symbolic level" takes place in the chapter on Scientific language and symbols: formulas and reaction symbols are derived as abbreviated models from the mental and concrete models of Chapt. 7 Many hints on appropriate scientific language have been added to avoid incomprehensible formulations of "lab jargon" for learners. All chapters have five exercise tasks that test the understanding of the contents of the respective chapter.

As it seems important to the authors that student teachers also learn about the development of many concepts and theories from the history of chemistry, an electronic appendix of more than 300 German written pages has been added to the German book (http://www.springer.com/de/book/978-3-662-56491-2), which documents significant milestones in the history of chemistry in brief form. There are 15 chapters, each of which also includes the subchapter "Relevance to Chemistry Education" and establishes the connection to chemistry education and chemistry teaching for student teachers. These 15 chapters refer to and reflect on findings from the following well-known researchers in the history of chemistry: Boyle, Cavendish, Scheele, Priestley, Lavoisier, Richter, Dalton, Gay-Lussac, Avogadro, Galvani, Volta, Davy, Faraday, Berzelius, Liebig, Wöhler, Kekulé, van't Hoff, Meyer, Mendeleev, Arrhenius, Brönsted, Werner, Röntgen, von Laue, Bragg, Watson and Crick. In Chap. 10 of this book, Chaps. 11–25 are outlined with photos and drawings, so that readers can estimate what content to expect in the electronic appendix of the German book..

The authors hope that students, trainee teachers, chemistry teachers, lecturers, both the considerations on chemistry didactics in Chap. 2–9 and the stations on the history of chemistry in Chap. 10 will reflect upon, successfully apply in classes and seminars, and acquire a good gain in didactic competence.

Hans-Dieter Barke
Günther Harsch
Simone Kröger
Annette Marohn

References

1. Safety concept and hazard symbols (Textbook "Chemie heute", NRW, 2016, Schroedel Publishing)
2. Disposal concept and waste bin symbols (Textbook "Chemie heute , NRW, 2016", Schroedel Publishing)

Contents

1	**Introduction to the "Pie Chart" for Chemistry Didactics**.........	1
	References..	9
2	**Student Conceptions**.......................................	11
	2.1 Scientific Focus: Theories from the History of Natural Sciences...	13
	2.1.1 Primal Matter Theories	13
	2.1.2 Transformation Concepts of the Alchemists	14
	2.1.3 The Phlogiston Theory...........................	15
	2.1.4 "Horror vacui" and Air Pressure	16
	2.1.5 Theories on Atomism and the Structure of Matter......	17
	2.2 Learners: Empirical Evidence of Student Conceptions.........	19
	2.2.1 Substances as Property Carriers....................	19
	2.2.2 Mixture Concept	20
	2.2.3 Destruction Concept.............................	21
	2.2.4 Energy Concept	22
	2.2.5 The Combustion Process	23
	2.2.6 Air and Other Gases.............................	26
	2.2.7 Structure of Matter..............................	27
	2.2.8 Horror vacui...................................	29
	2.2.9 Spatial Ability	29
	2.2.10 Homemade Misconceptions.......................	32
	2.3 Teaching Processes: Consideration of Student Conceptions.....	32
	2.4 Social Reference Fields: Student Conceptions and Colloquial Language	38
	2.5 Exercise Tasks	39
	2.6 Experiments..	40
	References..	54
3	**Motivation**..	57
	3.1 Learners: Developmental Status, Attitudes and Original Ideas ...	58
	3.1.1 Developmental Stage............................	59
	3.1.2 Attitudes......................................	60
	3.1.3 Original Student Conceptions	61

	3.2	Teaching Processes: Opportunities for Building Subject-Related Motivation....................................	61
		3.2.1 Comprehensible Teaching	62
		3.2.2 Introduction According to Wagenschein	63
		3.2.3 References to Everyday Life and Living Environment....................................	64
		3.2.4 Generation of Cognitive Conflicts..................	65
		3.2.5 Striking Experimental Effects	67
		3.2.6 Active Handling of Experimental or Model Building Materials...............................	69
	3.3	Scientific Focus: Experimental Skills for Demonstration Experiments...	70
	3.4	Social Reference Fields: Motivation through Everyday Language and Media....................................	70
	3.5	Exercise Tasks	73
	3.6	Experiments...	74
	References...		83
4	**Teaching Objectives**		85
	4.1	General Didactic Introduction	86
		4.1.1 Educational Objectives and Their Dimensions	86
		4.1.2 Didactic Models.................................	88
		4.1.3 Lesson Planning and Analysis	88
	4.2	Social Reference Fields: Educational Standards and Curricula	90
	4.3	Learners: Cognitive development, student conceptions, attitudes, interests......................................	96
		4.3.1 Learning Objectives and Developmental Psychology...	96
		4.3.2 Student Conceptions	97
		4.3.3 Attitudes and Interests	98
	4.4	Scientific Focus: Chemistry Lessons as a Spiral Curriculum	100
	4.5	Teaching Processes—Teaching Methods for Achieving Educational Goals	103
		4.5.1 The Research-Developing Teaching:................	104
		4.5.2 The Historically Problem-Oriented Teaching	106
		4.5.3 ChiK – Chemistry in Context	108
		4.5.4 choice^{2}learn	111
		4.5.5 The teaching method oriented towards student conceptions	124
		4.5.6 choice^{2}reflect	125
		4.5.7 The socio-critical problem-oriented teaching	130
		4.5.8 choice^{2}explore	131
		4.5.9 Further Teaching Concepts	137
	4.6	Exercise Tasks	142
	References...		142

Contents

5 Media 147
- 5.1 Teaching Processes: Media and their Functions in Teaching. 149
  - 5.1.1 Textbook. 151
  - 5.1.2 School Board 152
  - 5.1.3 Presentation Slides (via Overhead Projector or Computer/Tablet and Projector). 154
  - 5.1.4 Newspaper Report 155
  - 5.1.5 Videos, Films, Online Appearances. 156
  - 5.1.6 Computer, Tablet 156
  - 5.1.7 Multimedia. 158
  - 5.1.8 Interactive Whiteboard. 159
  - 5.1.9 Experiments 160
  - 5.1.10 Backgrounds and Light Wall 160
  - 5.1.11 Camera Use 161
  - 5.1.12 Projections 161
  - 5.1.13 Magnetic Whiteboards. 161
  - 5.1.14 Computer Use 162
  - 5.1.15 Data Acquisition Systems, Handheld Devices. 162
  - 5.1.16 Models 163
  - 5.1.17 Experiment Kits. 163
- 5.2 Scientific Focus: Objective Adequacy Appropriateness of Media. 163
- 5.3 Learners: Media Competence and Media Production 166
- 5.4 Social Reference Fields: Mass Media 168
  - 5.4.1 Webquest 169
  - 5.4.2 Movie Scenes. 170
- 5.5 Exercise Tasks 171
- 5.6 Experiments. 171
- References. 177

6 Experiments. 181
- 6.1 Scientific focus: Experiment, Experimental Skills, Safety 184
  - 6.1.1 Experiment and Process of Knowledge Acquisition 184
  - 6.1.2 Data Acquisition. 186
  - 6.1.3 Synthesis of New Substances. 188
  - 6.1.4 Experimental Skills and Abilities. 188
  - 6.1.5 Safety and Disposal 191
  - 6.1.6 Disposal 193
- 6.2 Teaching Processes: Functions, Selection Criteria and Forms of Experiments 194
  - 6.2.1 Functions of the Experiment 195
  - 6.2.2 Selection Criteria for Experiments. 200
  - 6.2.3 Implementations of the Experiment. 201
  - 6.2.4 Organizational Procedure of Experimental Teaching. 202

	6.3	Learners: Playfulness and Curiosity, Experimental Skills.......	203
	6.4	Social Reference Fields: Experiments on Everyday Life and Environment	204
	6.5	Exercise Tasks	206
	6.6	Examples of Experiments	207
	References...		226

7 Models and Model Representations

			227
	7.1	Scientific Focus: Models and their Functions	229
		7.1.1 Model concept and knowledge in the natural sciences	229
		7.1.2 Thought Models in Chemistry.....................	233
		7.1.3 Visual Models in Chemistry.......................	235
	7.2	Teaching Processes: Models and their Didactic Functions	237
		7.2.1 Conveying Chemical Facts through Model Ideas	240
		7.2.2 Adaptation and Extension of Models in Chemistry Education......................................	248
		7.2.3 Further Functions of Models and Model Ideas	250
	7.3	Learners: Experiences with Models.......................	252
		7.3.1 Toys	252
		7.3.2 Fun with Models	253
		7.3.3 Models from Other School Subjects	253
	7.4	Social Reference Fields: Interdisciplinary Model Ideas	255
	7.5	Exercise Tasks	255
	7.6	Model Building Workshop: Structures of Metals and Salts......	256
		7.6.1 Tasks and Construction Instructions	257
		7.6.2 Solutions and Drawings for the Tasks	261
	References...		264

8 Scientific Language and Symbols

			267
	8.1	Scientific Focus: Terms, Symbols, Quantities, Units...........	268
		8.1.1 Système Internationale and derived units	268
		8.1.2 School-Relevant Sizes and Units...................	269
		8.1.3 School-Relevant Terms...........................	277
	8.2	Teaching Processes: Everyday Language → Scientific Language → Symbolic Language	288
		8.2.1 Linking Everyday Language and Scientific Language	288
		8.2.2 The Chemical Symbol Language...................	292
		8.2.3 Derivation of First Chemical Symbols in Teaching.....	297
	8.3	Learners: Student Conceptions of Chemical Structures and Symbols	300
		8.3.1 Conceptions of Combustion.......................	300
		8.3.2 Concepts of the Ion Idea	300
		8.3.3 Concepts of Stoichiometry........................	301
		8.3.4 Laboratory Jargon and Misconceptions..............	303

	8.4	Social Reference Fields: Laypeople and Chemical Jargon	309
	8.5	Exercise Tasks	309
	References		312

9 Everyday Life and Chemistry 315
    9.1 Learners: Curiosity and Interest 316
        9.1.1 Student Interests 318
        9.1.2 Household Chemicals and Interest 320
        9.1.3 Attitudes towards Chemistry and Chemistry Education ... 320
    9.2 Scientific Focus: Subject Systematics versus Everyday Chemistry ... 322
        9.2.1 Everyday Phenomena and Chemistry 324
        9.2.2 Technical Interpretations, Experiments 324
    9.3 Teaching Processes: Subject Systematics plus Everyday Chemistry ... 332
        9.3.1 Methods for Mediation Processes 332
        9.3.2 Complete Curricula Based on Everyday Chemistry 333
        9.3.3 Chemistry in Context 337
        9.3.4 Chemistry for Life 338
        9.3.5 NRW Curricula and New Textbooks 339
    9.4 Social Reference Fields: Role-Playing and Environmental Education ... 342
    9.5 Exercise Tasks 344
    9.6 Experiments 345
    References 351

10 History of Chemistry for Student Teachers 353

Introduction to the "Pie Chart" for Chemistry Didactics

The Russian educator Itelson [1] once remarked very ironically about educators:

> If engineers in bridge construction, doctors in the treatment of people, and lawyers in the passing of judgments were to show such a tendency towards superficial justifications as we sometimes encounter in pedagogy, all bridges would have collapsed, the patients would have died, and the innocent would have been hanged.

At first, one will chuckle heartily about this quote, but then admit that the thoroughness of educators in preparing and arguing for their field of activity is usually not given to the same extent as in the fields of activity of other professions: The negative impact of mistakes in the profession of the educator does not become as obvious as mistakes in the professional field of an engineer, a doctor, or a lawyer.

The professionalism of student teachers or of teachers can be enhanced if the basics in pedagogy, didactics, and subject didactics are taught to an appropriate extent. Definitions related to this should be at the beginning before specific aspects of chemistry didactics are added. It must be clear: There cannot be *the* didactics or *the* subject didactics! The corresponding discussion takes place under a specific worldview, and every teacher has to lead it anew in his time and in his environment.

Pedagogy
This term is derived from "pedagogue" (Greek: child or boy leader) and is a collective term for different philosophical and psychological disciplines, whose common subject is social action. Roth [2] characterizes pedagogy in his *Handbook of Educational Science* with the following subject areas:

They concern

- the area of educational research (and concretize in its methodology and practices),
- the area of the school (in its historical and current references),
- the area of teaching (in its diverse conditionality as a complex interplay of effects)
- the area of vocational pedagogy (in its diverse expression and socio-political significance) [2].

Didactics
The Greek philosophy schools in antiquity created this central idea: It can be derived from *didaskein* (Greek: to teach, to prove) or from *didaktos* (Greek: teachable). These terms had an impact far beyond teaching and school. Only in the 17th century does Comenius increasingly refer to the teaching situation and lays down in his *Didactica Magna* of 1657 *didactic principles* such as relevance to life, topicality, and vividness. He understands didactics as the justified selection of content for the art of teaching (Latin: *docendi artificium*). Today, important *didactic models* such as educational theory [3], learning theory [4], information theory [5] or critical-communicative didactics [6] are discussed. A comprehensive overview is provided by Blankertz [7] and Ruprecht [8].

Aschersleben [9] defines quite generally:

> Didactics is the total of learning aids that the student gives himself or that the teacher gives him. The learning aids refer both to the selection of teaching subjects and to the methods of learning. In this context, teaching is the didactic situation in which learning takes place, and school is the corresponding institution. The teaching as a didactic situation includes everything that constitutes it: the participants, teachers and students, learning objects, media, working materials, and so on [9].

One is inclined to distinguish didactics and methodology from each other by assigning the question of content, the "what", to didactics and the question of content delivery, the "how", to methodology. However, the interdependence of these questions [4] and the task of didactics to also answer the "why" (justification question) and the "what for" (goal question) has led to the subsumption of all these questions under the term didactics. Two corresponding definitions are quoted:

"Didactics deals with the question,	"The following questions arise:
who	With what goals should
what	which contents under
when	which prerequisites and
with whom	under what conditions
where	at what level
how	with what methods
with what	in what time

1 Introduction to the "Pie Chart" for Chemistry Didactics

"Didactics deals with the question,	"The following questions arise:
why and	with what success
what for	taught by whom or
should learn" [10].	learned by whom" [11]?

The concept of *general didactics* becomes more understandable when one lists *specific didactics* and establishes connections:

- *School-specific* didactics: primary school, secondary school, intermediate school, grammar school or comprehensive school didactics
- *Grade-specific* didactics: didactics of primary level, orientation level, secondary level I and secondary level II (advanced level),
- *Subject-specific* didactics or *subject didactics*: all school subjects or also groups of school subjects (area didactics)
- *Vocational* didactics: subjects of vocational school and technical colleges, didactics for the vocational school year etc.

Subject Didactics
At first glance, the compound word may suggest that subject didactics is composed additively of the elements of the corresponding subject and those of didactics. A second look makes it clear that a direct union makes no sense and it is difficult to oversee all corresponding contents.

Subject didactics, however, refers to the content of the subject science on the one hand and to the content of general didactics on the other hand: These are the *reference sciences*. Subject didactics as the actual *professional science* of teachers is an independent interdisciplinary science with its own goals, tasks, and methods, which reflects the content of the reference sciences and applies it to the questions of subject didactics. Fig. 1.1 illustrates the subject didactics as independent bridge subjects between pedagogy and general didactics on the one hand and the subject sciences on the other hand [12]. Fig. 1.2 shows the linkage of educational science, subject science, and subject didactic components in the training of teacher students in the 1st phase of training and the independent reflection in the 2nd phase or in their own teaching [13].

Chemistry Didactics
To make the intertwining of educational science, didactic, and subject science aspects clear using an example and to illustrate the argumentation in chemistry didactics exemplarily, an interesting experiment is used as a basis, carried out in various subject didactic ways, and evaluated with alternative concepts.

Example "Iron Wool on the Balance Beam"
If you ignite a tuft of gray-shining iron wool, which was attached and balanced on one side of the balance beam, you observe a red glow front that slowly moves

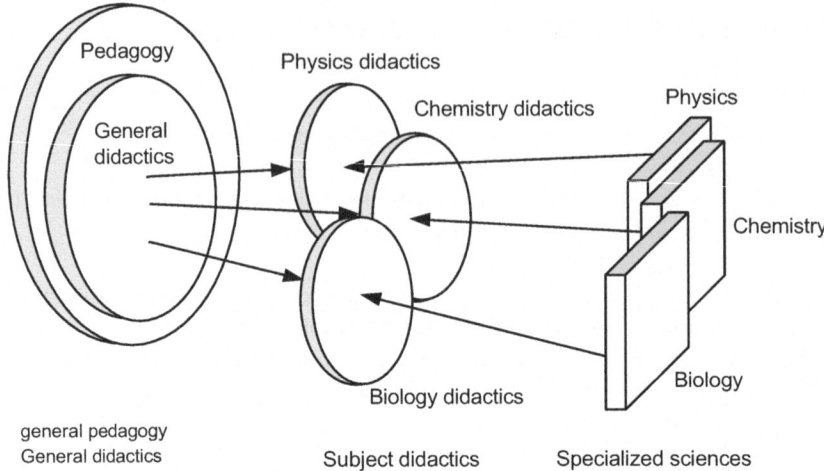

Fig. 1.1 Subject didactics as independent bridge subjects between the reference sciences, pedagogy/didactics, and the subject sciences [12]

through the wool. During the reaction, this side of the balance beam lowers, and a black product remains. The chemist describes this product as iron oxide with the composition Fe_3O_4, i.e., the composition of Fe^{2+}-, Fe^{3+}- and O^{2-}-ions: $(Fe^{2+})_1(Fe^{3+})_2(O^{2-})_4$. Thus, a redox reaction has taken place during the glowing of the iron: Fe atoms each give off 2 or 3 electrons and become cations, O atoms each take up 2 electrons and become anions; the types of ions link in the ionic lattice.

This correct procedure is feasible with students who already know the redox reaction as electron transfer in grade 10 or 11 — this interpretation is of course not possible in the initial teaching of grade 7 or 8. It is recognized that a *subject didactic reduction* is necessary in initial teaching to recognize and reflect the involvement of oxygen in the iron wool experiment and that the reacting oxygen from the air causes the observed increase in mass and forms iron oxide. In a next experiment, it is also to be shown that a sealed test tube, in which iron wool and oxygen are weighed together and brought to reaction, shows the same mass before and after the reaction, that the law of conservation of mass applies in chemical reactions.

Accordingly, this school experiment can be carried out in the form of various alternatives as well as evaluated in very different ways — depending on the pedagogical objectives set by the teacher, depending on the abilities and skills on the part of the students. Some ways are outlined in Fig. 1.3.

On **Path 1**, the experiment is carried out as described and observed by the learners: It will probably be surprising to find that "the black iron wool" is heavier

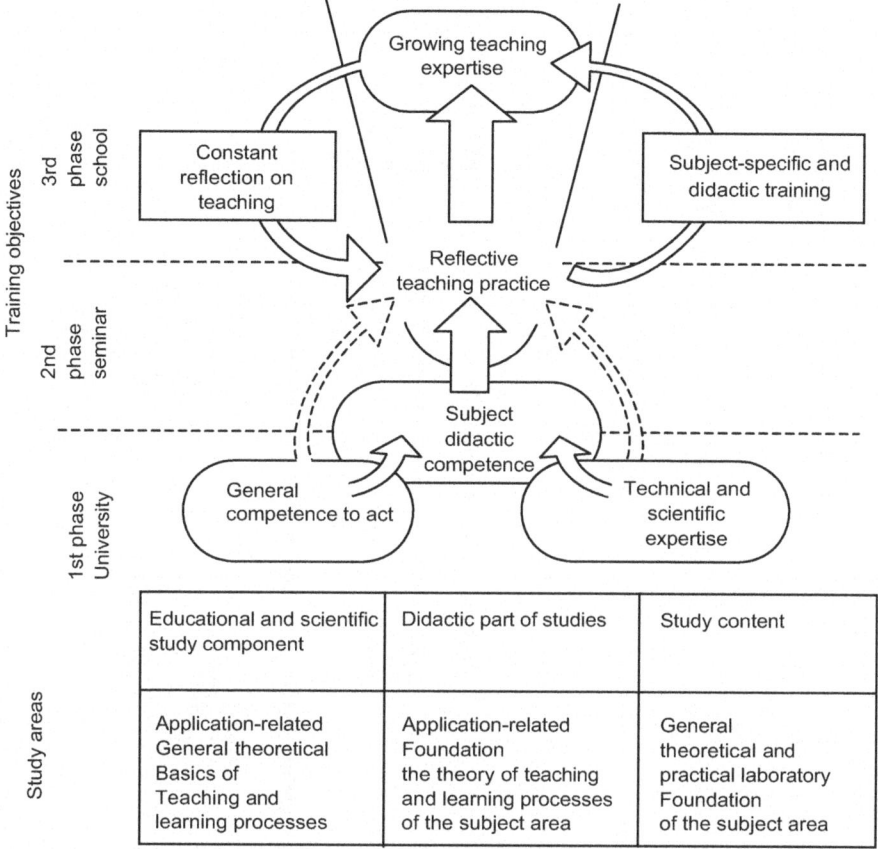

Fig. 1.2 Linkage of educational science, subject science, and subject didactic components in the professional development of teacher students [13]

than the gray iron wool before. In view of the evaluation of this increase in mass, the teacher must already decide:

- Should the learners themselves find the role of the incoming oxygen, or does the teacher name the oxygen as a reaction partner?
- Should they work out the reaction symbol in words, or are simplified element and compound symbols introduced?
- Is "+ heat" written for the indication of the energy turnover in the reaction scheme or separated by a semicolon "exothermic" or "$\Delta H < 0$" used?
- Do you show or draw structure models for the rearrangement of Fe and O atoms? Or do you even let suitable models be built by the students themselves — or have them make corresponding model drawings?

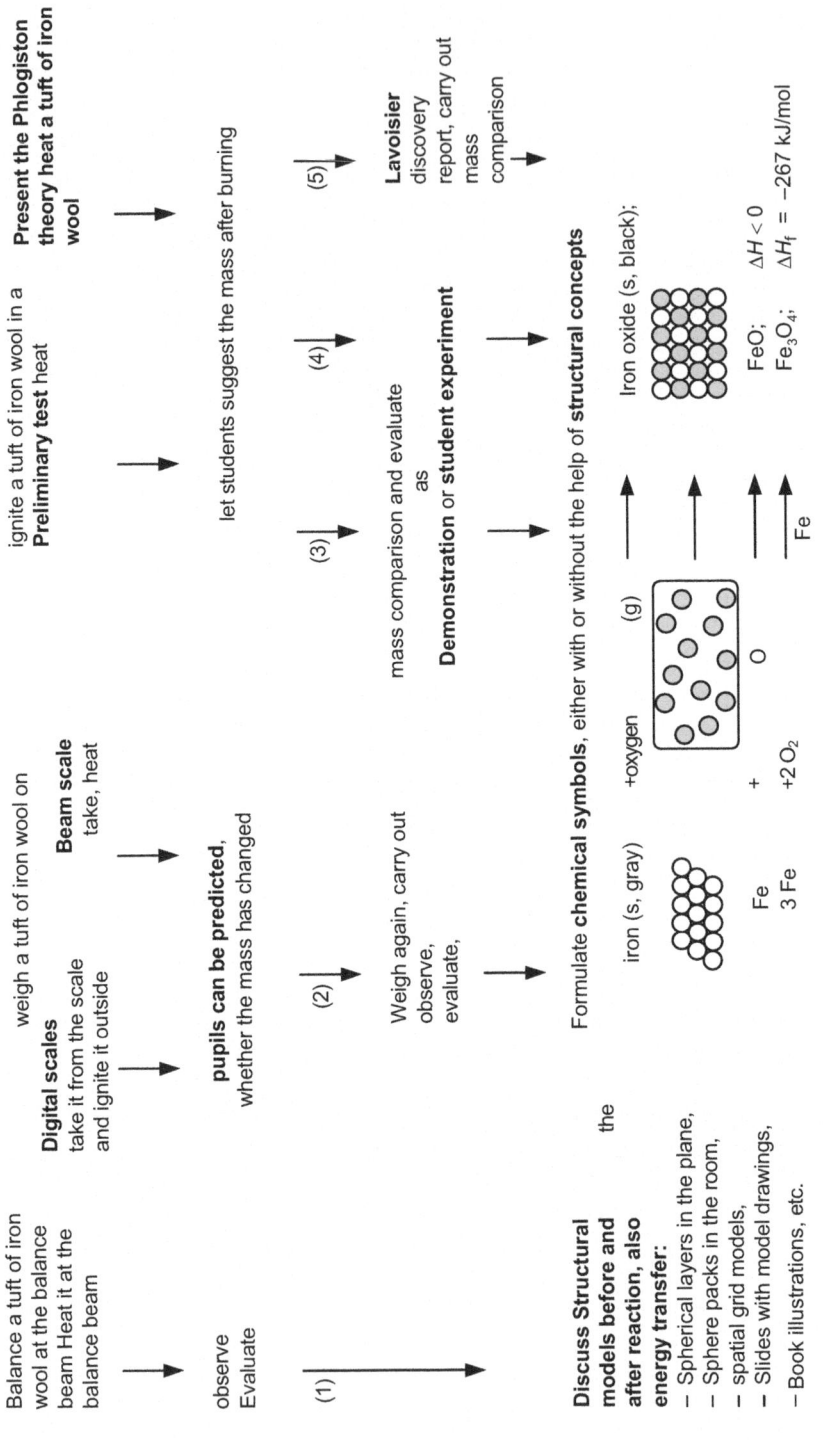

Fig. 1.3 Chemistry didactic alternatives using the example of the implementation and evaluation of the experiment "Iron Wool on the Balance Beam"

- Do you formulate a reaction symbol as a shortened model based on such structural ideas, or do you stick with the structural representation?
- Do you use the atom symbol for O atoms or appropriately the symbols for O_2 molecules for initial reaction symbols? Do you initially use the symbol FeO for a reduced structure of the iron oxide?
- Do you bring the different iron oxides into play and point out the black iron oxide formed in the experiment?

The same questions for evaluation also arise on Paths 2, 3, 4, and 5.

On **Path 2**, the tuft of iron wool is weighed — either with a digital scale or a beam balance. After removing it from the scale and igniting the iron, however, the teacher asks the students to predict whether the black product is lighter, heavier, or the same weight. He thus allows them to formulate their own hypotheses and develop solutions in which they are personally involved. This problem-oriented approach has the great advantage in terms of learning psychology that the experiment becomes the student's own problem and thus the motivation to think about solutions is much greater. Many students bring original ideas about the combustion process from their everyday life, will probably predict a "lightening" and be surprised that the product is heavier than the tuft before: A cognitive conflict contributes significantly to motivation (Chap. 3).

On a **Path 3**, the problem is presented even more openly. The teacher lets the students discuss their experiences with combustion and puts the combustion of the iron wool into this context with a short preliminary experiment. Depending on their ability and habit in their lessons, the students express their ideas and possibly suggest the investigation of the mass comparison themselves. The teacher carries out mass comparisons as demonstration or student experiments and evaluates them as described.

However, she can ask the students on **Path 4** to independently plan and carry out such an experiment as a student experiment. Depending on their own ideas and experimental skills, the student groups receive various experimental equipment and realize different solutions to the problem. The advantages of problem orientation are added to the advantages of action orientation and the related motivation — another didactic criterion.

The teacher can open the topic in a completely different way on **Path 5**. She asks the students about their ideas about the combustion process and expects the argument to come from the students: "Something goes into the air from the fuel." She reports that the same view was also held by scientists a few centuries ago, who called this something "phlogiston". The teacher can refer to the phlogiston theory of Stahl and the refutation of the theory by Lavoisier herself or suggest the reports to students. In discussions about experiments on the oxidation of mercury and the reduction of mercury oxide, as they were in the foreground with Lavoisier, the iron wool-oxygen reaction can be brought into focus.

Conclusion
Decisions regarding this large number of experimental and evaluation alternatives for a single issue can be more easily made and justified by teachers if they

are familiar with these or similar situations from their subject-didactic training. Especially in the initial chemistry instruction of secondary level I, a plethora of didactic decisions need to be made to achieve optimal learning success: Therefore, the following chemistry didactic reflections refer to the basic instruction of the subject chemistry [14]. This instruction begins in public schools in grades 7 or 8 depending on the federal state and ends with grades 9 or 10, before basic or advanced courses in the subject of chemistry on specific topics of secondary level II commence.

For these basic reflections, a thorough *subject-specific education* is assumed, which must have progressed to the extent that in-depth knowledge and a good understanding of the most important content for chemistry instruction at secondary level I are present. A possible list of such subject-specific content is contained in the publication of the German Chemical Society (GDCh): "Memorandum on Teacher Education for Chemistry Instruction in the Age Groups of Ten to Fifteen Year Olds" [15].

The same memorandum [15] also proposes a path for *chemistry didactic education*:

> The contents assigned to the goals of a didactic education of chemistry teachers are so diverse that they cannot all be dealt with adequately even with a favorable proportion of didactics in the total study volume. For this reason, a selection of important subject areas is made, each primarily assigned to one of four focus areas: the learners, the mediation processes, subject-specific focuses, and societal reference fields.

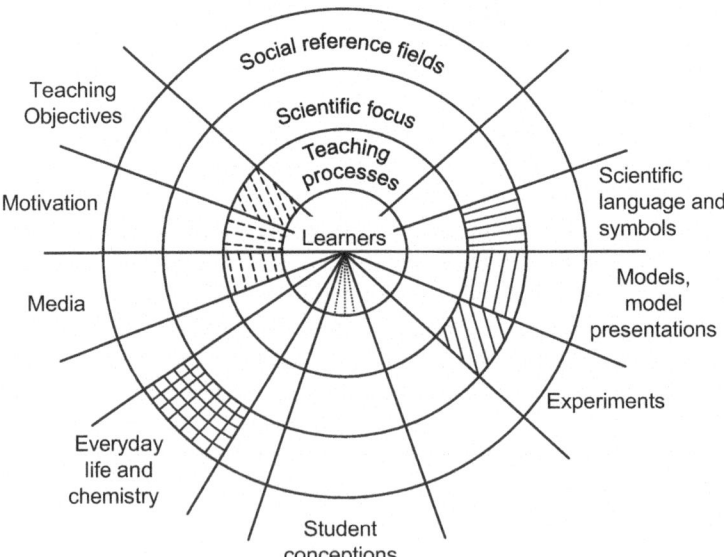

Fig. 1.4 Selection of topics and levels of reflection in the form of a pie chart [15]

The corresponding *four levels of reflection* can be found in Fig. 1.4 in the form of concentric circles. This suggests dividing each subject area into the mentioned four sections.

If the *learner* is at the center of every didactic consideration, the arc spans to the *societal reference fields*, which influence the learner as an individual and as a member of a community. The *subject-specific focuses* of the content are brought closer to the learner through *mediation processes*. Given the close interconnections of didactic questions and content references among each other, learners, forms of mediation, subject-specific focuses, and societal reference fields are not rigidly assigned for each individual subject area. Rather, the sections within the concentric circles are largely interchangeable, so that the possible alternating assignment of the contents allows the concentric rings to be understood as "rotating rings" [15].

The **pie chart** in Fig. 1.4 exemplarily represents *eight subject areas* that are suitable for special chemistry didactic reflection: the classic topics such as *experiments, models, model presentations, scientific language and symbols*, which are particularly assigned to the level "Scientific focus". Modern fields of didactics are *student conceptions* (assigned to the level "Learners") as well as *everyday life and chemistry* (assigned to the level "Social reference fields"). The reflection topics suitable for "mediation processes" are finally *teaching processes, motivation* and *media*.

The scheme deliberately contains empty segments without a topic designation: They are intended to make clear that there are many more topics for chemistry didactic reflection that can be included in the scheme. The topic "student conceptions" should be at the beginning, because there are original conceptions of students about substances and reactions that result from their everyday and life experiences and have been present for a long time before their chemistry instruction begins.

The explanations in this textbook are supplemented in various ways by other chemistry didactics textbooks that have already been published before this edition [16–26]: The study of chemistry didactics appears infinite and diverse to students — they must decide for themselves, based on their knowledge and attitudes, which argument they will follow!

References

1. Itelson L (1967) Mathematische und kybernetische Methoden in der Pädagogik. Volk und Wissen, Berlin
2. Roth L (1976) Handlexikon zur Erziehungswissenschaft. Ehrenwirth, München
3. Klafki W (1969) Didaktische Analyse als Kern der Unterrichtsvorbereitung. Schroedel, Hannover
4. Heimann P, Otto G, Schulz W (1970) Unterricht – Analyse und Planung. Schroedel, Hannover
5. Cube F (1968) Kybernetische Grundlagen des Lernens und Lehrens. Klett, Stuttgart
6. Winkel R (1980) Die kritisch-kommunikative Didaktik. WestPädBeitr 1:202
7. Blankertz B (1973) Theorien und Modelle der Didaktik. Juventa, München
8. Ruprecht H (1972) Modelle grundlegender didaktischer Theorien. Schroedel, Hannover

9. Aschersleben K (1983) Didaktik. Kohlhammer, Stuttgart
10. Jank W, Meyer H (1991) Didaktische Modelle. Cornelsen, Frankfurt
11. Vossen H (1979) Kompendium Didaktik der Chemie. Ehrenwirth, München
12. Riedel W, Trommer G (1981) Didaktik der Ökologie. Aulis, Köln
13. Hammer HO (1981) Fachdidaktik der Chemie an der Hochschule. CU 12:5
14. Scheible A (1966) Gedanken zum Einführungsunterricht in die Chemie. MNU 19:1
15. Barke H-D, Bitterling D, Gramm A, Hammer HO, Hermanns R, Leibold R, Lindemann H, Wambach H (1983) Denkschrift zur Lehrerbildung für den Chemieunterricht in den Altersstufen der Zehn- bis Fünfzehnjährigen. GDCh, Frankfurt
16. Schmidt HJ (1981) Fachdidaktische Grundlagen des Chemieunterrichts. Vieweg, Braunschweig, Wiesbaden
17. Christen HR (1990) Chemieunterricht. Eine praxisorientierte Didaktik. Birkhäuser, Basel, Boston, Berlin
18. Becker HJ, Glöckner W, Hoffmann F, Jüngel G (1992) Fachdidaktik Chemie. Aulis, Köln
19. Pfeifer P, Häusler K, Lutz B (1992) Konkrete Fachdidaktik Chemie. Oldenbourg, München
20. Lindemann H (1999) Einführung in die Didaktik der Chemie. Staccato, Düsseldorf
21. Harsch G, Heimann R (1998) Didaktik der Organischen Chemie nach dem PIN-Konzept. Vieweg, Braunschweig
22. Barke H-D, Harsch G (2001) Chemiedidaktik Heute. Lernprozesse in Theorie und Praxis. Springer, Berlin, Heidelberg
23. Heimann R (2005) Das Experiment – Ein Instrument zur Förderung des selbständigen Denkens. In: Rossa E (Hrsg) Chemie Didaktik – ein Praxisbuch für die Sekundarstufe I. Scriptor, Berlin
24. Rossa E (2005) Chemie Didaktik – ein Praxisbuch für die Sekundarstufe I. Scriptor, Berlin
25. Barke H-D (2006) Chemiedidaktik – Diagnose und Korrektur von Schülervorstellungen. Springer, Berlin, Heidelberg
26. Harsch G, Heimann R, Benmokhtar S, Wagner A (2014) Das START-Konzept – Teilchenmodelle und Formelsprache im Chemieanfangsunterricht. Aulis, Hallbergmoos

Student Conceptions 2

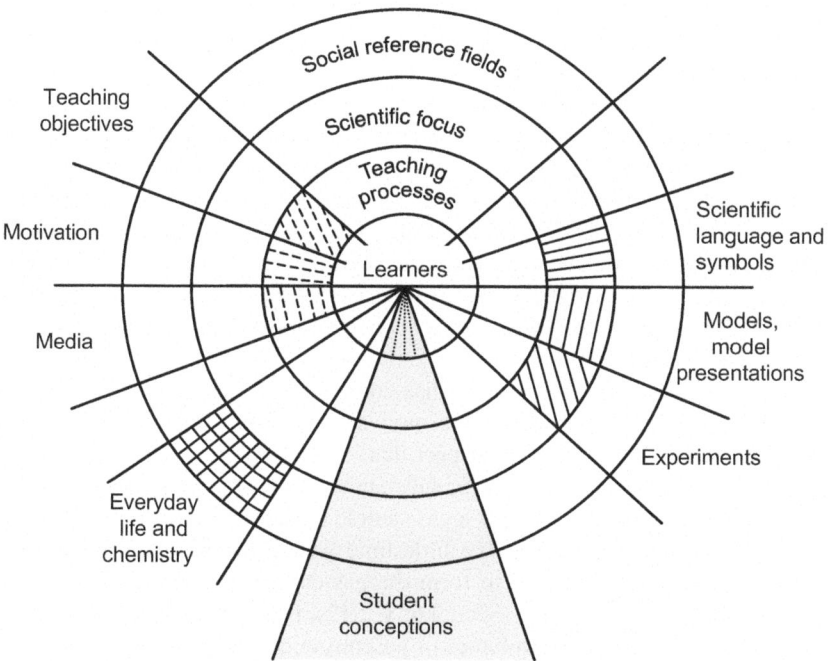

"The acid has eaten up the magnesium"—this is often the spontaneous reaction to the phenomena of metal-acid reactions. Young people cannot be blamed for this, as long as in everyday language "eating the metal" by acids or by rust is a common way of speaking. Such ideas of young people should be known to teachers in order to be able to address them in class before switching to the appropriate scientific language.

When planning lessons a few decades ago, it was assumed that students would have no scientific knowledge and that good lesson preparation would only have to decide in which order which new terms should be introduced with which didactic aids. However, subject-didactic surveys show that learners do bring their own ideas to many topics, which usually do not coincide with today's scientific ideas. For this reason, chemistry didactics should empirically determine which ideas exist for which facts, and which teaching is advantageous for their correction.

Student conceptions are often called *"false" conceptions*—without considering that students observe correctly and create their own, logical world of ideas. Therefore, the following designations are more accurate:

- Everyday conception or lifeworld conception,
- original or pre-scientific conception,
- Student pre-understanding or pre-concept,
- alternative conception.

Coal combustion as an example
An example may underline this view. One fills a round flask, which contains a few grains of coal, with oxygen, seals it with a balloon and weighs it (Sect. 2.6: V2.11). If you then heat the coal grains until they glow brightly and burn with a dazzling light when the round flask is swirled, the young people suspect that the mass of the round flask has decreased: "The coal grains are gone, they have completely burned." The observers have therefore observed correctly and correctly classify the observation in the series of their experiences with the combustion of wood and paper—they resort to everyday conceptions, which are usually also present in parents and siblings.

If you weigh the flask again after it has cooled down, the scale shows the same mass as before—and the observers are motivated to think about the reason. With the help of the teacher, they then suspect that, as with other reactions, a new substance has been formed. Since no new substance is visible, it can only be an invisible colorless gas: carbon dioxide. If a gas sample is taken from the round flask with the flask probe and passed through a little lime water, the suspicion is confirmed: carbon and oxygen have reacted to form the gas carbon dioxide. If the density of gases, especially carbon dioxide (Sect. 2.6: V 2.12), is also demonstrated, the learners may be convinced that the volumes of gaseous combustion products also have a certain mass. If the model conception of the reaction of C atoms with O_2 molecules to CO_2 molecules is additionally drawn on the board (Sect. 2.6: picture to V2.11), the reaction can finally be clarified and possibly already here the law of the conservation of mass in chemical reactions can be introduced. Of course, further combustion reactions have to be reflected and discussed in the light of the new experiences.

A discussion based on these experimental experiences softens the original conceptions and brings the scientific conceptions sustainably to the fore. This learning process cannot be completed in a single lesson, but must be expanded through continued problem-oriented teaching on the combustion process. It is advantageous to always first talk to the learners about their original conceptions before introducing the scientific conception (such as the historical oxidation theory of oxygen transfer according to Lavoisier or the redox reaction with electron transfer).

Surprisingly 'students' conceptions of combustion contain elements of the historical Phlogiston theory: Apparently, there are parallels between students' conceptions and the course of historical knowledge processes in the subject science. Therefore, it makes sense to study the transition from Stahl's Phlogiston theory of 1697 to Lavoisier's Oxidation theory of 1787 and to derive insights for chemistry teaching from it. There are also other historical theories that have undergone profound changes over the centuries—it is worth investigating which parallels exist in students' thinking. Such theories include:

- Theories of primal matter of the Greek philosophy schools in antiquity,
- Transformation concepts of the alchemists,
- The Phlogiston theory or theories of heat substance,
- The "Horror vacui" and the recognition of air pressure,
- Theories on atomism or the structure of matter, etc.

Since these reflections are of a scientific-historical nature, they will be presented in the following under " Scientific Focus". Subsequently, empirical experiences with students' conceptions are to be reported under the category "Learners" and compared with the historical theories. Under " Teaching Processes", recommendations for teaching are to be reflected, and under "Social Reference Fields", finally, influences of everyday language and advertising in the media on students' conceptions and thus on chemistry teaching.

2.1 Scientific Focus: Theories from the History of Natural Sciences

The Greek philosophers of antiquity have thoroughly thought through many facts of human life and created recognized theories for many areas: Our current cultures and basic values are largely based on Greek philosophy.

2.1.1 Primal Matter Theories

For the Greek natural philosophers, the fundamental question arose about the primal matter: What does the world consist of? What is "the primal ground, the primal matter, the primal substance, the element? Equally important was the second basic idea that this basic substance should be of eternal existence, that nothing (from nothing) arises nor (into nothing) perishes, that only the forms of appearance change. This drew attention to the following problems:

1. the materiality of the world,
2. the indestructibility and uncreatability of matter,
3. the transformability of matter 'while preserving the primal substance'"

As primal substances or "elements", the Greek philosophers initially discussed "earth, water, air, and fire". In particular, Aristotle taught the categorical separation of thing and property. This distinction of the thing as a "carrier of properties" on the one hand and "properties of the things themselves" on the other hand, was not yet fully understood by the Greek philosophers before Aristotle. Based on this insight, Aristotle came to the thesis that "development and change, becoming and perishing is nothing else than the transition from one way of being to another way of being" [1].

2.1.2 Transformation Concepts of the Alchemists

The age of alchemy extends from about the 4th to the 16th century, with the Arabs particularly involved in the development of alchemy. For them, the term alchemy was just another word for chemistry, formed from the Arabic "al" and the Greek "*chyma*": metal casting. This term shows the great importance of metals to people and their desire to make first household items, first swords and other weapons from metals—and above all, to transform base metals into gold.

Many Arab writings already contain instructions for artificial gold production with the help of the "ferment of ferments", the "elixir of elixirs". The correct mixture of the four elements is necessary, and the "spirit" (the heated, volatile mercury) must penetrate the "bodies" (lead, copper, etc.). "The mysterious 'elixir' itself only comes about through the correct union of the four elements, the body (metal) and spirit (mercury), the male and the female. It assimilates the bodies and colors them (hence also called 'tincture'), by turning them into gold, up to a thousandfold amount" [2]. Even the great scholar Albertus Magnus "believes in the possibility of artificial gold production, but says that he has not found any alchemist who has completely succeeded in metal transformation" [2].

Even "practical" and "obvious" proofs were not lacking in alchemy until the 18th century. Coins were shown that were supposed to be minted from alchemical gold, or nails, one half of which was supposed to be made of iron, the other half of iron transformed into gold. Even court judgments were made in favor of alchemical operations, and not least, clever fraudsters ensured that the belief in the possibility of metal transformation was repeatedly revived by their "successful transmutations" [3].

Nevertheless, in 1923, science was once again excited when a Berlin university professor announced that he had succeeded in transforming mercury into gold by treating it with electric currents. The correctness of the finding was initially confirmed not only from various sides, but several "researchers" (even from Japan) also reported that they had found the same thing earlier. Upon thorough experimental verification, it was only discovered after two years that the small traces of gold came from the electrodes. With this, the dream of gold, which has captivated fantastic minds for over a millennium, is probably finally over [2].

2.1.3 The Phlogiston Theory

It has always been observed that during combustion, the fuel, such as coal or sulfur, "disappears". Georg Ernst Stahl published his interpretation of these observations in 1697 and introduced the term phlogiston (Greek: *phlox*, the flame): He started from the combustion of sulfur and believed that "the sulfurous acid produced by combustion was sulfur deprived of its combustible principle" [4].

Stahl had also observed in Thuringian smelters that from highly heated mixtures of yellow "zinc lime" (today: zinc oxide) and pieces of coal, the silver-shining zinc melt flowed out, and developed his theory on this basis: He claimed that phlogiston would be contained in all combustible substances and would leave the fuel during combustion. The air played a role insofar as it was necessary to absorb the phlogiston. Thus, the phlogiston got into leaves, woods, and coal, which should contain the largest proportion of phlogiston. By heating metal lime on a piece of charcoal, the phlogiston was transferred and the pure metal could be produced. The "proof" was provided by the experimental "calcination of a metal" by heating and the reduction of the metal lime by coal in the laboratory [3]:

$$\text{Metall} \rightarrow \text{Metallkalk} + \text{Phlogiston}$$

$$\text{Metallkalk} + \text{Phlogiston (Kohle)} \rightarrow \text{Metall}.$$

However, Stahl had to regard the pure metal as a *compound* of metal and phlogiston, and the metal lime—that is, today's metal oxide—as an *element*. And experiments with the balance contradicted his theory, because the "calcined metals" increased in mass. Stahl wanted to save his theory by attributing a "negative weight" to phlogiston.

The viewpoint of the phlogistonists becomes more understandable if one does not consider it from the standpoint of material processes, but from that of energetics. During combustion, one not only sees a flame, but *heat* is also constantly generated. If one replaces phlogiston with this heat energy, one does justice to the facts in a certain way [2].

The phlogiston theory was replaced 100 years later by Lavoisier's oxidation theory. Lavoisier found oxygen through weighings using the synthesis and decomposition of mercury oxide as an example, defined the *oxidation theory* in terms of oxygen transfer, and formulated the law of conservation of mass in chemical reactions . The *law of conservation of mass* was also found by Mikhail Lomonosov in Russia at the same time. Until its discovery, oxygen was also referred to as "fire substance" (Empedocles) or "fire air" (Scheele). "A remnant of these terms remained in use as *'heat substance'* for a few decades until the cause of heat was recognized as the movement of smallest particles" [3].

2.1.4 "Horror vacui" and Air Pressure

Experiments with pipettes and siphons already alerted natural philosophers in antiquity to the fact that there are no air-free spaces on earth, that as soon as one substance leaves a space, another substance occupies it. In this context, a formulation by Canonicus has become known, which states that "nature has a horror vacui, an abhorrence of the vacuum" [5].

Galileo Galilei was also familiar with this phenomenon and knew from well builders that it is not possible to suck water from a depth of over 10 m upwards with a pump. He considered this measure to be the utmost force with which "nature could prevent a vacuum", and in 1643 invented a thought experiment to measure this force, the "resistenza del vacuo" [5] (Fig. 2.1a): A A large cylinder should be filled half with water, a large mass has to be attached to the movable piston until a vacuum is formed in the cylinder. It is not known "whether this experiment is only described on paper or whether it has actually been carried out" [5]. This apparatus inspired his student Evangelista Torricelli to replace the difficult-to-realize dense piston with liquid mercury, which is specifically heavy and slides like a piston in the one-sidedly closed glass tube (Fig. 2.1b). With this experiment first described in 1643 [5], he was able to prove the normal air pressure of 760 mm mercury column and the existence of a matter-free space above the mercury in the glass tube. To do this, Torricelli filled an approximately 1 m long, one-sidedly melted glass tube with mercury, sealed it and opened it under the surface of a tub filled with mercury (Fig. 2.1b). He observed that each time the mercury column in the glass tube drops to about 76 cm height, and showed that when the glass tube is tilted, the metal fills the tube in its entire length, while in a vertical arrangement

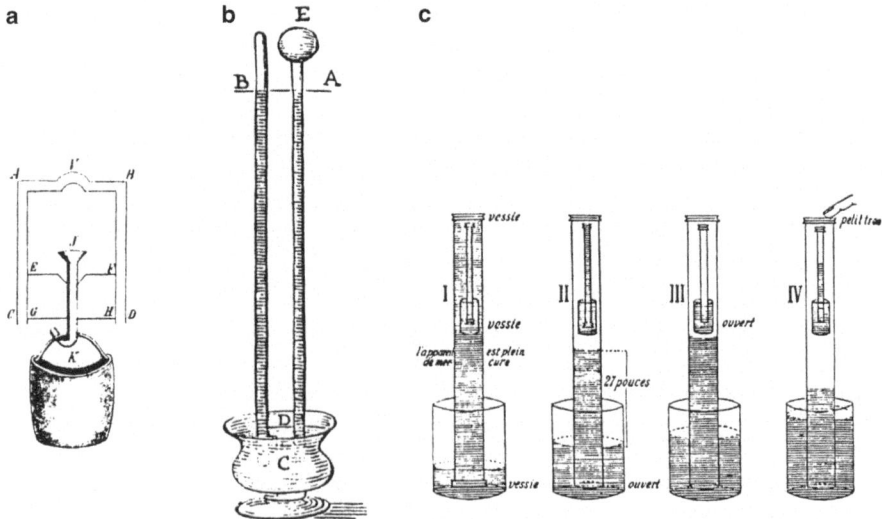

Fig. 2.1 Historical experiments to overcome the concept of "horror vacui" [5]

the 76 cm column reappears. The interpretation led to the proof of air pressure: The approximately 20 km high air layer of our atmosphere balances a 76 cm high mercury column. At high air pressures, the column can also be 77 cm high, at low pressure it drops to 75 cm. Torricelli had thus constructed the first barometer, in his honor the new pressure unit Torr was defined: 1 mm mercury column = 1 Torr, the normal air pressure thus corresponds to the value of 760 Torr.

Torricelli interpreted the empty space above the metal column as a vacuum—otherwise the metal could not completely fill the glass tube below a height of 76 cm when tilted. This empty space caused considerable difficulties for most colleagues: They were still subject to the "horror vacui" and demanded a special "ether" in this empty space. To demonstrate a matter-free space to learners as well, a simple experiment is suggested (Sect. 2.6: V2.1)

In 1647, Blaise Pascal provided the final proof through the experiment "du vide dans le vide" (Fig. 2.1c) that a "Torricelli apparatus in the Torricellian vacuum" does not indicate gas pressure and that there is no special "ether" in the free space above the mercury column [5]: In Fig. 2.1cI it can be seen that all openings are still closed by a bladder membrane ("vessie"); if the lower seal is removed in Fig. 2.1cII, the mercury drops to the usual 76 cm. Now in Fig. 2.1cIII the seal of the small Torricelli apparatus is removed and in this case the metal drops completely to the mercury surface in the beaker, thus indicating zero pressure. Only in Fig. 2.1cIV does air enter the entire apparatus—normal air pressure is displayed again.

For final certainty, Pascal had the barometer experiment carried out on a mountain about 900 m high near Paris in 1648 and found a mercury column of about 690 mm. This finally proved the existence of atmospheric air pressure and the theory of the "colonne d'air" [5], i.e. the pressure of the more than 20 km high air column above sea level, which balances a 760 mm high mercury column.

Further attempts to prove the existence of air pressure were planned by Otto von Guericke based on existing knowledge. After he invented various techniques for creating a vacuum—that is, a series of technically advanced air pumps—he conducted the experiment with the "Magdeburg Half-spheres" as the mayor of the city of Magdeburg (Fig. 2.2):

> I placed these half-spheres on top of each other and quickly pumped out the air. I saw the force with which the two shells were joined. Compressed by the pressure of the external air, they were so firmly connected that sixteen horses could not or only with difficulty tear them apart. But when they were finally separated with all their forces, it caused a noise like a gunshot [6].

2.1.5 Theories on Atomism and the Structure of Matter

These theories also have their origins in interpretations of Greek philosophical schools. There were essentially two directions of thought: A group around Democritus and Leucippus was convinced that a repeated division of a portion of matter has an end and matter is made up of indivisible particles, the atoms (Greek: *atomos*, indivisible). This concept assumed particles and empty space around them

Fig. 2.2 The Guericke experiment of the "Magdeburg Half-spheres" [6]

and is today also called the discontinuity hypothesis. Aristotle and other philosophers claimed that the repeated division of bodies has no end. In particular, the then unthinkable impossibility of empty space, which had to separate individual particles from each other—the "horror vacui"—convinced them of the continuous structure of matter: Since then, the *continuum hypothesis* has been discussed. Due to the great influence of the Aristotelian school, it was taught everywhere and thus Democritus' discontinuity hypothesis was suppressed for almost two millennia.

After the "horror vacui" had been overcome in a macroscopic sense with Torricelli's experiments and the vacuum became conceivable, Pierre Gassendi transferred these findings to the existence of the vacuum in a submicroscopic sense, disregarded the Aristotelian views and revived Democritus' idea in 1649: "*The atoms and the empty space* are the only principles of nature, as the absolutely full, and the absolutely empty space, nothing third is conceivable" [7]. After an almost 2000-year interruption, scientists now based their work on the discontinuity hypothesis and began to think about the structure of matter from smallest particles. The English philosopher John Dalton came up in 1808 with the idea of different types of atoms that should build up the different elements, and proved the existence of atoms by calculating atomic masses and creating a first table of atomic masses. In 1912, the German physicist Max von Laue discovered X-ray structure analysis, which could be used to investigate the spatial arrangement of atoms, ions or molecules in crystals.

2.2 Learners: Empirical Evidence of Student Conceptions

Students of introductory in teaching chemistry are predominantly assigned to the stage of concrete thinking operations according to Piaget, they are predominantly fixated on the concrete object in their conceptions. As a result, they describe phenomena in a *concrete-visual* and *magical-animistic* manner. The following examples are known from experience:

- The wood does not want to burn, the flame wants to go out, the flame consumes the candle,
- Chemicals attack materials, acids eat metals, rust devours iron etc.

The interpretations of students often correspond to simple *analogies*, causes are *personified*:

- Sodium reacts with water "like a fizzing tablet".
- When copper sulfate dissolves, it is "like the spreading of red cabbage broth in water".
- The grain grows in the fields so that people can feed themselves.
- The wood burns so that one can warm oneself etc.

Particularly, parallels can be seen in the thinking of students to the historical course of knowledge processes in chemistry. For this reason, the concepts mentioned in Sect. 2.1 from the history of natural sciences should be taken up and compared with empirically observed examples in the statements of our students today.

2.2.1 Substances as Property Carriers

In the imaginations of the students, new substances do not consistently emerge in chemical reactions, but new properties are assumed:

- Copper sheet turns black when heated, copper roofs turn green, silver tarnishes dark, a colorless solution turns deep blue, among others.

It seems—just like for the philosophers of antiquity—there is a primal substance or property carrier that somehow remains and only changes its outer appearance again and again. Therefore, students should be made aware that

- the black layer on the copper sheet can be removed: The black substance copper oxide is formed by the reaction of reddish-brown copper with colorless oxygen in the air (Sect. 2.6: V2.1);

- a green layer forms on copper roofs: The substance copper carbonate is formed on the copper over decades by the reaction of copper with rainwater, which is saturated with carbon dioxide from the air;
- the black layer proves to be a different substance than silver: black silver sulfide is formed by reactions of silver with the gas hydrogen sulfide (H_2S) and other sulfur compounds that are mixed in air;
- the colorless alkaline solution turns blue because added universal indicator causes a specific dye to form.

In an experiment, the metal can be recovered from black silver sulfide or from black copper oxide (Sect. 2.6: V2.2) and it can be shown that in some way "the copper is contained in copper oxide", and that by a new reaction of the copper with oxygen or sulfur, the black oxides or sulfides are formed again, meaning the reactions are reversible. However, this leads to the question of what remains in the compounds and can be brought out again. The answer is difficult and can only be formulated at the level of the core-shell model of the atom: The cores of metal and non-metal atoms remain in the reactions. The question of how red copper oxide (Cu_2O) differs from black copper oxide (CuO) can only be answered with the concept of atoms and ions. This discussion will be taken up again later (Chap. 7).

2.2.2 Mixture Concept

In chemical reactions, new substances are formed. These new substances are often described by chemistry teachers and authors of textbooks with the elements "as their components":

- Copper sulfide *contains* copper and sulfur,
- Water *consists of* hydrogen and oxygen,
- a hydrocarbon *is made up of* carbon and hydrogen.

These formulations suggest to young learners a mixture concept: "Compounds are mixtures of elements". In this regard, it is very important to demonstrate the *heterogeneous* mixture of two substances, such as that of copper shavings and sulfur powder. It is to be compared with the powder of the *homogeneous* copper sulfide: With the best microscope, no two types of crystals can be recognized anymore. The formulation "Copper sulfide is a compound of the elements", or "Hydrogen and oxygen are chemically bound in water" are preferable to the above, but do not help the learners further. Only at the level of Dalton's atomic model can one speak of a water molecule "consisting of" or "containing" two hydrogen atoms and one oxygen atom.

The **dissolving of salt in water** can be compared more to mixing processes of two types of substances: The salt is mixed withwater into salt solution, and by

2.2 Learners: Empirical Evidence of Student Conceptions

heating the solution and evaporating water, the salt can be recovered. However, learners also develop notions that

- after the dissolving process, the salt no longer exists—only the salty taste of "salt water" is present,
- from 100 g of water and 10 g of salt, a solution is formed that has a mass of 100 g: "the salt is gone",
- in general, substances cease to exist when they are no longer visible.

Therefore, one should firstly show in experiments that the salt can be recovered from the salt solutions at any time (unfortunately, this does not work so well with sugar from sugar solutions, because when the sugar solution is heated, a sugar syrup is formed, which decomposes). Secondly, one should omit the terms "salt water or sugar water" and choose the appropriate names "salt solution or sugar solution": This makes learners aware that a substance is dissolved in another substance.

2.2.3 Destruction Concept

Children and adolescents observe many phenomena in their first decade of life where substances disappear: wood or paper "are gone when you light and burn them", the puddle is soon "disappeared in the sunshine, the water is gone", the sugar disappears when dissolved in water, "the sugar is gone". A destruction concept can be identified in learners, which is not only created by their own accurate observations, but is also reinforced by familiar phrases in everyday language:

- Candles, spirits or gasoline "burn" up, charcoal "glows" away, wood is destroyed by burning
- Ethanol or acetone "evaporate" quickly, afterwards they are away,
- Plants "rot" away, animal carcasses "decompose", food is "digested",
- Rock "weathers" away, sandstone figures "crumble",
- Limestone in a water boiler is "dissolved", metals are "corroded",
- Grease stains are "removed", residues are "destroyed" etc.

These phrases should be reflected upon with the learners, in particular, it should be illustrated through experiments that no "destruction" of matter takes place. In the case of combustion processes, reference should be made to the colorless gases that the observer cannot see: The fuel reacts with the oxygen in the air, the reaction products are carbon dioxide and water vapor, in the case of charcoal only the carbon dioxide (Sect. 2.2.5). The evaporation of acetone can indeed be made clear by using a sensitive scale, but it should also be shown that a large volume of acetone vapor is formed (Sect. 2.6: V2.3): The acetone is thus not "gone". The reaction of lithium or sodium with water should demonstrate the resulting hydrogen gas and the portion of a new substance remaining after the water has evaporated (Sect. 2.6: V2.4), the removal of stains should be interpreted by dissolving fat in gasoline: The fat is not "gone", but remains in the cleaning cloth after the cleaning process (Sect. 2.6: V2.5).

The interpretation of the experiments should lead to the **development of a conservation concept**: The water in the puddle remains as water vapor, as does the acetone as acetone vapor, the alkali metal pieces may no longer be visible, but they have reacted with the water and metal compounds have formed, the fat is only dissolved and can be recovered. The actual understanding of the preservation of the elements in the compounds can only be conveyed when the types of atoms are known and chemical reactions are understood as a rearrangement of the atoms or ions (Chap. 7 and 8).

2.2.4 Energy Concept

There is also no conservation concept for energy among students, their ideas are mostly based on everyday language:

- Energy is "used up", electricity*consumption* and fuel*consumption* have to be paid for,
- Batteries are "empty", the battery is "empty" and needs to be recharged,
- Energy "is gone, one has no more energy, chocolate brings back used energy".

It should be made clear when observing energy conversions that in all cases the energy is conserved, that there is a conversion from one form of energy to another: Thus, electrical energy is converted into heat energy (water boiler), into mechanical (electric motor) or into light energy (incandescent lamp). The *chemical energy* is the least illustrative of all types of energy: For a corresponding view, every exothermic or endothermic reaction is to be interpreted as the chemical energy of the reaction products becoming smaller or larger. Only then can the exothermic conversion of fuels in the car engine be understood and the fuel consumption correctly interpreted: From the energy-rich system gasoline/oxygen, an energy-poor system carbon dioxide/steam is formed, the differences in the chemical energies of both systems are the cause for the release of energy, for the exothermic reaction.

To make the unit Joule (J) or kilojoule (kJ) for energy clear and to distinguish it from the unit °C for temperature, the following experiment is possible (Sect. 2.6, V2.6): A beaker with 100 g of water is provided on a tripod and wire mesh with an immersed thermometer, a portable butane burner is weighed accurately. The temperature of the water is measured accurately. The burner is ignited, placed under the beaker with a roaring flame for about a minute, then immediately extinguished; it is weighed again. The thermometer is stirred until the temperature of the water remains constant and the temperature difference is determined. If the difference is 20 °C, then with the knowledge that the energy of 4.2 J raises the portion of 1 g of

2.2 Learners: Empirical Evidence of Student Conceptions

water by 1 °C, the following is calculated: The presented 100 g of water take from the burner flame the energy E = 4.2 J / (g °C) · 100 g · 20 °C = 8400 J = 8.4 kJ.

Through the known mass difference of the butane burner, the amount of energy released by the butane-oxygen reaction can be estimated from the mass of the missing butane and compared with the amount of energy absorbed by the water: The amounts of energy released to the environment and thus the efficiency can be determined (Sect. 2.6: V2.6).

2.2.5 The Combustion Process

Both ideas of the property carrier ("copper turns black when heated in the flame") and those of the destruction concept ("wood is burned, the wood is gone") refer to original ideas of combustion. Years of observing the always fascinating flames of fuels such as paper, wood and charcoal or alcohol and gasoline lead to the statements that during combustion *something* is released into the air, that *something* is lost and little *ash* remains. This "something" is probably the "substance" that the scientist Stahl called phlogiston in the 17th century.

The teacher must make it clear to the students in chemistry lessons that the colorless gas oxygen reacts with the fuel, and in the case of candles, paper or wood, the colorless gases carbon dioxide and water vapor are produced as reaction products. Even in the case of metals, oxygen reacts, but solid reaction products are formed, which the learner can see and examine. If this combustion process is not accompanied experimentally and only a diffuse reaction equation is formulated for learners, the ideas of the "something" are not corrected for a long time.

A group of 9th grade students was asked to provide the reaction equation for the combustion of magnesium, and up to 95% of the learners noted appropriate symbols such as "Mg + O → MgO". However, they were also asked what happens to the particles of magnesium during combustion in the air. Many students argued—contrary to their correct reaction equation—that some of the magnesium particles go into the air and another part of the particles remain as white ash [8]. A German student's statement and corresponding drawing are shown in Fig. 2.3. This figure exemplifies that formulating a correct reaction equation does not guarantee that the context has been understood—not only the preconceptions of the learners about combustion need to be discussed more clearly and carefully compared with the observations from experiments, but also spatial models or model drawings of the structures before and after the reaction need to be conveyed (Chap. 7).

Other subject didacticians like Rosalind Driver [9] observe similar student ideas (Fig. 2.4). She investigates—in addition to the well-known reaction of iron wool on a balance beam—the reaction of phosphorus in a round-bottom flask, which is deliberately sealed with a stopper and remains sealed during the thought experiment. She asks what mass the round-bottom flask has after the reaction of the phosphorus. Despite the sealed stopper, many respondents believe that "phosphorus and white phosphorus smoke are gone", and in their opinion, the flask is lighter afterwards.

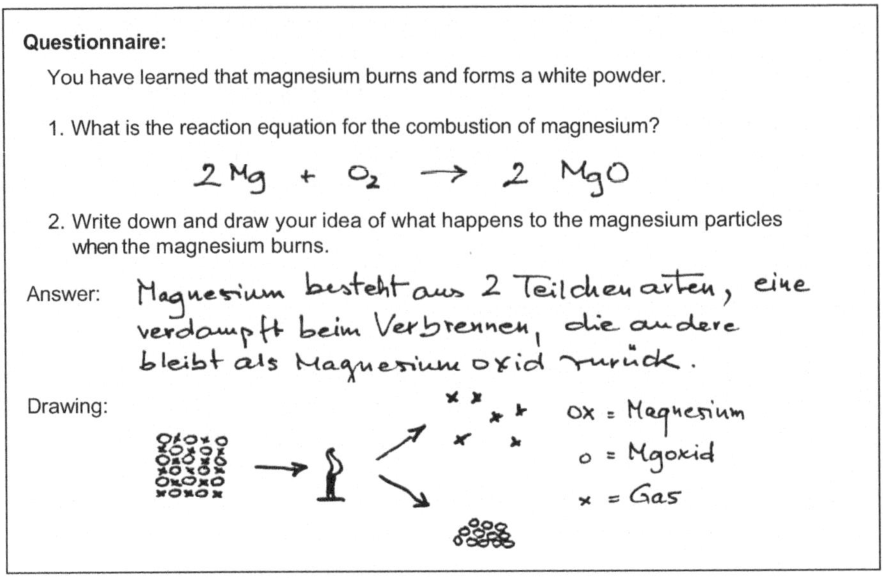

> **Questionnaire:**
> You have learned that magnesium burns and forms a white powder.
>
> 1. What is the reaction equation for the combustion of magnesium?
>
> $$2\,Mg + O_2 \rightarrow 2\,MgO$$
>
> 2. Write down and draw your idea of what happens to the magnesium particles when the magnesium burns.
>
> Answer: Magnesium besteht aus 2 Teilchenarten, eine verdampft beim Verbrennen, die andere bleibt als Magnesiumoxid zurück.
>
> Drawing:
>
> ox = Magnesium
> o = Mgoxid
> x = Gas

Fig. 2.3 9th grade German student's ideas about combustion [8]

Even older students who have already received several years of chemistry instruction often cannot separate themselves from the idea of annihilation. Helga Pfundt [10] reports of a 10th grade student who claimed: "According to the formula, carbon should be obtainable from CO_2, but of course it is impossible to obtain a black solid from a colorless gas."

The original ideas should be discussed with the teenagers in any case; attention should be drawn to two important facts: the conversion of the air or the oxygen as a part of the air, and the formation of solid reaction products in combustion reactions of metals. Experimentally, it can be shown that during combustion the mass increases when an open apparatus is used: iron wool becomes heavier on the balance beam, the mass of the combustion product of magnesium is larger than that of the magnesium before (Sect. 2.6: V2.7). To show this effect of "becoming heavier" also for reactions that produce gaseous combustion products like carbon dioxide and water vapor, a candle on a balance beam should be ignited in such a way that the resulting gases are chemically absorbed and weighed (Sect. 2.6: V2.8). From the colorless carbon dioxide gas, black carbon is actually obtained—contrary to the student's idea—by immersing burning magnesium in a cylinder filled with carbon dioxide (Sect. 2.6: V2.9).

However, learners should not remain with the idea that all portions of a substance become heavier during combustion—it is the *law of conservation* that needs to be demonstrated for these reactions in the experiment. It should be assumed

2.2 Learners: Empirical Evidence of Student Conceptions

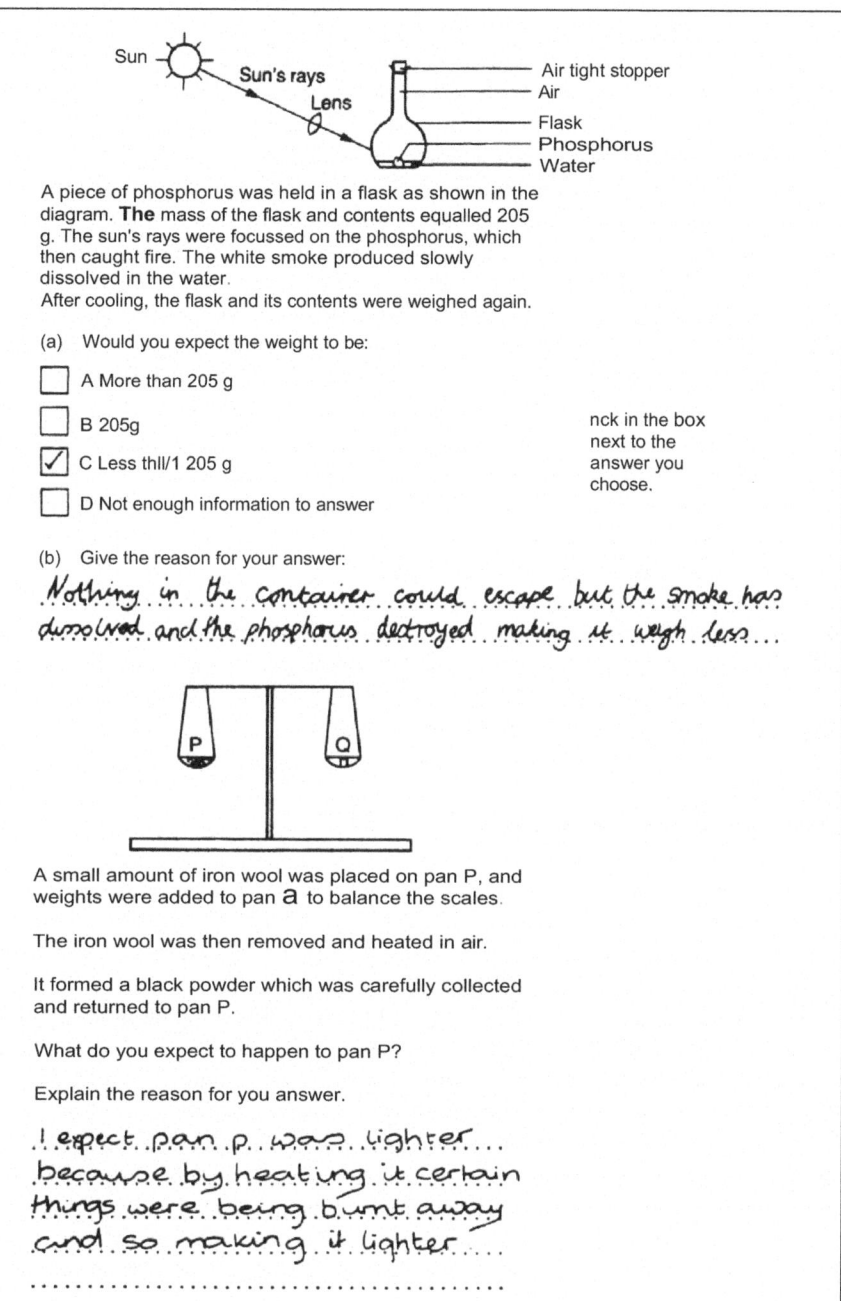

Fig. 2.4 Statements by English teenagers about the combustion of iron wool and phosphorus [9]

that one can choose *open apparatus* for combustion reactions as in V2.7 and V2.8 (Sect. 2.6), but also closed apparatus. For example, if a test tube containing a tuft of steel wool and air is sealed with a balloon, weighed, and heated with a burner, the metal in the test tube glows and black iron oxide forms. If you weigh it again after cooling, you will find the same mass as before (Sect. 2.6: V2.10a). To show the same experiment with an organic fuel, some matches can be heated in the same way and mass comparisons can take place as before (Sect. 2.6: V2.10b): The experiment with closed apparatus shows the law of mass conservation in chemical reactions. It applies, of course, in all cases—both for open and closed systems.

To understand the role of oxygen in the combustion of carbon and the "disappearance" of the coal, Petermann, Friedrich and Oetken suggested the "Boyle experiment" in a closed glass apparatus [11]: Some grains of activated carbon are placed in a round flask flushed with oxygen, the sealed flask is weighed (Sect. 2.6: V2.11). By heating with the burner, the carbon grains are ignited, they glow when swirled in the glass flask and then react without any visible residue. The subsequent weighing of the cooled closed flask shows mass constancy, a lime water test of the colorless gas mixture in the flask shows the presence of carbon dioxide. The layman naturally expects a lower mass after the "disappearance of the carbon grains"—the surprising observation of mass constancy can bring the weighed oxygen and the coal-oxygen reaction or the formation of carbon dioxide of the same mass into focus. If the Dalton atomic model is available, the reaction can be interpreted by the rearrangement of the C atoms and O_2 molecules to CO_2 molecules (see picture for model idea at V2.11).

2.2.6 Air and Other Gases

Many experts of past centuries neither recognized air as a substance nor distinguished other colorless gases from air. This is similarly difficult for young people today. Since air constantly surrounds us seemingly weightless and warm air is known to rise, a portion of air in children's imagination has no mass and is therefore not seen as a substance. Thus, Münch [12] was able to show in an empirical survey that about half of the adolescents between 10 and 16 years believe that a football—inflated in front of them with a common air pump—is lighter than the same ball that has been inflated only slightly.

The mass of a certain portion of air and thus the density of air can be quickly and convincingly demonstrated to students. If a glass sphere is evacuated with a water jet pump, weighed accurately with an analytical balance, and weighed again after filling with 100 mL of air from the piston sampler, they can read off a mass of 0.13 g (Sect. 2.6: V2.12). If the *densities* of other gases are determined in the same way, air and other gases can be clearly distinguished from each other by the new measurement values. Well-known gases such as oxygen, nitrogen, hydrogen, and carbon dioxide can also be presented experimentally and distinguished by the test with the burning or smoldering wooden stick (Sect. 2.6: V2.13). Finally, the *oxygen content of the air* of about 20 vol.-% can be demonstrated by showing the reaction of steel wool or phosphorus with oxygen in the air to form solid oxides in closed apparatuses (Sect. 2.6: V2.14).

Many everyday conceptions of gases come from *everyday language*. Thus, Weerda [13] receives the following statements from students:

- Fresh air is "good" air, air without oxygen is "bad".
- A fireplace needs "inlet air" and "exhaust air". Cars emit "exhaust gases" into the air.
- Colorless gases are "air" or similar to air. Water evaporates "into air".
- Gases are flammable, are used for cooking and heating.
- Gases are dangerous, are explosive, are toxic.
- Gases "can be liquid", there is "liquefied gas" in lighters.

To clarify the term *"liquefied gas"*, the following experiment can be carried out. Butane gas is filled into a gas liquefaction pump by displacing the air, this is put under high pressure by a piston: A large drop of liquid butane forms in the presence of gaseous butane (Sect. 2.6: V2.15). It should be made clear that in lighters and camping gas cartridges, liquid and gaseous butane coexist, and that the gas pressure in the cartridge is maintained as long as liquid butane is present or can evaporate.

2.2.7 Structure of Matter

It is possible to resume the discussion of the natural philosophers from two thousand years ago: Does repeated division of a substance sample come to an end and to smallest particles or not? Since students have no personal experience with the smallest particles of matter, the continuum hypothesis is naturally closer to them. For example, one student answered the question of whether he could imagine the smallest water particles: "No, you can spread a drop of water as wide as you want."

Even if the adolescents accept the particle concept in a discussion, difficulties still arise in the ideas: The particle concept is not applied consistently. Pfundt [14] showed the dissolution of a blue copper sulfate crystal in water and asked students about their ideas. She differentiated pre-defined answers not only in terms of a continuum or discontinuum concept, but also with regard to the possibility that particles may first form during the dissolution process (Fig. 2.5) or during crystallization existing particles from the solution reassemble into a continuous substance, so to speak "merge". In this context, Pfundt called them *"non-preformed particles"*: They can emerge or disappear again. In the other case, there are forever existing *"preformed particles"*.

The surveys showed that adolescents in grades 7, 8 and 9 predominantly chose answers to the continuum hypothesis and considered "non-preformed particles"

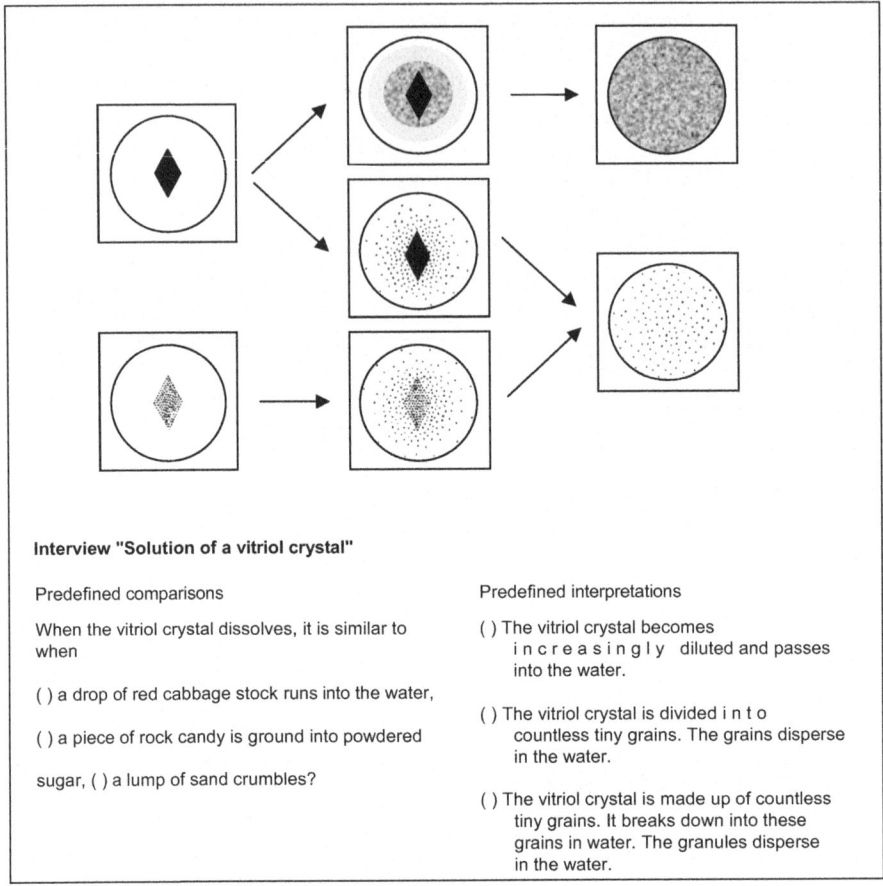

Fig. 2.5 Questionnaire on the particle concept of students for the dissolution process [14]

possible. Only a few students ticked off ideas about preformed particles and consistently argued with the particle concept in changes of state, dissolution and crystallization processes. For the successful teaching of the particle idea, it is therefore necessary to consistently work with this model in class and to deepen it in many different examples.

Marohn [15] also found numerous misconceptions about the dissolution process through surveys in the secondary level II of the grammar school. For example, students express the idea of dissolving ethanol in gasoline, that C_2H_5OH molecules disintegrate into atoms or "dissolve down to the atoms", or they think of molecule fragments like C_2H_4 and H_2O, i.e. fragments that "taken together again form the original molecules" [15].

2.2.8 Horror vacui

In model drawings for the structure of substances, Pfundt found that students prefer squares as models for particles over the usual circles. When asked why, the answers were that the models "must fit, they must be able to be seamlessly joined together" [14]. When circles are drawn together, cavities would probably form, which, from the students' point of view, should not exist: The "horror vacui" in the learners' conceptions accordingly leads to a preference for squares over circles.

Novick and Nussbaum [16] also conducted surveys on particle understanding and found that the majority of teenagers or students in the USA have the idea that there is air or other matter between the particles of gases. Subsequent investigations were carried out to what extent the "horror vacui" regarding the spaces between particles of a gas is present among German learners [17]. An experiment was conducted using the gas butane as an example (Sect. 2.6: V2.15), the model drawing was requested and questions were asked about the spaces between the butane particles (Fig. 2.6).

The result of the survey in grades 9, 10 and 11 showed that almost all participants reproduced the model drawing correctly, but in fact only about 50% of all subjects ticked the alternatives "nothing" or "empty". This means that the other half of the participants assume that the spaces between the butane particles are filled with butane, air or other matter: These teenagers are subject to the "horror vacui"!

The justifications for the answers showed this in particular: "The space between the particles cannot be empty or there is not nothing present", "I cannot imagine that there is nothing there", "if there was no air, there would have to be a vacuum, and I cannot imagine that", "something must be there, there is no place where there is absolutely nothing", "the space cannot simply contain nothing", "something must be there" [17].

In teaching the particle model, it is therefore necessary to point out and discuss the matter-free space between the particles in addition to introducing the smallest particles themselves or various arrangements of the particles: Teaching on this subject is proposed in detail elsewhere [17].

2.2.9 Spatial Ability

In order to understand the three-dimensional structure of matter from the smallest particles, adolescents should have sufficient spatial ability. Studies using tests of spatial ability have shown that from grade 8 onwards, this ability is present in most students [17]. Spatial ability is presented as a factor of intelligence, the test of spatial ability (Fig. 2.7) and test results are compared [17]. In particular, it was confirmed what has been observed in many international surveys on spatial ability, that girls generally show weaker results (Fig. 2.8).

> **3. Models for vaporizing butane (camping gas)**
>
> The casing of a hand pump is completely filled with butane gas.
> The gas is converted into a liquid under pressure using a mild hand pump.
> When the piston of the pump is released, liquid butane evaporates again.
> *Draw your idea of particles:* 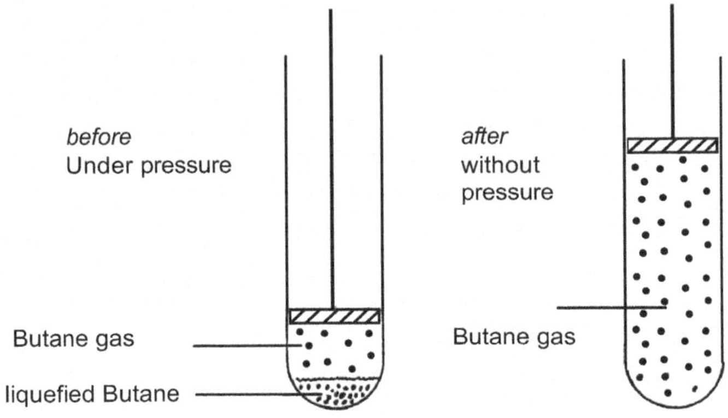
>
> before
> Under pressure
>
> Butane gas
>
> liquefied Butane
>
> after
> without
> pressure
>
> Butane gas
>
> How do you imagine the space *between the particles in the butane gas*? I imagine that
>
> ☐ there is also *butane gas* between the particles,
> ☐ the space between the particles is *empty*,
> ☐ there is *nothing* between the particles,
> ☐ there is *air* between the particles,
> ☐ is a special invisible substance between the particles.

Fig. 2.6 Excerpt from the questionnaire on "horror vacui" in the particle concept for gases (particle representation of a student already drawn in)

Cosima Schwöppe [18] was also able to confirm this result in studies of grades 5 and 6 of all types of schools. In her "Recommendations for School," she suggests: "Since the spatial ability of participants of the examined age group is in a developmental phase, the initial science instruction should contribute to the promotion of this ability. ... With regard to chemical teaching content, spatial abilities can be promoted through the targeted use of simple models for the structure of matter" [18]. The work with sphere packing and spatial molecule models proposed in the later Chap. 7 is intended to promote not only spatial abilities, but also the understanding of chemistry. In particular, girls would benefit in both aspects.

However, regarding gender differences, an empirical comparison of adolescents in Germany and Ethiopia showed that children at Ethiopian private schools initially also show the differences between boys and girls, but not the children at

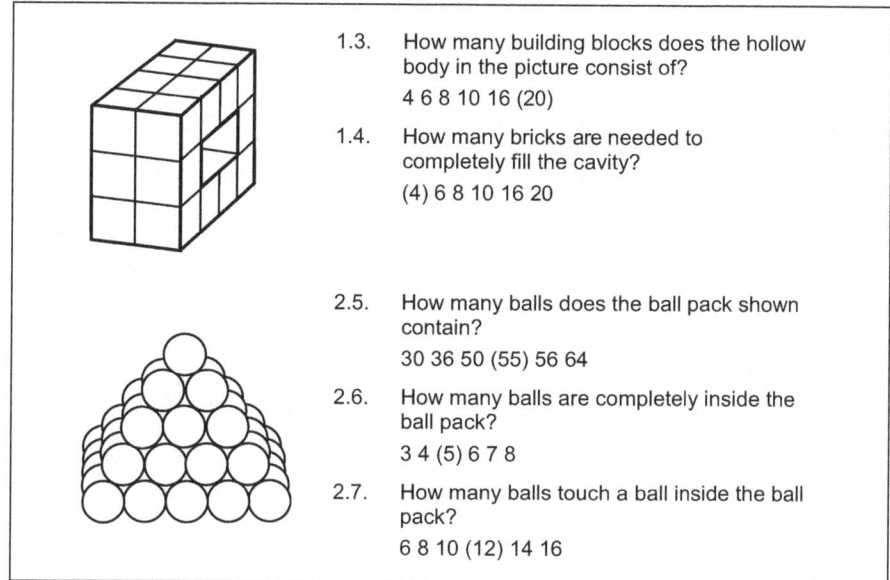

Fig. 2.7 Examples of tasks from the test of spatial ability, solutions in brackets [17]

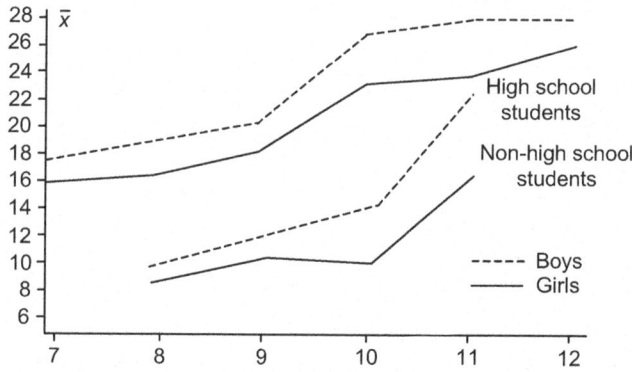

Fig. 2.8 Averages of a spatial ability test depending on gender and grade level [17]

the state Government Schools. This finding was interpreted to mean that children of wealthy parents have usual technical toys and boys develop their spatial ability with them. The children of poor parents do not have this toy, both boys and girls play with objects in nature, form spatial figures from mud or from woven branches of trees. In this way, they develop a comparable spatial ability and do not significantly differ in empirical surveys [17]. Further studies would need to provide certainty in this regard.

2.2.10 Homemade Misconceptions

Many other empirical surveys show [19] that in addition to the preconceptions brought from the real world, there are also student conceptions of subject areas that have been taught by teachers but do not originate from the real world. Since they have been generated in school, they are to be called "school-made" or "homemade" misconceptions. For example, the concept of ions is introduced in grade 9 or 10 and ions are discussed as the smallest particles of salts, acidic and alkaline solutions—yet surveys on the smallest particles in acidic or alkaline solutions do not yield the desired hydronium or hydroxide ions, but usually H-Cl molecules in hydrochloric acid or Na-O-H molecules in alkaline solutions, also Na-Cl molecules in solid sodium chloride and related solution [19].

Such misconceptions could be avoided with better teaching. Therefore, these false concepts should be known to chemistry teachers so that they can already build strategies in the preparation of lessons that reduce these misconceptions. For this purpose, these often occurring homemade misconceptions have been compiled for the following subject areas [19]:

- Structure-Property Relationships
- Chemical Equilibrium
- Acid-Base Reactions
- Redox Reactions
- Complex Reactions
- Energy

2.3 Teaching Processes: Consideration of Student Conceptions

All teaching must start with the children's experience (Dewey).

All new experiences that students make in class are organized with the help of already existing ideas (Ausubel).

Without explicitly dismantling false ideas, no sustainable new ideas are acquired (Piaget and Inhelder).

The teaching must not merely guide from ignorance to knowledge, it must also replace existing knowledge with different knowledge (Pfundt).

Chemistry teaching must build a supporting bridge from the students' original ideas to the currently valid ideas (Pfundt).

These statements impressively demonstrate that teachers must not compare their students with "blank slates" that are "just to be filled". A lesson that does not take into account existing ideas usually leads to students following the lesson only until the next test is written, then gradually forgetting the new ideas and returning to their old, long-acquired ideas.

Today, subject didacticians and teachers agree that one must know or determine students' ideas on a topic before "the bridge from the original ideas to the scientific ideas" can be successfully built. An important goal of the mediation process is therefore to show the students their own contradictions in classroom discussions or to make them aware of the contradictions of their world of ideas to scientific interpretations when interpreting new content in terms of subject science. This should motivate them to want to overcome these contradictions. Only when they have realized that they cannot get any further with their own explanations are they ready to follow the teacher's lesson and thus build new thought structures.

For the mediation process it is therefore important that in connection with the students' level of development, reflection is given to

- existing contradictions within the learners' own explanations,
- contradictions between preconceptions and scientific ideas,
- contradictions between preconceptions and accurate explanations of experimental phenomena,
- possibilities for dismantling original student ideas,
- possibilities for building sustainable and subject-appropriate interpretations.

It is also important to consider that according to the *constructivist theory* a concept change from original to scientific ideas is only possible if

- each individual builds their own learning structures individually,
- activity and self-activity of each learner is given,
- a "Conceptual Growth" (according to the assimilation by Piaget) or
- a "Conceptual Change" (according to the accommodation by Piaget) takes place [20].

The previous text has referred to students' original interpretations and contradictions on some topics and also offered suggestions on how a *Conceptual Change* or a *concept change* can take place. In particular, Fig. 2.3 has shown that such a change to the scientific explanation of combustion processes cannot be brought about solely by reaction equations like "$Mg + O \rightarrow MgO$"—weeks and months of problem-oriented teaching using structural models and model drawings are required (see also Chaps. 7 and 8).

In addition, many subject didacticians agree that original ideas or preconceptions must be addressed in class in order to successfully correct them, to successfully achieve a *Conceptual Change*. Various ways have been described on which this discussion with the students is possible.

Petermann, Friedreich and Oetken [11]
In chemistry class, they chose the combustion of coal as a topic and created a "student-oriented teaching method". It requires that the learners first express *their*

ideas about coal combustion and interpret the Boyle experiment (Sect. 2.6: V2.11) *from their perspective*. Only after this discussion is the scientific concept conveyed with further experiments and model ideas (Sect. 2.6: Model drawing at V2.11). The next step is a confrontation with expressed student ideas to ensure that learners absorb the new interpretation of combustion in a factually appropriate manner. Even if they initially add this new interpretation to their preconception acquired over many years and use both ideas in parallel, the experimental teaching that follows over months and years can consistently build up the scientific explanation convincingly and lead to the final *Conceptual Change*.

Dörfler [21]
Dörfler has developed another example of success in the "student-oriented teaching method". He found good results in a test after planning and conducting the teaching of a class 11 based on the known misconceptions about acids and bases and their neutralization. In a double lesson after the test, he formed student groups and gave each group a poster with a known misconception. For example, one group received the statement "Hydrochloric acid consists of HCl molecules" or another group the statement " Salt is the product of every neutralization". After group work, he discussed misconceptions in comparison with the scientific interpretation in class and thus recorded for all students which mistakes they should not make. A test after this double lesson showed a further increase in performance because the misconceptions had been addressed. In other words, the opportunity was created to dismantle original ideas about neutralization and to build up scientific ideas instead.

Marohn [22]
In her concept *Choice2learn*, Marohn is guided by the approach of the "cognitive conflict" and the related insight that a sustainable change in the cognitive structure can only be successfully realized if learners become aware of their own ideas and deal with them. Marohn explains this concept in detail (Sect. 4.4).

Temechegn and Sileshi [23]
Both Ethiopian colleagues discovered the teaching tool of *Concept Cartoons* by Naylor and Keogh [24] and suggested them for various topics of chemistry teaching to let the learners discuss misconceptions disguised in speech bubbles in a motivating way (Fig. 2.9). Essentially, there are *two strategies* on how to counter both preconceptions and homemade misconceptions using Concept Cartoons.

Preconceptions of learners are determined, discussed, and compared using the corresponding Concept Cartoon at the beginning of a teaching unit; based on the knowledge of such preconceptions, reflective teaching is planned and carried out towards the correct idea. Afterwards, the Concept Cartoon is to be offered again: Now the learners recognize the scientific idea and defend it against the erroneous ideas listed in the cartoon.

2.3 Teaching Processes: Consideration of Student Conceptions

Fig. 2.9 Concept Cartoon on the mass of a rusting iron nail [23]

Homemade misconceptions are noted by the teacher using corresponding Concept Cartoons; on this basis, reflective teaching is planned and carried out towards the scientific idea. At the end, the Concept Cartoon is presented to the students; in the discussion, the scientific idea is recognized. The presented misconceptions of the cartoon are then to be compared with the newly acquired scientific idea, corresponding errors are each to be recognized and corrected by the new idea.

In both ways, the misconceptions are included in the teaching and addressed for students. In this way, they can clearly recognize at the end of a teaching unit which scientifically valid idea exists and to what extent the alternative ideas are erroneous.

Examples of Concept Cartoons [23]
A selection of questions regarding observed misconceptions is attached, which can be worked on either as multiple-choice questions or as Concept Cartoons with the provided faces (Fig. 2.9), correct answers are marked with an arrow:

1. *Evaporation of water*: Imagine a small amount of pure water is completely evaporated in a closed and evacuated glass vessel by heating. What does the glass vessel contain then?
    a. Hydrogen and oxygen
    b. Water vapor ←
    c. Water vapor, hydrogen, and oxygen
    d. Air
2. *Masses when dissolving in water*: 1 kg of table salt is added to 20 kg of water and completely dissolved by stirring. How heavy is the salt solution?
    a. 19 kg
    b. 20 kg
    c. 21 kg ←
    d. More than 21 kg
3. *Combustion*: Charcoal burns on the grill. Is it irretrievably destroyed?
    a. Yes, of course—the charcoal has burned out.
    b. No, when charcoal burns, the gas carbon dioxide is produced. ←
    c. No, there is still the ash.
    d. Yes, of course: burned is burned.
4. Removal of grease stains. The grease stain from the pants is removed with a cloth and gasoline, the stain is no longer visible. Is the grease irretrievably gone?
    a. No, it's in the cloth. ←
    b. Yes, the stain was removed.
    c. Yes, the stain is gone.
    d. No, the grease has chemically reacted, a new substance has been created.
5. *Smallest water particles or H_2O molecules*. After the rain, a puddle of water is visible, later it has disappeared again. Explain!
    a. New particles are formed from water particles: water vapor particles.
    b. Water particles have mixed with air particles. ←
    c. H_2O molecules are split into H atoms and O atoms.

d. H$_2$O molecules react with the molecules of the air.
6. *Smallest particles in ethanol vapor.* Ethanol (C$_2$H$_5$OH) is heated until gas bubbles form in the liquid. What are they made of?
   a. from C atoms, H atoms, and O atoms
   b. from H$_2$ molecules, O$_2$ molecules, and CO$_2$ molecules
   c. from C$_2$H$_5$OH molecules ←
   d. from C$_2$H$_4$ molecules and H$_2$O molecules
7. *Structure of solid sodium chloride (NaCl).* Sodium chloride crystals are structured in a certain way. Which model assumption is correct?
   a. Na-Cl molecules, covalent bonding
   b. Na → e$^-$ → Cl, transfer of an electron
   c. Na$^+$Cl$^-$ ion pairs
   d. Na$^+$Cl$^-$ ion lattice ←
8. *Composition of mineral water.* Mineral water contains dissolved carbon dioxide and various minerals such as sodium chloride and calcium sulfate. Which model assumption is correct?
   a. CO$_2$(aq), NaCl(aq), CaSO$_4$(aq)
   b. Na-Cl, O=C=O, Ca=SO$_4$
   c. Na$^+$(aq), Cl$^-$(aq), CO$_2$(aq), Ca^{2+}(aq), SO$_4^{2-}$(aq) ←
   d. Na$^+$Cl$^-$(aq), CO$_2$, Ca^{2+}SO$_4^{2-}$(aq)
9. *Chemical equilibrium.* Amounts of substance and concentrations in equilibrium—which statement is correct?
   a. The amounts of all involved substances are equal.
   b. The concentrations of all types of particles are equal.
   c. The concentrations of reactants and products are equal.
   d. The speeds of the forward and reverse reactions are equal. ←
10. *Particles in acidic and alkaline solutions.* What particles does diluted hydrochloric acid contain?
    a. H$^+$(aq) ions, Cl$^-$(aq) ions and H$_2$O molecules ←
    b. HCl molecules and H$_2$O molecules
    c. HCl(aq) particles
    d. H$^+$Cl$^-$ particles and H$_2$O molecules
11. *Neutralization.* Hydrochloric acid and sodium hydroxide are mixed together, creating a neutral solution. Which model representation applies to the neutral solution?
    a. HCl and NaOH are present in equal amounts
    b. H$^+$(aq), Cl$^-$(aq), Na$^+$(aq), OH$^-$(aq)
    c. Na$^+$(aq), Cl$^-$(aq), H$_2$O ←
    d. Na$^+$Cl$^-$, H$_2$O
12. *Precipitation of metals from their solutions.* An iron nail is dipped in copper sulfate solution: a reddish-brown coating forms. Explain!
    a. Copper sulfate is reduced; copper deposits on the nail.
    b. Cu^{2+} ions are reduced to Cu atoms; copper forms. ←
    c. Iron takes oxygen from the sulfate; rust deposits.
    d. Copper leaves the copper sulfate and deposits on the nail.

13. *Metal-acid reactions.* A piece of magnesium ribbon is added to hydrochloric acid: gas development occurs. Which statement is correct?
    a. An acid-base reaction occurs.
    b. Cl^-(aq) ions of hydrochloric acid are oxidized to chlorine.
    c. Chlorine gas is produced in a redox reaction.
    d. A redox reaction occurs; hydrogen is produced. ←
14. *Energy in coal combustion.* Burning charcoal provides large amounts of heat—which statement is correct?
    a. Coal transforms into energy; ash remains.
    b. Coal reacts with the oxygen in the air; energy is released. ←
    c. When coal burns, the mass decreases, but the energy increases.
    d. C atoms of coal react directly to energy.
15. *Gasoline explosions in car engines.* In the car's carburetor, gasoline is vaporized and exploded in the cylinder. Which statement is correct?
    a. Gasoline molecules transform directly into energy.
    b. Gasoline molecules explode and release energy.
    c. Gasoline molecules release H atoms; oxyhydrogen explodes.
    d. Gasoline vapor reacts with the oxygen in the air; energy is released. ←

All the answers given appeared in empirical surveys. Cartoons 1–6 correspond more to the *preconceptions* of young students in initial instruction, while cartoons 7–15 represent *homemade misconceptions* of learners from secondary level II, which were given in surveys. If in some cases there is no clear answer, weighing the different alternatives should also lead to the correct model representation. Once learners finally recognize which ideas are definitely incorrect, they can purposefully build up the scientific interpretation for themselves and realize the desired *Conceptual Change* in relation to the preconceptions. Regarding the homemade misconceptions, the teacher can plan the chemistry lessons in such a way that possible misconceptions are reflected upon—but in the end, the scientific representation results.

2.4 Social Reference Fields: Student Conceptions and Colloquial Language

One must be clear that newly acquired concepts are not sustainable for all time and can be significantly impaired soon after the lesson: Everyday life conceptions, which have been acquired over many years, are more deeply rooted than novel explanatory concepts that are taken up after a few weeks of lessons or even just a few hours of lessons.

So, on the one hand, it is important to repeatedly query scientific conceptions in specific teaching situations and to deepen them with new contexts in order to achieve their firm *rooting* among the learners. On the other hand, we must be aware that conversations with friends and relatives about scientific topics can unsettle students with their not yet deeply rooted newly acquired conceptions.

The *colloquial and everyday language* remains opposed to these new conceptions in many aspects, students still have to deal with statements in everyday life such as "the puddle is gone", "electricity is consumed" or "the battery is empty". One would have to achieve that students start reflecting on colloquial expressions and offer this reflection in conversation to relatives and friends—then these students would acquire a competence that also greatly promotes the universally desired ability to criticize. Such a competence could then have a positive influence on society by appropriately describing and passing on scientific facts.

Finally, the *influence of the media* on the newly acquired conceptions of students should be pointed out. On the one hand, there are commercials on radio and television that convey very diffuse conceptions of scientific phenomena: "No chemistry comes to my skin" it says—however, a strongly exothermic reaction of pyrophoric iron and the incoming oxygen takes place after opening the heat pad recommended in the advertisement. On the other hand, factual consumer advice can usually bring out positive aspects and support newly acquired conceptions in the desired direction. In this regard, Becker offered "consumer questions in the RIAS telephone studio" and proposed them as a subject of didactic reflection [25].

2.5 Exercise Tasks

A2.1
The inappropriate conceptions in some subject areas among learners were often also conceptions of scientists of past centuries. Specify three subject areas and show parallels between historical conceptions and today's preconceptions. Explain each fact from the currently valid scientific point of view.

A2.2
Chemical compounds are often described as "containing" certain elements or as "consisting" of certain elements, such as: "Water consists of hydrogen and oxygen". What technical and didactic problems are hidden behind such statements? Suggest formulations that are more appropriate.

A2.3
In didactic discussions on the structure of matter, terms such as continuum, discontinuum, preformed and non-preformed particles or the "horror vacui" are argued. Explain this discussion. What kind of teaching, experiments and models do you suggest for a better understanding of these facts?

A2.4
The experiment is a very convincing instrument to make misconceptions conscious and to motivate students to dismantle them in favor of accurate conceptions. Describe a possible experimental approach in three contexts and sketch corresponding teaching paths to dismantle inappropriate conceptions.

A2.5
To address preconceptions and misconceptions in chemistry teaching, there are teaching methods oriented towards student conceptions, related concept cartoons or the Choice2learn program. Choose a topic and one of the three ways and explain how you would implement the inclusion of student conceptions in chemistry teaching.

2.6 Experiments

V2.1 Reaction of Copper in Vacuum and in Air

Problem
Like scientists of past centuries, learners are subject to the "horror vacui" and often express the idea that the surrounding air fills any empty space, as is the case for an empty bottle or glass. To make it clear that there are also air-free or gas-free spaces, a test tube with a valve containing a blank piece of copper sheet is evacuated. If it is heated strongly, the black copper oxide does not form as usual, but the metal sheet remains unchanged. Only when the valve is opened and air flows onto the still hot sheet, the black substance can be observed. You can also hear the familiar whistling sound of the incoming air.

Material
Large sealable test tube with stopper, side tube and tap, water jet pump, burner; Copper sheet

Procedure
A strip of copper sheet is placed in the test tube, the tube is sealed. It is evacuated with the tap open using the water jet pump, the tap is closed. To test the vacuum, the tap can be briefly opened (whistling sound!), then the air can be pumped out again. Now the test tube is strongly heated at the point where the metal is located. After a short time, the burner is removed and the tap is opened.

Observation
Upon first opening of the tap, a whistling sound of the incoming air can be heard. When heating the copper sheet, it remains unchanged, upon subsequent opening of the tap, both the familiar whistling sound can be heard and the sudden appearance of black copper oxide can be seen.

V2.2 Reaction of Copper Oxide with Hydrogen

Problem
This experiment is intended to exemplify that the shiny reddish-brown copper can be recovered from black copper oxide using hydrogen. It is possible to discuss "property carriers" versus "new substances" using this example: There is no "black copper metal", but the black substance is a different material called copper oxide. Furthermore, the simultaneous oxidation and reduction can be introduced vividly and the redox reaction in terms of oxygen transfer: It just needs to be clear that free oxygen does not occur in any case.

Materials
Combustion tube with stopper and outlet tube, porcelain boat, test tube; black copper oxide (wire form), hydrogen

Procedure
Hydrogen is passed through the combustion tube, which contains a porcelain boat filled with copper oxide. After a negative result of the bang test, the hydrogen is ignited at the outlet tube and the copper oxide is strongly heated. As soon as the reaction starts, recognizable by the formation of metallic copper, the burner can be removed. After the exothermic reaction has taken place, the reaction tube is allowed to cool in the hydrogen stream and only then is the hydrogen supply interrupted.

Observation
Shiny reddish-brown copper is formed under glowing, and water droplets can be clearly seen in the outlet tube.

Disposal
The copper can be used for further experiments or it can be oxidized in the air stream to reuse the oxide for the same experiment.

V2.3 Mass Change During the Evaporation of Acetone

Problem
Students often interpret evaporation as "disappearance" of the substance. To show that the substance transitions from the liquid state to the gaseous state, the evaporation should first be observed with a scale, and then the formation of plplacetone vapor should be followed with a piston sampler.

Materials
Digital scale with display, watch glass, Erlenmeyer flask with glass beads and perforated stopper, piston sampler; acetone or ether

Procedure
The watch glass is provided with a few drops of acetone or ether, placed on the scale pan and the scale display is observed. A few drops of the volatile substance are added to the Erlenmeyer flask, the piston sampler is connected, the Erlenmeyer flask with the glass beads is strongly shaken.

Observation
The scale shows decreasing masses until the liquid has completely evaporated and the vapor has mixed with the air. The piston sampler gradually fills up until the volume of gaseous acetone or ether finally remains constant.

V2.4 Reaction of Alkali Metals with Water

Problem
The popular reaction of sodium with water leads students to state that the sodium has "disappeared". On the one hand, gas formation and streaking or the color reaction with phenolphthalein can be shown during the reaction, on the other hand, the newly formed reaction product sodium hydroxide can be obtained from the solution after the water has evaporated.

Materials
Large glass tub, stand cylinder with cover glass, test tubes, beaker, tweezers, knife, filter paper; Lithium, Sodium, Phenolphthalein and universal indicator solution

Procedure
The glass tub is filled halfway with water and placed on the overhead projector. A small piece of sodium is brought to the water surface, the path of the sodium ball is closely observed. The experiment is repeated with lithium.

The stand cylinder is filled with water, the cover glass is removed underwater and a piece of lithium is placed in the cylinder with tweezers (this experiment must not be carried out with sodium). After the reaction of the lithium piece, the stand cylinder is closed with the cover glass, removed from the tub and placed upright. The gas produced is ignited.

A sample of the solution from the tub is tested in the test tube with the indicator solutions, a small part of the solution is evaporated in the beaker.

Observation
During the reaction, streaks can be clearly seen in the projection, the hissing sound can be heard and the gas formation can be observed. The gas formed burns in the air, briefly showing a red-colored flame. The indicators are colored red in one case or blue by the other solution. A solid white substance remains in the beaker.

Disposal
The solution in the glass tub is so diluted that it can be poured into the wastewater. Residues of sodium or lithium are to be reacted with ethanol, the ethanol solutions are to be diluted and disposed of in the drain.

V2.5 Comparison of gasoline and a solution of fat in gasoline

Problem
When "removing" grease stains from clothing, the grease "disappears" in the students' imagination: "It's gone." To trace back the stain removal to the dissolution process of fat in gasoline, it should be made clear that the fat is no longer in the clothing, but in the wiping cloth. Solvents can evaporate there without leaving any residue, but when a fat solution evaporates, the fat remains.

Materials
Light gasoline, solution of olive oil in light gasoline, filter paper

Procedure
A few drops of gasoline are placed on a filter paper, a few drops of fat solution are placed on a second filter paper at the same time. Both papers are observed.

Observation
The spot of the pure solvent becomes smaller and smaller and is finally no longer visible: The solvent has turned into a gaseous state. The spot of the fat solution also becomes smaller, but a fat stain clearly remains on the paper.

V2.6 Energy transfer when heating water

Problem
To illustrate chemical energy and in particular to make clear the unit Joule (J) for energy and to distinguish it from temperature measurements in the unit °C, a specific amount of a fuel should be used to heat a certain portion of water to a desired temperature. If you determine this mass of the converted fuel by, for example, weighing a portable butane burner before and after heating, you can account for the transferred energy amount of the butane portion and compare it with the energy absorbed by the water. For quantitative evaluation, you only need to know that

a. to increase the temperature of 1 g of water by 1 °C, 4.2 J of energy is required and
b. when 1 g of butane is burned, approximately 50 kJ of heat energy is released:

$$\text{Butan (g)} + \text{Sauerstoff (g)} \rightarrow \text{Kohlenstoffdioxid (g)} + \text{Wasser (g)};$$
$$\Delta H = -50 \, \text{kJ/g}$$

Materials
Beaker, tripod with wire mesh, thermometer (0–100 °C), portable butane burner, digital scale; Water

Procedure
A beaker with 100 g of water is provided on the tripod with an immersed thermometer, the butane burner is weighed accurately. The temperature of the water is measured precisely. The burner is ignited and placed under the beaker with a roaring flame for about a minute, the burner is immediately turned off and weighed again. The water is stirred with the thermometer until the temperature remains constant, the temperature difference is determined.

Observation
If the temperature difference $\Delta t = 20$ °C, then it is calculated: 100 g of water take from the burner the energy $E = 4.2 \, \text{J} / (\text{g} \, °\text{C}) \cdot 100 \, \text{g} \cdot 20 \, °\text{C} = 8400 \, \text{J} = 8.4 \, \text{kJ}$.

If the mass difference of the butane burner is exactly 1 g, then it is known that the energy amount $E = 50$ kJ is released. Since the water has absorbed the energy $E = 8.4$ kJ, the difference of $E = 50 \, \text{kJ} - 8.4 \, \text{kJ} = 41.6 \, \text{kJ}$ has been absorbed by the surroundings: from the beaker, the thermometer, tripod and wire mesh, from the surrounding air. The efficiency, based on the heating of water, is therefore only about 20 %.

V2.7 Combustion of metals on the scale

Problem
Based on his everyday experiences, the student believes in a "loss of mass" or the "lightening" of the burning substances when burning spirits, paper or candles. To demonstrate initially using the example of metal combustion that a mass increase occurs due to the bound oxygen content in the solid metal oxides, corresponding experiments are carried out on the scale.

Material
Beam balance, digital scale, porcelain crucible with lid; steel wool, magnesium ribbon

Procedure
a. The steel wool hanging on one side of a balanced beam balance is ignited; if necessary, one blows lightly against the steel wool to accelerate the reaction and to better see the glowing.
b. In the porcelain crucible, a roll of about 10 cm magnesium ribbon is weighed accurately. The crucible is strongly heated with the roaring burner flame with the lid initially closed and the magnesium is ignited, during the reaction the crucible is alternately briefly uncovered and covered again. The cooled crucible is weighed again.

Observation
a. The beam with the red-hot glowing steel wool lowers, a black reaction product is formed.
b. A white combustion product is formed from the magnesium under bright glowing. The scale shows a larger mass afterwards than before the combustion. When the product is broken down, a green substance becomes visible: magnesium nitride. It clearly develops the gas ammonia when moistened with a little water.

V2.8 Burning candle on the scale

Problem
The students understand the mass increase during the combustion of metals and the formation of solid metal oxides, but will object that this cannot apply to spirits, paper or candles. To convince the students also for these cases, a candle should burn on the beam of the scale, only carbon dioxide and water vapor as invisible gaseous combustion products must be bound: by soda lime, a mixture of sodium

hydroxide and calcium oxide. In the device for this (see picture), the combustion gases are absorbed: Their mass is larger due to the bound oxygen content than that of the candle material before.

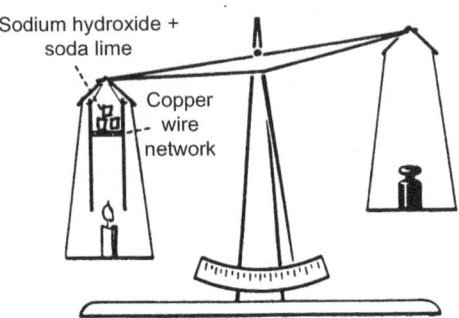

As a preliminary experiment, it can be shown that tealights suffocate on their own combustion products, that they go out under a beaker after a certain time due to the formed carbon dioxide:

Paraffin (s) + Sauerstoff (g) → Kohlenstoffdioxid (g) + Wasser (g)

Materials
Tea lights, three differently sized beakers, beam balance or digital scale, glass cylinder with copper mesh and soda lime (sodium hydroxide can also be used)

Procedure
a. Three burning tea lights are placed under three differently sized beakers.
b. A tea light is placed on the scale and lit.
c. A tea light is positioned under the glass cylinder, which contains absorbing chemicals (see picture).

The soda lime is loosely filled about 2 cm high so that the resulting gases can flow through. The scale is balanced and the candle is lit. If smoke develops in the glass cylinder (too dense packing of the absorbent!), the experiment must be repeated.

Observation
a. All three tea lights gradually go out, the one under the largest beaker goes out last.
b. The mass of the tea light slowly decreases by a few milligrams per minute.
c. The scale pan with the burning candle tilts downwards, a digital scale shows a mass increase of up to 200 mg.

2.6 Experiments

V2.9 Reaction of Carbon Dioxide with Magnesium

Problem

Students initially accept that the corresponding metals can be recovered from metal oxides (see V2.2). However, they usually cannot imagine that carbon can be released from the colorless gas carbon dioxide as a black solid. To convince them, the reaction of carbon dioxide with burning magnesium is carried out in a standing cylinder.

Materials

Standing cylinder with cover glass, crucible tongs; magnesium ribbon, carbon dioxide, sand

Procedure

A small amount of sand is added to the standing cylinder to protect the bottom of the cylinder. It is filled with carbon dioxide by displacing the air and covered. A magnesium ribbon about 10 cm long is to be ignited and immersed deep into the cylinder using crucible tongs.

Observation

The flame does not go out, but continues to burn with a crackling sound. White magnesium oxide is formed. After the reaction, black spots can be observed on the inner wall of the cylinder, which prove to be soot when wiped off with a finger.

V2.10 Combustion Reactions—The Mass Remains Conserved

Problem

In order to prevent the "heavier metals" from becoming entrenched in the minds of the students, the law of conservation of mass must be derived experimentally immediately afterwards. For this, in contrast to the open arrangement as in V2.7 and V2.8, a closed apparatus must be chosen. The metal reaction in a closed apparatus can be repeated and the constancy of the mass before and after can be demonstrated by weighing. On the other hand, matches can be enclosed in the test tube to also experimentally show the conservation law on wood, a fuel known from everyday life.

Materials
Test tubes, balloon, digital scale; steel wool, matches

Procedure
a. A large test tube is half-filled with steel wool, sealed with a balloon, and weighed accurately. The test tube is heated with the roaring flame of a burner where the steel wool is located, and after cooling, it is weighed again.
b. The experiment is repeated with 5-8 matches: After weighing the test tube with matches and balloon, the match heads are ignited by strongly heating the test tube, and after cooling, it is weighed again.

Observation
The same mass is observed before and after the reaction: In the first case, a black reaction product has formed; in the second case, charred remains of the wood can be seen after a flash flame. The balloon briefly inflated.

V2.11 Coal combustion—the mass is preserved here too [11]

Problem
Since grilling with charcoal is particularly well known from everyday life and laypeople often believe that the "coal is irretrievably gone afterwards", this experiment should be shown additionally to eliminate doubts about the law of conservation of mass. Such experiments are also called Boyle experiments, because the English chemist Boyle first conducted this experiment in history.

Material
1-L-round flask, stopper (with glass tube and balloon), burner, 100-ml syringe ; activated carbon, oxygen, test tube with lime water

Procedure
The round flask is flushed with oxygen and provided with 5–6 grains of activated carbon, sealed with a stopper and weighed. The part of the round flask where the grains are located is strongly heated until the grains ignite. The stopper is to be pressed firmly with the thumb or the balloon is to be used to balance the pressure. The burner is removed and the flask is swirled until all the charcoal grains have burned out. After cooling, the flask is weighed again. A gas sample is to be taken with the syringe and connected glass tube and passed through few mL of lime water.

2.6 Experiments

Observation

The carbon grains light up and burn out in pure oxygen, without leaving any ash residue. The mass afterwards is the same as before the reaction. The extracted gas sample precipitates a white precipitate from the lime water: carbon dioxide in a higher concentration than usual air.

▶ **Note** As soon as the atomic model according to Dalton has been introduced, the rearrangement of the C-atoms and O_2 molecules to CO_2 molecules can be interpreted and illustrated by a model drawing (see picture). The law of conservation of mass in chemical reactions is thus very vivid and easy for learners to understand.

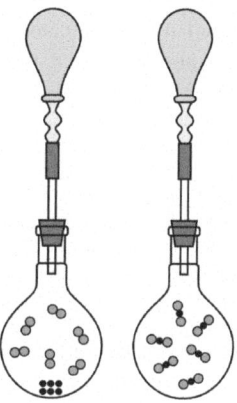

V2.12 Density of air and carbon dioxide

Problem

Students are certainly aware of the existence of the earth's atmosphere, but to a much lesser extent they identify air as a space-filling substance or as a mixture of substances with a characteristic and measurable density. This density is to be determined and compared with that of another gas.

Material

Digital balance, 250-mL round flask with tap (see picture), 100-mL syringe, water jet pump, hose; carbon dioxide

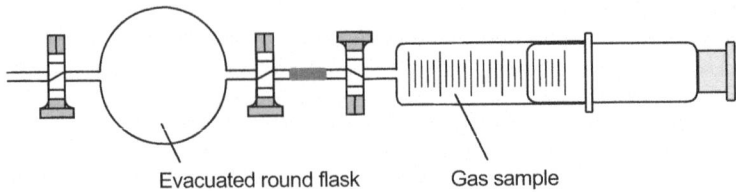

Evacuated round flask   Gas sample

Procedure
The syringe is filled with 100 ml of air and sealed (see picture). The round flask is evacuated with the pump, the tap is closed, the ball is weighed accurately. The syringe is connected and the portion of air is transferred into the glass ball by opening the taps. It is weighed again. The density of the air is to be calculated from the difference in mass and the given volume. The experiment is to be repeated with carbon dioxide.

Observation
100 ml of air weigh 0.13 g, 100 ml of carbon dioxide 0.2 g. The densities are calculated for air to be 1.3 g/L (table value 1.29 g/L), for carbon dioxide to be 2.0 g/L (table value 1.97 g/L).

▶ **Note** The experiment can also be carried out using an empty plastic bottle (distilled water bottle) with a stopper and tap: The bottle is weighed accurately, the portion of 100 ml of gas is quickly pumped to the existing air with the syringe, the tap is closed and the bottle is weighed again.

V2.13 Properties of Hydrogen and Other Colorless Gases

Problem
Students often uncritically identify colorless gases with air. For this reason, some colorless gases and corresponding detection reactions should be introduced, which clearly highlight the differences in the properties of different gases. Since the properties of hydrogen are particularly new to students, these should be demonstrated in detail.

Materials
5 standing cylinders with cover glass, wooden stick, balloon, combustion spoon, glass tube, beaker, empty tin can with concentric hole of about 1 mm diameter; hydrogen, oxygen, nitrogen, carbon dioxide, methane, lime water

2.6 Experiments

Procedure

The mentioned gases are filled into the cylinders by air displacement, covered and labeled. A burning wooden stick is first dipped into all cylinders, then a smoldering wooden stick. To distinguish between nitrogen and carbon dioxide, both cylinders are mixed with a little lime water and shaken.

Observation

Hydrogen ignites with a gentle bang and burns with a colorless flame. In oxygen, the stick burns very brightly and a smoldering stick even ignites (glowing stick test). In nitrogen and carbon dioxide, both flame and glowind stick go out, in carbon dioxide a white substance precipitates milky from the colorless lime water (lime water test), not in nitrogen. Methane is ignited and burns quietly with a yellow flame.

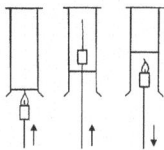

Conducting Further Hydrogen Experiments

a. A balloon is filled with hydrogen, it is tied at the mouthpiece and released.
b. A candle, which is attached to the combustion spoon, is brought burning to the balloon until the reaction starts (be careful, bang).
c. Hydrogen flowing from the steel bottle is ignited at a glass tube, a small flame is set and a dry beaker is held over it.
d. Hydrogen is filled into an upside-down hanging cylinder by air displacement, a burning candle is introduced, which is attached to a combustion spoon (see picture). The candle is slowly pulled out and reintroduced, the procedure is repeated several times.
e. A cylinder is filled with hydrogen (opening downwards!), and placed on an equally large cylinder filled with air, both gases are mixed by turning. They are separated from each other with cover glasses and tested with the burning wooden stick (bang!).
f. An empty tin can is provided centrally on the bottom plate with a hole of diameter $d = 1$ mm. It is set up with the large opening facing down and filled with hydrogen from below by air displacement. The gas flowing out of the small hole is to be ignited: beware, loud bang after about 20 s.

Observation

a. The balloon rises to the ceiling of the room.
b. The gas in the balloon burns quickly with the balloon bursting loudly.
c. The pure hydrogen burns very quietly, the beaker fogs up due to water droplets.
d. The candle goes out in the cylinder, but each time it reignites on the burning hydrogen when it is pulled out.
e. The mixture of hydrogen and air burns very quickly with a gentle bang (bang gas!).

f. The hydrogen initially burns completely quietly (you can hold a paper over the hole for control: it ignites). After about 20 s a soft humming can be heard and shortly afterwards a very violent bang (spectators must definitely be warned!).

▶ **Note** The steel bottles needed for these experiments must each be secured against falling by the transport trolley or by a chain attached to the table. If they fall onto the valve and the valve is knocked off, the steel bottle will go through walls like a rocket.

V2.14 Composition of Air

Problem
Students use the terms "good air" and "used air" from everyday language, but do not imagine the oxygen content of the air. Therefore, it is important for this reason alone to conduct experiments on the composition of air. The question of why either a metal or phosphorus is used should be explained by the fact that in these cases a solid is formed that binds the oxygen in the air and thus removes it from the air volume. With a 20 vol.-% oxygen content, it is expected that 20 mL of oxygen from 100 mL of air will be bound to the solid and about 80 mL of nitrogen will remain.

Materials
Two 100-mL syringes, combustion tube with suitable stoppers (see picture), glass tubes, small stand cylinder with cover glass, wooden stick, glass bell, water tub, combustion spoon with stopper, ruler; steel wool, phosphorus (red).

Procedure
A combustion apparatus is set up (see picture). The enclosed air portion of 100 mL is pushed several times over the heated steel wool and the volume of the cooled residual gas is determined. The residual gas is pneumatically collected in the small cylinder and tested with a burning wooden stick.

2.6 Experiments

A glass bell with a tube is placed in sealing water of the glass tub (see picture). A small portion of phosphorus is picked up with a combustion spoon and ignited, the burning phosphorus is introduced into the glass bell and it is closed with the stopper of the combustion spoon. The rise of the liquid level in the glass bell is observed, the remaining proportion of residual gas is estimated (ruler).

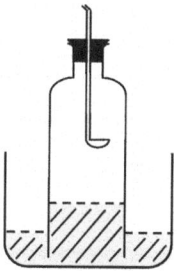

Observation
a. The steel wool glows and turns into a black product, the gas volume reduces to 80 mL, this residual gas extinguishes a burning wooden stick.
b. The phosphorus burns for some time, white smoke forms, after the flame goes out, the water level of the glass bell rises, the volume of the residual gas is also about 80 vol.-%.

V2.15 Condensation of Butane Gas Under Pressure

Problem
Students know butane lighters and the term "liquefied gas". Perhaps they have once observed the liquid butane phase and the gaseous butane phase above it in a transparent lighter: Despite such observations, the term "liquefied gas" usually remains in the imagination and should be reflected by the following experiment. The experiment also shows the properties of this gas: It can be condensed by pressure and thus become a visible liquid with a specific boiling point.

Materials
Gas liquefaction pump (see picture), hose; Butane (camping gas cartridge)

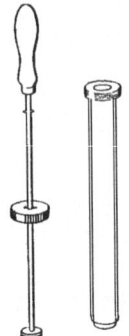

Procedure

The pump is opened and filled with butane from the cartridge by displacing air (use a deep immersion hose). The piston is put on, pressed into the gas pump with strong pressure and locked. The lock is released and the piston is observed. This process can be repeated as often as desired.

Observation

When the gas is compressed, a large drop of liquid forms, the gas volume is only about a tenth. If the lock is released, the piston moves out of the pump on its own, the liquid drop completely evaporates with cooling, the same gas volume as before can be determined.

References

1. Reuber R et al (1972) Chemikon—Chemie in Übersichten. Umschau, Frankfurt
2. Lockemann G (1950) Geschichte der Chemie. De Gruyter, Berlin
3. Strube W (1976) Der historische Weg der Chemie. VEB Grundstoffindustrie, Leipzig
4. Bugge G (1955) Das Buch der Grossen Chemiker Bd. 1. Verlag Chemie, Weinheim
5. Dijksterhuis FJ (1956) Die Mechanisierung des Weltbildes. Springer, Berlin
6. v Guericke O (1894) Neue "Magdeburgische" Versuche über den leeren Raum. In: Ostwald's Klassiker der exakten Naturwissenschaften. Engelmann, Leipzig
7. Lasswitz K (1890) Geschichte der Atomistik. Bände 1 und 2. Voss, Hamburg
8. Barke H-D (1995) Strukturorientierter Chemieunterricht und Teilchenverknüpfungsregeln. ChemSch 42:49
9. Driver R (1985) Children's ideas in science. Open University Press, Philadelphia
10. Pfundt H (1975) Ursprüngliche Vorstellungen der Schüler für chemische Vorgänge. MNU 28:157
11. Petermann K, Friedrich J, Oetken M (2008) Das an Schülervorstellungen orientierte Unterrichtsverfahren. Inhaltliche Auseinandersetzung mit Schülervorstellungen im naturwissenschaftlichen Unterricht. CHEMKON 15:110
12. Münch R et al (1982) Luft und Gewicht. NiU-P/C 30:429
13. Weerda J (1981) Zur Entwicklung des Gasbegriffs beim Kinde. NiU-P/C 29:90
14. Pfundt H (1981) Das Atom—Letztes Teilungsstück oder Erster Aufbaustein. Chimdid 7:75

References

15. Marohn A (2008) Schülervorstellungen zum Lösen und Sieden—auf der Suche nach "elementaren" Vorstellungen. MNU 61:451
16. Novick S, Nussbaum J (1981) Pupils' understanding of the particulate nature of matter. ScEd 65:187
17. Barke H-D (2011) Der "Horror vacui" in den Vorstellungen vom Teilchenkonzept. Raumvorstellung zur Struktur von Teilchenverbänden. In: Barke H-D, Harsch G (Hrsg) Chemiedidaktik kompakt. Lernprozesse in Theorie und Praxis, 1. Aufl. Springer, Heidelberg
18. Schwöppe C (2007) Das Raumvorstellungsvermögen von Münsteraner Schülerinnen und Schülern der Jahrgangsstufe 5/6 im Hinblick auf chemische Unterrichtsinhalte. Dissertation. Schüling, Münster
19. Barke H-D (2006) Chemiedidaktik—Diagnose und Korrektur von Schülervorstellungen. Springer, Berlin, Heidelberg
20. Duit R (1996) Lernen als Konzeptwechsel im naturwissenschaftlichen Unterricht. In: Lernen in den Naturwissenschaften. IPN, Kiel
21. Dörfler T, Barke H-D (2009) Das an Schülervorstellungen orientierte Unterrichtsverfahren. Das Beispiel Neutralisation. Chemkon 15:141
22. Marohn A, Egbers M (2011) Vorstellungen verändern—Lernmaterialien zum Thema "Verdampfen" im Rahmen der Unterrichtskonzeption "choice2learn". Pdn—Chem 60(3):5
23. Temechegn E, Sileshi Y (2004) Concept cartoons as A strategy in learning, teaching and assessment chemistry. Addis Ababa University, Addis Ababa
24. Naylor S, Keogh B (2000) Concept cartoons in science education. Millgate House Publishers, London
25. Becker H-J (1988) Verbraucherfragen im RIAS-Telefonstudio: Gegenstand fachdidaktischer Forschung? Chimdid 14:69

Motivation 3

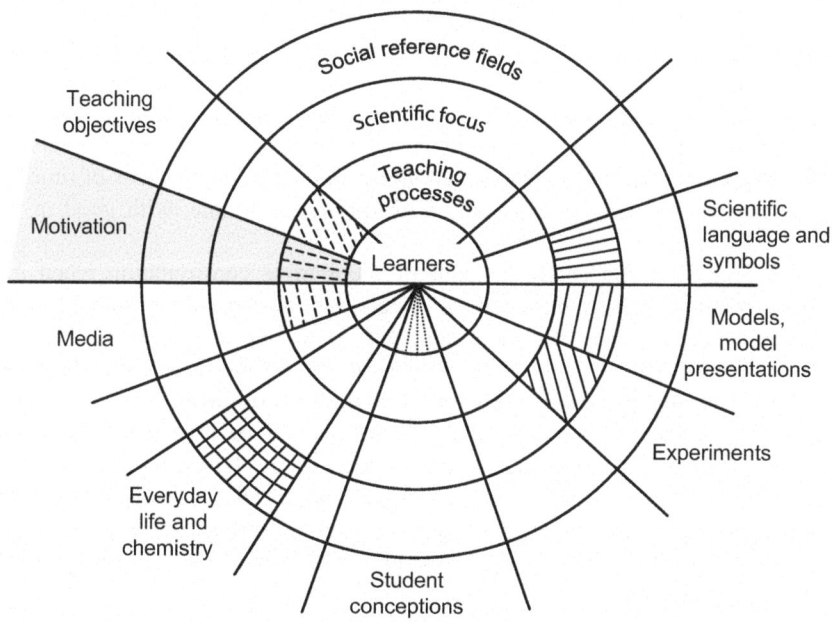

Janna, please go through all the tasks for the test again. Anyone who can't do them should expect to fail the class.
Lynn is the only one who has handed in correct homework. I will enter an A in the notebook.
Jonathan and Tilmann—if you don't hand in all the missing protocols by Friday, I will call your parents.

Such and similar statements are quite common for teenagers to hear at school. In particular, the great importance of grades for student performance in society often prompts teachers to threaten with poor grading and thus discipline the students.

Some teachers and parents believe they can motivate with this grade pressure and see it as the only way to get their protégés "to learn". They do not consider that the students, based on the weak foundation of this short-lived *extrinsic motivation*, only cooperate until the next test, until the reward by teachers or parents, until the achievement of the class goal, only to forget almost everything afterwards: Desired, long-term learning does not take place!

An important task of the teacher is therefore to think of other measures that stimulate students to learn and do not simply force them. Therefore, when preparing lessons, there are no limits to the creativity of teachers to motivate students in a subject-related way over a longer period of time, in order to generate *intrinsic motivation*. It is the challenge of every subject didactics in this respect to identify the possibilities to motivate intrinsically and to arouse interest in the subject.

In the natural science subjects, it is relatively easy to motivate through natural or laboratory phenomena and to ignite *curiosity* or *interest* in students. If the short-term interest in individual contexts even leads to a longer-lasting interest in the school subject, i.e. if one thus guides the learner from "situational to personal interest" [1], one achieves subject-related motivation over long periods of time and does not need to motivate extrinsically with praise and blame, with good or bad grades.

Interest unfolds more easily in favor of a cognitive confrontation when it is accompanied by positively experienced affects: Humans strive to establish "consistency between affection and cognition" [2]. So if one also addresses positive *emotions* of the learners in the desired generation of long-term motivation, one will be able to evoke positive *attitudes* beyond interest. In this context, the teacher in the subject of chemistry does not have a hard time: Experiments and even more so student experiments carried out by the students themselves usually evoke positive emotions and are excellently suited to generate the desired long-term intrinsic motivation and even a positive attitude towards the subject through the related positive emotion. The following explanations are intended to explain this in detail and with many examples.

3.1 Learners: Developmental Status, Attitudes and Original Ideas

The task of chemistry didactics in this respect is first of all to reflect on the conditions that must be observed in order to build up the desired intrinsic motivation:

- Status of the mental development of the learners
- Existing attitudes towards chemistry or chemistry lessons
- Original ideas about natural phenomena or chemical processes

3.1.1 Developmental Stage

According to the theory of Piaget [3], learners at the secondary level I (ages 10 to 16 years old) are in the stage of concrete or formal thought operations in terms of their cognitive abilities. However, the age limits to be assumed in this respect can vary considerably: For example, it has been found that only 25% of 16-year-old adolescents in grade 10 reach the stage of formal thought operations [3].

Measures for motivation must be based on such developmental stages: For adolescents at the concrete operational thought stage, individual phenomena are therefore preferable to laws. For example, the constancy of the melting temperature of an ice-water mixture can be investigated (Sect. 3.6: V3.1) or the lowering of the boiling temperature of water with decreasing pressure (Sect. 3.6: V3.2) can be used and evaluated in a motivating way. On the other hand, it is hardly possible to show the vapor pressure curve of water (Fig. 3.1) and hope that learners in grades 7 or 8 will understand this curve—they will not be motivated to think about this graph. Even learners at the stage of formal thought operations in grades 10 and 11 will initially like to see or observe the individual phenomena in student experiments before they are then more motivated to take note of the vapor pressure curve of water and discuss the dependence of the boiling temperature on pressure.

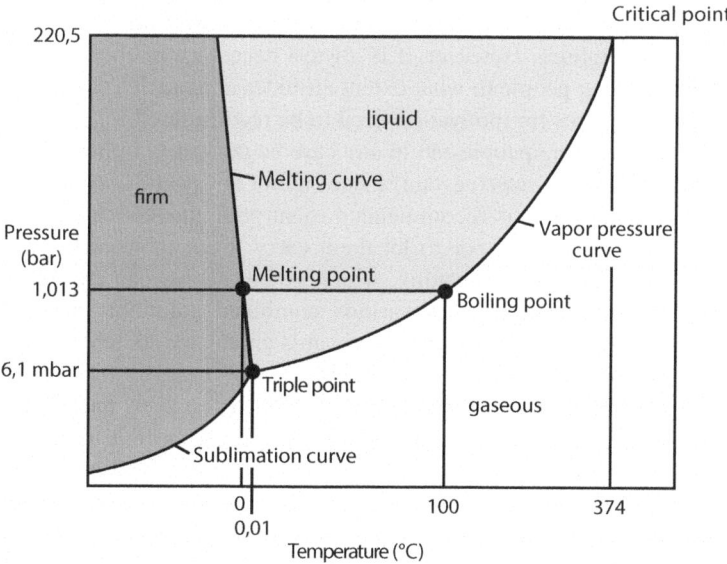

Fig. 3.1 Vapor pressure curve of water [4]

3.1.2 Attitudes

Motivation can only make sense if there is a neutral or even positive attitude towards the corresponding school subject or the matter to be learned. With a negative attitude, young people would not be willing to turn to the desired subject matter or think about it.

For the empirical collection of attitudes, Heilbronner and Wyss [5] asked many Swiss young people aged 11–15 years at the beginning of the 1980s to "draw their image of chemistry". The pictures showed smoking chimneys of factories, polluted rivers, containers of toxic chemicals with skull symbols, animals dying from animal experiments. The authors [5] found an extremely *negative attitudetowards chemistry* due to the dominance of these images (Chap. 9). At that time in Switzerland, a disaster had occurred in a chemical plant and not only the students, but all other citizens were shaped by the attitude: "Chemistry is evil, it is a threat to everyone." Since such chemical accidents can always surprise us, it is a special challenge for the teacher who is teaching chemistry to engage in discussion with the young people and prevent them from acquiring a generally negative long-term attitude towards chemistry.

Barke and Hilbing [6] tested this result for the end of the nineties on young people of the same age group from North Rhine-Westphalia and Brandenburg, also expressed attitudes of young people through pictures to be drawn and questioned these pictures through a questionnaire. The evaluation shows that about 75% of the pictures contain at least one motif that represents a rather positive attitude towards chemistry (Chap. 9). In this respect, attitudes have improved compared to the seventies and eighties. However, it is always necessary to check through conversations with young people to what extent attitudes change due to environmental influences and measures for motivation need to be reconsidered.

The attitudes of young people can in any case be positively influenced by *beautiful and attractive experiments*, even if these cannot always be used for evaluation in class. For this reason, it is recommended to surprise the students from time to time with a show experiment, or to let them carry it out themselves. For example, color pictures are very motivating, which young people can create in student experiments by precipitation from various combined solutions on filter paper [7]—they will develop a positive attitude towards chemistry and chemistry lessons through aesthetically appealing "Runge pictures".

However, spectacular experiments alone are not enough to motivate students in the long term: The most beautiful effects quickly fizzle out and are quickly devalued in our media-saturated world due to sensory overload and habituation. It is therefore necessary to make connections tangible through simple, but interconnected experiments that can be discovered by the students themselves: The Phenomenological-Integrative Network Concept (PIN concept) is very well suited for this [8].

Often, students come to the introductory lessons of the subject chemistry and develop a initially positive attitude towards chemistry lessons due to the beautiful experiments. However, if formulas and reaction equations are introduced after the first half-year and not immediately understood, the positive inclination usually fades. Many studies have shown that linking the formulas with illustrative models of the structure of corresponding chemical compounds can help maintain a positive attitude (Chap. 7 and 8): On the one hand, young people understand the formulas and equations much better, on the other hand, they are particularly motivated by building such sphere packs and molecule models themselves [9, 10].

3.1.3 Original Student Conceptions

As outlined in Chap. 2, teenagers prefer their own explanations for many facts, which do not coincide with today's scientific conceptions: material change, conservation law, energy conversion, gas concept, combustion, etc. It was pointed out that it is beneficial for a good understanding of the facts to incorporate these original conceptions into the development of scientific conceptions (Chap. 2).

This approach in teaching is particularly advantageous for building subject-related motivation : Through the *cognitive conflict* between one's own pre-existing, original conceptions and the experimentally demonstrated phenomena that the teacher presents regarding the scientific facts, curiosity and interest can arise. The students recognize the problem as significant for themselves and are motivated to find the solution to the problem—and finally to realize the **Conceptual Change** to the scientific conception. A series of experimental examples will specify this in the following.

3.2 Teaching Processes: Opportunities for Building Subject-Related Motivation

Motivation is not to be understood as a one-time act that takes place at the beginning of a lesson: Often teachers try to make a problem appealing to their students by initially relying on the students' experiences, but then using this reference solely as a vehicle to convey content that is not comprehensible to the students. For example, the teacher may ask what the students know about the rusting of an iron nail, only to deal with the difficult relationship between electrochemical standard potentials and local elements for the corrosion of metals after a few minutes. Such an approach is soon seen through and rejected by the students, it counteracts an attempt at motivation in a counterproductive way.

Successful opportunities for motivation are based on other measures:

1. Designing chemistry lessons in a comprehensible way for learners (Sect. 3.2.1)
2. Implementing introduction and genetic learning according to Wagenschein (Sect. 3.2.2)
3. Establishing continuous references to the everyday life and world of the teenagers (Sect. 3.2.3)

4. Generating and productively using cognitive conflicts in the students' conceptions (Sect. 3.2.4)
5. Demonstrating striking experimental effects (show experiments) (Sect. 3.2.5)
6. Enabling hands-on interaction with experimental or model building materials (Sec. 3.2.6)

3.2.1 Comprehensible Teaching

Regarding the subject of chemistry, one often hears the argument, "one would not understand chemistry, the constant work with formulas and reaction equations was incomprehensible, the calculation of material conversions using the concept of moles was opaque". As long as students cannot comprehend the teacher's approach and thus do not understand the lesson, no motivation for further participation is built: learners then only participate to survive the next test or to demonstrate sufficient performance in chemistry by memorizing theorems.

To achieve long-term motivation, it is therefore the first and most important commandment to follow a comprehensible path in teaching: students should feel that they understand a chemical fact through the lesson, i.e., they are successful in learning. It is not enough to formulate a reaction equation at the end of an interpretation of phenomena, but—as soon as the teaching conditions allow—to work out explanations regarding the "submicro level" (Fig. 7.9): With the question, "which atoms, ions or molecules react in the considered reaction", an acid-base reaction is distinguished from a redox reaction. An example: First, the dissolution of calcium carbonate in hydrochloric acid is described with the known reaction equation

$$CaCO_3 + 2HCl \rightarrow CaCl_2 + H_2CO_3 \quad (H_2CO_3 \rightarrow H_2O + CO_2).$$

The actual understanding of the reaction is now achieved by asking about the smallest particles responsible for the reaction. In calcium carbonate, these are obviously the carbonate ions, in hydrochloric acid the hydronium ions, which by transferring one proton each cause the carbonate ion to react to form a carbonic acid molecule:

$$CO_3^{2-} + 2H_3O^+(aq) \rightarrow H_2CO_3(aq) + 2H_2O \quad (H_2CO_3 \rightarrow H_2O + CO_2).$$

If one emphasizes that the calcium ions and the chloride ions are not changed as accompanying ions and remain in solution, then one has a better understanding of the acid-base reaction than through the simple gross reaction equation alone. This rather conveys the impression that "2 HCl molecules" react and creates corresponding misconceptions. If beaker models for the particle associations before and after the reaction are additionally drawn (Fig. 7.15, 7.16, 7.17), then the learners understand the reaction optimally and recognize that it is completely irrelevant whether sodium or potassium carbonate react with hydrochloric acid, diluted sulfuric acid or nitric acid: It is always the same proton transfer from hydronium ions to carbonate ions that determines the reaction.

Through this optimal understanding at the "submicro level" (Fig. 7.9), learners are motivated to continue to enjoy working for the subject of chemistry. The following chapters on experiments, models, and symbols offer further concrete aids to design a comprehensible, i.e., understandable introduction to chemistry.

3.2.2 Introduction According to Wagenschein

When reading the writings of Wagenschein [11], one can sense the unusually motivating way in which a posed question is used as an introduction to stimulate long-lasting thought—to encourage learning. An introduction should "not penetrate the problem too complexly and not too little complexly", the mental work should be "exemplary", leading "down to elements and up to complicated questions" [11].

An example from school physics may explain this in Wagenschein's own words: "An introduction to mechanics that I have often tested is the innocuous-looking question: Where does a stone fall that is held out of the window of a high tower and then let go? It initially appears trivial. But it immediately confuses in a most captivating way when one gradually thinks of the curvature of the earth and the—alleged—rotation of the earth, and when one then initially takes a lag to the west for granted, then doubts the rotation of the earth, blames the rotating air for the stone's movement. But why does it go along? Why isn't there a constant east wind? Analogous experiences in a railway carriage, open and closed, come to mind. These questions can trigger hours of bitter discussions. They end with the discovery of the law of inertia, and finally—and this is now a sensation—with the east deviation. In the end, the students really believe that the earth is rotating. I have tested this topic in 1946 in graduate courses for returning young soldiers and also as a topic of weeks-long breath with upper secondary students. One can break open all of mechanics with it and then enter into it" [11].

This *exemplary* (i.e., starting from concrete experiences), *Socratic* (i.e., kept going by a constant question-answer game) or *genetic* (i.e., psychologically gradually building up, developing) way of questioning and doubting the supposedly self-evident, promoting the confusion of students and thereby creating a productive tension, characterizes Wagenschein's approach. He not only wants to convey knowledge in a motivating way, but also the path of gaining knowledge: "There are two completely different teaching styles: whether the teacher only wants to prove to the student that it is as it is, or: whether he also wants to let him experience how man, mankind, could and had to come up with something like this. Only the second, the genetic kind has to do with the, in the strict sense understood, exemplary teaching " [11]. However, it should not be underestimated that this teaching method places high demands on teachers and students.

3.2.3 References to Everyday Life and Living Environment

Typical chemistry lessons—especially at grammar schools—usually favor a conceptual structure: This concept orientation is often brought by teachers from their university lectures and, based on these experiences, they apply the finished subject systematics to their chemistry lessons. The young people who dutifully come to school and perhaps think about learning something from their everyday life, recognize little or no connection to their living environment on the concept-oriented path. They are forced to learn the new subject quite formally in order to achieve the desired good grades: Motivation in this way remains extrinsic in many cases, only in a few cases do young people perceive this approach as advantageous.

To convey the required conceptual structure in an intrinsically motivated way, it is helpful for chemistry lessons to integrate everyday life and the living environment of the students: In addition to the usual laboratory substances, substances and reactions from the kitchen, bathroom, garden and garage should be included, with regard to the living environment from school, hobby, travel or sport. Chapter 9 "Everyday Life and Chemistry" offers many experimental examples in this regard, establishes everyday references for motivation at the beginning of a teaching unit or at its end for deepening and repetition. A new curriculum in the state of North Rhine-Westphalia [12] has virtually forced textbook authors to put everyday references at the beginning of each chapter. Thus, the new textbooks "Chemie heute" [13], "elemente chemie" [14] or "Chemie 2000+" [15] offer such special introductions and excursions to questions of everyday life and environment much more intensively. Schwedt [16] has published on this topic, featuring supermarket products and their examination by school means.

With some topics, it is possible to conduct the lessons consistently with substances from everyday life—in such cases, the conditions for motivating chemistry lessons are optimal. Wanjek [17] was able to show in a scientific accompaniment of the teaching unit "Acids and Bases" that food and cleaning substances increase interest in chemistry and thus motivation. Especially girls, who expressed much less interest in chemistry in a questionnaire before this unit than boys, increased their interest after this lesson to the level of boys [17]. This everyday-related teaching consciously and successfully linked the conceptual structure of chemistry and references to the living environment.

In the UK, the "Salters Chemistry" program has designed everyday references throughout the entire chemistry curriculum (Chapter 9). Ilka Parchmann got to know this program on site and, together with Reinhard Demuth and Bernd Ralle, created the chemistry work "Chemie im Kontext" for secondary level II (grade 11–13 ending with the A-levels), which also links each topic with references to the living environment. It will be presented in detail later (Chapter 9).

However, it will not always be possible to bring everyday substances and references to the living environment so that a viable factual structure can be developed from them alone. This is by no means necessary. Harsch and Heimann [8] show in the context of the PIN concept with many examples how everyday references can be integrated into a genetically growing subject systematics in a way that increases motivation: detection of basic organic substances in food and household products, relationships between test tube syntheses and biochemical metabolic processes, model experiments for understanding chemical recycling including material and energy balances.

3.2.4 Generation of Cognitive Conflicts

The classic possibility of intrinsic motivation is called by Piaget "equilibration process by generating a cognitive conflict", Copei describes it as "fruitful moment by shaking self-evidences", Roth calls it "original encounter through an anomalous starting point".

Lind [18] formulates the same context with the *Incongruity Theory*: The difference between a perceived stimulus and the stimulus expected by the individual is called incongruity. This incongruity can be the cause of subject-motivated behavior, because students strive to eliminate the anomaly experienced in this way, i.e., to close the gap between expectation and actual observation.

In the respective lesson preparation, the teacher must

- know or correctly assess the prior knowledge or ideas of the students,
- select the event to be presented in such a way that it is incongruent with the students' expectations,
- determine the presentation mode of the anomaly,
- determine the type and strength of the incongruity [18].

In chemistry lessons, it is possible at many points to create incongruities through observations of certain natural phenomena or laboratory experiments and thus to motivate the students in a special way. It is up to the creativity of each teacher to select suitable phenomena or experiments for a topic—some examples are outlined.

Melting Temperature of Ice
Students express their idea that heating a substance should lead to a certain temperature increase. They heat a mixture of ice and water, stir well with the thermometer and continuously read the temperature (Sect. 3.6: V3.1). They do not expect the observation that the temperature remains at 0 °C despite heating, and are motivated to think about it—with the help of the teacher. In this context, the melting heat of ice is discussed and it is pointed out that the energy amount of 335 J is required to dissolve the water molecules from the molecular lattice for 1 g of ice (Fig. 3.2). It can also be pointed out at this point that—as soon as ice is present in the mixture with water at normal pressure—the temperature is always 0 °C, that a chemical equilibrium exists between ice and water:

$$\text{ice (s, 0 °C)} \rightleftarrows \text{water (l, 0 °C)}$$

Boiling Temperatures of Water

Students usually know the value of 100 °C as the boiling temperature of water, but do not pay attention to the fact that this value always applies only to normal pressure. To make this important condition clear, two incongruities can be created. Either the story of mountaineers who find a boiling temperature of 92 °C when boiling water at 5000 m altitude is told, or a corresponding experiment is conducted that shows the dependence of the boiling temperature on pressure: The "presentation mode of the anomaly, type or strength of the incongruity" is determined [18]. This is also possible in terms of experimental procedure: The pressure relationship can be demonstrated by directly connecting the water jet pump to a boiling apparatus for water, or by performing the "boiling by cooling" in a closed apparatus filled with steam—a further incongruity occurs for the students (Sect. 3.6: V3.2). In this context, attention can be drawn to the evaporation heat of water, which is more than 6 times the melting heat (Fig. 3.2), and possibly to the vapor pressure curve of water (Fig. 3.1). The boiling of water at normal pressure is again described by the following equilibrium:

$$\text{water (l, 100\,°C)} \rightleftarrows \text{vapor (g, 100\,°C)}$$

Solubility of Carbon Dioxide

Students are familiar with the effervescent appearance of "carbonic acid" when dissolving an effervescent tablet in water and have the idea that a certain volume of gas is produced per tablet. In a pneumatically standing measuring cylinder filled with water and placed in a water bath, an effervescent tablet is dissolved and a gas volume of approx. 70 ml is observed (Sect. 3.6: V3.3). At this point, the students are asked to predict the gas volume the second tablet will produce, and of course, they say: "the same volume". However, the second tablet develops almost 200 ml of carbon dioxide—the students' expectations are not met, they start thinking about how to eliminate the occurring anomaly: With the help of the teacher, they find out that the first tablet is put into pure water and produces a saturated solution

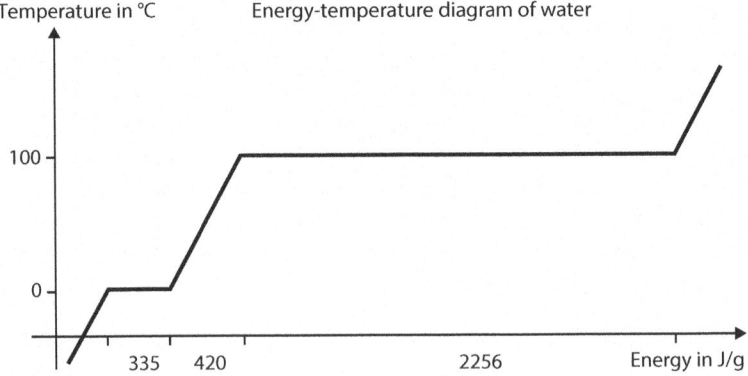

Fig. 3.2 Diagram of the melting heat and evaporation heat of water

of carbon dioxide in water, that the second tablet releases all the carbon dioxide as gas and produces the large volume because it hardly reacts with the already saturated solution of the gas in water. The equilibrium can also be formulated here:

$$\text{carbon dioxide (g)} \rightleftarrows \text{carbon dioxide (aq)}$$

Combustion

The experiment already described in Chap. 2 on "Iron wool on the scale" (Sect. 3.6: V2.7) classically shows a cognitive conflict of the adolescents, who in their experience have always observed the "lightening" of substances during combustion—such as charcoal, wood, and paper. A tuft of iron wool is weighed, ignited, and the students are asked to predict: "Has the tuft become heavier, lighter, or is it the same weight as before"? The expectation "lighter" is not met after the second measurement, the opposite is found: The tuft becomes heavier. The students are now highly motivated to eliminate this incongruity and come to the conclusion in the discussion that the increase in mass is due to the reaction of the added oxygen. If the teacher immediately follows up with the weighing of iron wool and air in a closed test tube and compares the masses before and after the reaction of the iron wool in the closed test tube, then—after two or three other examples—the law of conservation of mass in chemical reactions is inferred.

Extinguishing Fires

Students are well acquainted with extinguishing fires with water. When asked how, for example, burning fat from a deep fryer or burning metal shavings in a metal workshop are extinguished, the answer of most teenagers based on their experiences will be obvious: "with water". During the actual execution (Sect. 3.6: V3.4), the students are very surprised by the intense flare-ups: They did not expect this and are highly motivated to consider factually appropriate answers. With the help of the teacher, they find out that the very hot fat at about 300 °C immediately evaporates the extinguishing water and the spontaneously arising steam carries along fine droplets of fat, which burn explosively when dispersed in air. In the reaction of burning magnesium with water, hydrogen is even produced—a new fuel that reacts with oxygen to form water:

$$Mg\ (s) + H_2O\ (l) \rightarrow MgO\ (s) + H_2\ (g); \quad \text{exothermic,}$$
$$2H_2\ (g) + O_2\ (g) \rightarrow 2H_2O\ (g); \quad \text{exothermic}$$

3.2.5 Striking Experimental Effects

"No motivation without emotion" is the abbreviated understanding that interests and attitudes develop optimally with positive emotions. Especially in the natural sciences, such positive emotions can easily be triggered by experimental effects, and at all times there were events where show experiments were demonstrated. Liebig's evening lectures in the second half of the 19th century were also part of this, they were often even attended by the Bavarian royal couple in Munich.

An event exemplifies the emotions such experiments can even trigger in a queen: She was so surprised by the beautiful blue flash of the nitrogen monoxide-sulfur carbon reaction (Sect. 3.6: V3.5) that she wanted to see it again. Liebig repeated the experiment—however, the glass flask exploded due to an assistant's mistake, who handed the experimenter oxygen instead of nitrogen monoxide: Both the royal couple and Liebig were injured [19].

Our students also want to see beautiful experiments several times, so their (safe!) repetition should always be planned for the lesson. However, only a teacher who has gained familiarity with the experiments over a long time and can safely demonstrate them should plan and perform show experiments in front of young people.

Regarding these effects, there are two different concepts. On the one hand, the well-known *show experiments* such as the above-mentioned nitrogen monoxide-sulfur carbon reaction are perceived as aesthetically beautiful attempts and thus as exciting experiences, but usually no evaluation is based on them: The experiments do not serve for subject-related motivation, but rather represent a kind of extrinsic motivation for "Christmas lectures". They are presented in many experimental books, for example in the form of "fairground chemistry" by Krätz [19] or as "chemical cabinet pieces" by Roesky [20].

Another example should illustrate that show experiments can also lead to subject-related motivation and an evaluation. If, for example, the issue of "density of different substances" is to be introduced, usual weighings and volume measurements on metal pieces can be carried out and evaluated and the metals can be identified using the density table. This path is important if concept-oriented learning objectives are to be achieved directly or if student experiments are to be used to practice experimental skills. To introduce the topic of density in a more motivating way, an effect can also be shown that the learners probably do not know. A can of "Coca Cola" and "Diet Coke" of the same size (330 ml) are placed in ice water: the former sinks, the latter floats (Sect. 3.6: V3.6). If this effect is used to tell the little story that at the last party you always had to reach deep into the cold water when someone wanted "Coca Cola", while "Diet Coke" was simply taken from the water surface, then the students will be further motivated to think about this effect: It not only triggers subject-related motivation and perhaps also emotions, but also establishes a connection to everyday life. The discussion about the sugar content of both types of cola can finally explain the different densities.

Three more effects are listed that can trigger subject-related motivation: The experiment "Ice bursts a bottle" (Sect. 3.6: V3.7) is an attempt to show the anomaly of water and to lead to a discussion of the structure of ice. "Black carbon from white sugar" (Sect. 3.6: V3.8) may introduce the topic of sugar and the composition of carbohydrates, the astonishing effect "Electricity from a lemon" (Sect. 3.6: V3.9) motivates a discussion of the voltage series of metals. Similarly effective experiments can be found for almost any problem—the creativity of teachers knows no bounds!

3.2.6 Active Handling of Experimental or Model Building Materials

Especially for children, but also for teenagers, it is always interesting when they do not have to sit quietly in their chairs at school, but can move around, run or do something manually: motivation in the *psychomotor area*. For this reason, student-conducted experiments are ofgreat importance. Students are not only motivated to learn something through varied movement and their own actions, but they understand and retain chemistry in a more action-oriented way than through a teacher-led demonstration or even without any experiment. Student experiments are exemplified in Chap. 6 "Experiments".

If students in an action-oriented class even produce certain *products* that they can take home, then the motivational effect is particularly strong. If the students are asked to make a brass sign for their front door with their name on it and take it home when studying "Reactions of Acids", they are very motivated to make this sign. They coat the brass plate with the wax of burning candles, carefully write their name in the wax surface and etch the resulting free metal spots with nitric acid (Sect. 3.6: V3.10). They are happy to show such a sign to their family and friends and can explain exactly how to make it: The motivation extends far beyond the classroom.

Also, the *construction of structural models*, such as sphere packing or lattice structures for crystal structures and molecule models for the structure of molecules, is perceived as beneficial by students: The motivation regarding psychomotor skills can be used to make the structures of different substances clear to them. If the formulas of the corresponding substances become clear from such structural models, then in this case the psychomotor motivation even promotes the understanding of the chemical symbol language. Relevant examples can be found in Chap. 7 "Models, Model Concepts" and Chap. 8 "Technical Language and Symbols".

If students are asked to build a sphere packing as a model for the sodium chloride structure (Sect. 3.6: V3.11) and to take this, for example, as a paperweight for their own desk, they will not only carefully build the model, but also explain the corresponding structure to friends and acquaintances—in such cases the motivation extends into the private sphere of students and families. Even seasoned teachers who attended our further education courses "Structural Models and Understanding Chemistry" admitted that the driving motivation for attending the course was the offer to take the self-built models for their lessons!

Finally, you can offer various fruits or soft, differently colored candies to build molecule models. With two types of candies and toothpicks, models of H_2O, CO_2, or CH_4 molecules can be built, and with three types, models of C_2H_5OH or CH_3COOH molecules can be made. Once these candy models have been explained and the corresponding molecule symbols understood, you can even allow the learners to eat their models—a special form of motivation to build models for chemical structures!

3.3 Scientific Focus: Experimental Skills for Demonstration Experiments

Reflection at the level of "teaching processes" has shown the possibilities of intrinsically motivating students through experimental means, as well as allowing extrinsic aspects—such as show experiments—to come into play. In both categories, effective experiments are suggested, often involving dangerous experiments: Even an experimenter of Liebig's caliber made a mistake in the evening lecture that could have led to a catastrophe for the Bavarian royal couple and himself [19].

Therefore, teacher students must be taught *experimental skills and abilities* in order to be able to safely demonstrate rapid combustions (Sect. 3.6: V3.4) or explosive phenomena (Sect. 3.6: V3.5). In a practical course on school experiments, these skills must be demonstrated, preferably in the form of experimental lectures.

Teachers already working in the classroom should also try out new spectacular experiments before they can be safely demonstrated for the experimenter himself and for the audience. Only after eliminating potential hazards can one step in front of the class and demonstrate dangerous phenomena, taking into account all aids (safety glasses, protective screen, exhaust, splinter basket). To survive an entire "Christmas lecture" unscathed, a long experience of experimenting in front of school classes is required. Finally, the relevant regulations regarding the hazardous substances ordinance must be observed (Chap. 6).

To exploit many possibilities of motivation, in addition to experimental safety, a *scientifically sound basic education* is also required. One not only wants to be able to correctly assess spontaneous student comments in order to possibly create a suitable cognitive conflict for motivation, but also wants to be flexible and express motivating thoughts or carry out motivating experiments in as many teaching situations as possible. For example, if a learner claims that black carbon can never be obtained from the colorless gas carbon dioxide of the formula CO_2, the experienced teacher will fill a standing cylinder with the colorless gas, hold a burning strip of magnesium inside and show that black grains of soot form on the inner wall of the cylinder (Sect. 2.6: V2.9). The cautious young teacher will initially only demonstrate a previously tried out motivation experiment at planned points in the lesson or carry out a planned group work with student experiments.

3.4 Social Reference Fields: Motivation through Everyday Language and Media

Formulations of everyday language often obscure the factually correct context, but on the other hand provide motivating occasions to think about these facts. For example, in everyday language one says "the copper roof turns green"—and leads to the idea that copper can "appear red-brown at one time and green at another". If one starts from the knowledge about specific properties that characterize certain substances, it can be determined that only one of the colors can be specific to copper. In this way, the cognitive conflict or incongruity is established and learners may be motivated to think about the "change of copper color". The result of the reflection should be the finding that copper reacts with the weakly acidic carbon dioxide-saturated rainwater over decades and forms the green copper carbonate as a cover layer on the metal. An experimental verification of this assumption can follow and eliminate the incongruity. The media also provide—mostly unintentionally—statements that can lead to motivating discussions.

Newspaper Articles
When an eco-magazine claims that toxic benzene in unleaded gasoline "is not burned, but is largely released into the breathing air through the exhaust" (Fig. 3.3), then the author of the article has never seen in the laboratory how easily

Ein Wasch-Wunder blieb geheim – weil es Chemie ist
Dabei ist das neue Mittel von Hoechst nicht nur sparsamer, sondern bedeutend umweltfreundlicher als frühere

a

Dünnsäure steckt Fischen noch immer in den Gräten
Untersuchung: Krankheitsbefall im ehemaligen Verklappungsgebiet ist leicht erhöht

Weltweit einmaliges Vorhaben
Sprit fürs Auto aus der Luft

Strom soll aus der Pflanze kommen
Duderstadt plant bundesweit erstes Pilotprojekt mit Biomasse

Natrium explodierte
Unfall bei Degussa – Ein Arbeiter verletzt

Amerikaner drehen Hahn für Trinkwasser zu
Zuviel Chemie im kostbaren Naß

b

2 Feuerlöscher: 1 x mit Pulver, 1 x mit Sauerstoff gef. Dunstabzugshaube 150 cm lang. kpl. ncuw.
☎ ▓▓▓▓▓▓ 14–18 Uhr.

Gift brannte auf Autobahn
dpa Lausanne. Die Autobahn Lausanne-Genf mußte am Montag mehrere Stunden in beiden Richtungen gesperrt werden, nachdem sich der Inhalt mehrerer Fässer mit Salzsäure und Schwefelsäure auf die Fahrbahngossen hatte und in Brand geraten war. Ein Lastwagen, der die Fässer transportierte, war wegen eines geplatzten Reifens umgekippt.

27 Verletzte bei Chlorgasunfall
hi Ganderkesee. Nach einem Arbeitsunfall in einem Galvanisierungsbetrieb in Ganderkesee (Landkreis Oldenburg) wurden gestern mittag durch austretendes Chlorgas 27 Personen verletzt. Wie die Polizei mitteilte, geschah der Unfall beim Abfüllen eines Tanks. Aus bisher ungeklärter Ursache gelangte dabei Salzsäure in den Behälter, der zu einem Teil mit Natronlauge gefüllt war. Durch die Reaktion beider Stoffe entstand die Chlorgaswolke. Am Unfallort selbst wurden zwei Arbeiter verletzt, einer davon schwer. Auf dem Gelände eines benachbarten Betriebes atmeten 25 Beschäftigte das Gas ein, sie mußten wegen Verätzungen der Atemwege ins Krankenhaus eingeliefert werden. Während des Giftgasalarms wurde die Bevölkerung über Rundfunk und Lautsprecher dazu aufgerufen, in den Häusern zu bleiben. Gegen 15 Uhr konnte die Absperrung des Unfallortes aufgehoben werden.

Verpestet „Bleifrei" die Luft?
dpa. Bonn/Frankfurt. Bleifreies Benzin enthält nach Darstellung des Öko-Testmagazins (Oktober-Ausgabe) zur Zeit durchschnittlich doppelt soviel krebserregendes Benzol und zweieinhalbmal soviel Toluol wie verbleiter Kraftstoff. Benzol werde im Motor nicht verbrannt, sondern durch den Auspuff weitgehend an die Atemluft abgegeben. Ferner werde der benzolähnliche Stoff Toluol beim Verbrennen in Benzol umgewandelt, so daß sich die Benzolwerte in der Atemluft stark erhöhten, wenn bleifreies Benzin ohne Katalysator gefahren werde.
Das Bundesinnenministerium hat die Darstellung des Öko-Testmagazins als falsch zurückgewiesen.

Fig. 3.3 Newspaper articles with errors for motivation in class [21]

benzene is flammable and burns violently with a sooty flame. In chemistry lessons, such reports can be discussed to lead the young people to a critical approach to newspaper reports on the one hand, and on the other hand to use the motivation to uncover and correct errors in a report.

Haupt [21] has compiled a collection of newspaper clippings on many topics of chemistry, which contain examples of similar journalistic errors and provide motivation to find, discuss, and seek correct solutions to these errors (Fig. 3.3a). Haupt [22] uses a selection of newspaper articles with insufficient journalistic research as an opportunity to formulate suitable tasks for students based on them and to help them analyze the errors and find corrected formulations (Fig. 3.3b).

Television
For example, when a journalist comments on a factory fire with the words: "No chemicals were involved in the fire" (example from American television, the journalist may have meant hazardous substances with the word "chemical"), students can bring in their knowledge and correct, depending on their knowledge, that all fuels are always chemicals, as is the involved oxygen in the air a chemical involved in the fire. The incongruity between the journalist's statement and one's own knowledge can be motivating to uncover the error and as a young student to prove a faulty report to a seasoned reporter.

Even in regular television advertising, which teenagers often enjoy watching, one finds similar insufficient explanations, such as "chemical-free drugs". The program booklet argues that "chemical-free drugs offer hope", that a "special active ingredient from a medicinal plant that grows in North and South America combats joint pain as effectively as gently" [23]. So it is an active ingredient that is present as a chemical—and therefore, of course, no "chemical-free" medication is offered. Critical chemistry students should be motivated or be motivated by the teacher to think about such news and to recognize that there can be no "chemical-free" substances, that every substance is a chemical.

Movies
As is well known, teenagers like to go to the cinema and watch James Bond films, for example. They often encounter scientific content and are motivated to think about it. Ducci, Oetken, Friedrich, and Rubner [24] use this motivation to go into chemistry lessons with such content and reflect on it with the help of the teacher.

For example, they show the scene from the film "James Bond—Diamonds Are Forever" of a body being burned in a crematorium and the subsequent recovery of a diamond treasure by James Bond. As soon as the question is raised whether diamonds can withstand the high temperatures of a fire unchanged, the teenagers are motivated to think about it or even plan an experiment in this regard. If the teacher puts some cheap industrial diamonds in a combustion tube and heats them very strongly with the addition of oxygen, then the viewers see the dazzling white flame of the diamonds and can detect the gaseous combustion product carbon dioxide. At this moment, the lesson on the topic of diamond and graphite is perceived completely differently than without the motivation through the movie—and the teenagers can finally determine: "No, the diamonds would have burned under the conditions and would no longer be findable" [23].

The feature films "Das Boot" and "Apollo 13" are used as a basis to address problems with too high CO_2 concentrations in the breathing air [25], scenes from other films like "Dante's Peak" [26] and "Erin Brockovich" [27] can be reflected upon to make the problems of too high SO_2 concentrations or toxic chromium compounds the motivating content of certain lessons.

3.5 Exercise Tasks

A3.1
Provide examples of extrinsic and intrinsic motivation, discuss the differences between both types of motivation. Describe three teaching situations and corresponding examples of intrinsic motivation.

A3.2
Original student conceptions are particularly suitable for opening a motivating, subject-related discussion in class with incongruities and anomalies. Explain this connection using three examples of your choice, clarify the incongruity and show how you want to resolve it.

A3.3
Students can easily be motivated to watch attentively through striking experimental effects. Explain using experiments of your choice to what extent only an extrinsic motivation exists and in which cases a subject-related motivation can be established.

A3.4
The introduction to a new teaching topic should be motivating. Choose common textbook topics and show an introduction for each that is particularly motivating by a) connecting to the students' prior knowledge, b) through an incongruity, c) through a reference to everyday life, d) through the students' own activity.

A3.5
Everyday language contains phrases that are not always factually correct and therefore motivate to think and correct. Explain this using three examples of your choice and suggest as correct as possible formulations for the facts.

3.6 Experiments

V3.1 Constant Melting Temperatures

Problem

Students observe in their everyday life that every heating of a substance leads to its temperature increase. However, if a pure substance melts during heating, the temperature remains constant until the substance is completely melted: During melting, the energy supplied is used to destroy the crystal lattice of the solid substance (heat of fusion). This relationship should be recognized by the students with the following experiments.

Material

Thermometer (temperature probe and digital display), tripod and wire mesh, reagent and beaker glasses, wooden clamp; ice, naphthalene or stearic acid

Procedure

a. An ice-water mixture is heated several times in the beaker, after good stirring the temperature is to be read with the thermometer.
b. A test tube is filled with a quarter of naphthalene, carefully melted in the burner flame, the melt is stirred with the thermometer and observed (Caution: When immersing the thermometer, the readable maximum temperature must not be exceeded! Thermometer may break!). The temperature is noted every 30 seconds until the substance is completely solidified.

Observation

As long as the ice melts, the temperature of the ice-water mixture remains constant at 0 °C. As long as a mixture of naphthalene melt and solid naphthalene is present, the temperature remains constant at 80 °C.

Disposal

The test tubes with solid naphthalene are sealed with stoppers and stored in the collection until the next experiment.

V3.2 Boiling Temperatures of Water

Problem

Students often only know the shortened formulation "Water boils at 100 °C". To establish the pressure relationship, corresponding boiling temperatures can be determined at reduced pressures by connecting a vacuum pump or a water jet pump, and relationships between boiling temperature and pressure can be formulated.

There is also the possibility to replace the air in the round flask with water vapor, condense it by cooling, and measure the decreasing boiling temperatures at the corresponding negative pressure: For the students, there is the motivating incongruity that the water is not brought to a boil by heating, but by "cooling".

Material
Round flask with side tube and tap, stopper with thermometer (temperature probe), water jet pump, stand, boiling stones

Procedure
a. The round flask is filled with a quarter of water, the water is brought to a boil (boiling stones) and the boiling temperature at normal pressure is determined. The water jet pump is connected, the boiling temperature is measured again during air suction (Caution: Almost a vacuum, wear safety glasses!).
b. The water in the flask is heated to boiling and kept boiling for one minute until the air from the flask has been completely replaced by water vapor (image). The tap is then closed. The flask is to be rotated 180°, a wet cloth is to be placed on it and the thermometer is to be read. The process is repeated several times, finally the flask is upright and the tap is carefully opened (Caution vacuum, wear safety glasses!).

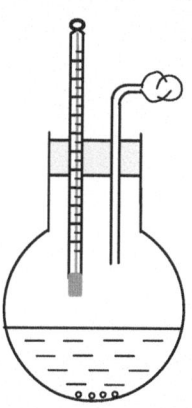
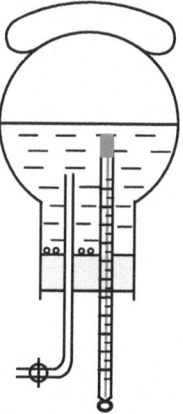

Observation
The thermometer shows boiling temperatures under 100 °C at reduced pressure. When cooling the vented, inverted flask, the water starts to boil again, temperatures down to 70 °C and below can be measured. Finally, air enters the flask with a whistling sound.

V3.3 Different Gas Volumes of the Same Effervescent Tablets

Problem
Students know that carbon dioxide gas ("carbonic acid") is dissolved in mineral water and recognize the gas when it is released in the form of small gas bubbles when effervescent tablets are dissolved in water. However, when they collect the gas pneumatically by dissolving an effervescent tablet, they overlook the fact that only a part of the released gas can be observed in the cylinder, while the other part dissolves in the water until saturation. If a second, identical tablet is dissolved, a much larger gas volume appears due to the already saturated solution. Since the learners expect the same volume, they are motivated by the incongruity with their preconceptions to think about the observations and independently discover the phenomena of the solubility of gases in water and in the saturated solution.

Materials
Measuring cylinder (250 ml) and suitable stopper or cover glass, glass pan; effervescent tablets (type "carbonate/citric acid")

Procedure
The measuring cylinder is completely filled with water and pneumatically placed in the half-filled glass pan using the stopper or the cover glass. A tablet is placed under the cylinder opening and the developed gas volume is marked at the end of the reaction. A second tablet is added, and this volume is also recorded.

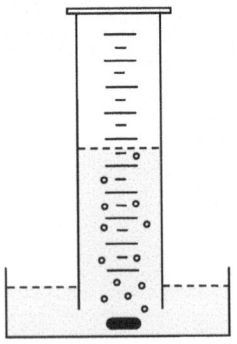

Observation

The gas volume in the first case is about 70 ml, in the second case 70 ml + 130 ml, so a total of 200 ml (the occurring volumes of the tablet types should be tested beforehand).

V3.4 Extinguishing Fat and Metal Fires

Problem

Due to appropriate experiences, extinguishing fires with water is such a common conception that people naturally resort to water even in the case of fat or metal fires—and thus risk serious accidents! The experiments can be demonstrated both for a discussion of these safety aspects and to create a cognitive conflict for students and let them solve it: They should recognize that the fat ignited at high temperature (approx. 300 °C) instantly vaporizes the extinguishing water and forms an explosive mixture in the air with the fat droplets carried along with the steam, which reacts violently with a yellow flame. In the case of a metal fire, the burning metal reacts with water to form metal hydroxides and combustible hydrogen: hydrogen and metal react with the oxygen of the surrounding air with a white, bright jet flame.

Materials

Triangular stand with wire mesh; tealight, magnesium shavings, spray bottle with water

Procedure

a. The wick of a tealight is cut off, the paraffin in the aluminum container is heated very strongly until the smoke produced by decomposition can be ignited. A jet of water is directed onto the paraffin burning without a wick (Caution: wear safety glasses!).

b. A cone of magnesium shavings is ignited on the wire mesh and heated from all sides until almost the entire cone glows red. A jet of water from the spray bottle is held directly into the embers (Caution: wear safety glasses!).

Observation

A yellow jet flame up to one meter high forms above the burning fat. The glowing magnesium powder reacts with a high white dazzling jet flame.

V3.5 Blue Flash of Carbon Disulfide in Nitric Oxide

Problem
A historical show experiment is intended to demonstrate that effective experiments are usually associated with great emotions: Even the Bavarian queen wished to see the "beautiful blue flash" a second time and Liebig wanted to perform the reaction again. However, he used carbon disulfide and oxygen in a bulbous bottle and triggered an explosion of the glass vessel. These reactions can be carried out safely in a standing cylinder.

Materials
Standing cylinder with cover glass, large test tube with delivery tube, pneumatic trough, plastic pipette, butane gas burner; ammonium nitrate, carbon disulfide, oxygen

Procedure
In the preparation, ammonium nitrate is decomposed in the test tube by careful heating, the resulting colorless nitrogen oxide gas is pneumatically filled into the standing cylinder (safety glasses, fume hood!). The pipette is to be filled with 2 ml of carbon disulfide and injected into the standing cylinder, the cylinder is to be covered with the cover glass. The mixture is to be ignited with the flame of the butane gas burner. The experiment is repeated with oxygen and carbon disulfide.

Observation
A bright blue flash appears, a sound reminiscent of a barking dog can be heard (hence the experiment is often called "Barking Dog"). In the second experiment, a white flash is seen, accompanied by a loud bang.

V3.6 Diet Coke is "lighter" than Coca-Cola

Problem
The concept of density can be introduced into the classroom through an effect that surprises most students and therefore challenges them to explain the effect: A can of Coca-Cola sinks in ice water, the can of Diet Coke floats. They will independently come to the high sugar content of the Coca-Cola solution and thus to the higher density in the discussion and conclude that sugar-free Diet Coke solution probably has a smaller density. By the way, with the addition of "light", the cola manufacturer does not mean the lower density, but the lower nutrient content compared to normal Coca-Cola.

Materials
Large standing cylinder, scale, hydrometer; Can of "Coca-Cola", can of "Diet Coke", ice water

Procedure
Both cans are weighed. The standing cylinder is filled three quarters with ice water, the beverage cans are added in the order indicated. The densities of the cola solutions are determined by weighing certain volumes or using the hydrometer.

Observation
The cola can weighs about 20 g more than the Diet Coke can at the same volume. The Coca-Cola can sinks in the ice water, the Diet Coke can floats on top. The density of Coca-Cola solution is greater than 1.00 g/mL.

▶ **Note** The manufacturing technique of the cans each results in an air inclusion in the can. This inclusion can vary, and it can therefore happen that the Diet Coke can also sinks.

V3.7 Ice bursts a bottle

Problem
Another density phenomenon is the density anomaly of water: Ice occupies a larger volume than the water portion of the same mass. The ice floating on water is very familiar to all of us and we do not think about the fact that in the normal case the solid sinks in its melt, so for example a candle in liquid wax or a metal object sinks in the melt of the metal. To make this water anomaly clear, a container can be filled to the brim with water, sealed and cooled below the freezing point: The container will be burst by the larger ice volume.

Materials
Small glass bottle with screw cap, thermometer (-20 to 100 °C); ice water, ice-salt cold mixture

Procedure
The cold mixture of ice and a certain amount of salt is prepared, the temperature significantly below the freezing point is demonstrated with the thermometer. The bottle is filled to the brim with ice water and tightly closed and immersed in the cold mixture. After a few minutes, it is taken out again.

Observation
The water freezes to ice, the bottle bursts.

V3.8 Black carbon from white sugar

Problem
A most astonishing phenomenon for students of any age or grade level is the reaction of white sugar and colorless pure sulfuric acid to black carbon and steam. It can be shown that sugar is a carbon compound, that carbon is not contained as a substance in the white sugar, but C atoms in the $C_{12}H_{22}O_{11}$ molecules of sucrose. Starting from sugar, this could be a motivating introduction to the chemistry of carbon compounds or organic chemistry.

Materials
Beaker (250 ml, tall form), glass rod; household sugar, pure sulfuric acid, water

Procedure
About 5 cm high of sugar is added to the beaker and mixed with a little water to form a paste. Approximately 3 cm high of sulfuric acid is layered on top, briefly stirred with the glass rod, the beaker is placed on a heat-resistant surface or on the tiled table and left to stand.

Observation
A vigorous hissing and swelling of the mixture indicates a reaction, during which it heats up very strongly. A black, porous substance forms in the shape of a sausage, which can be up to 20 cm long. A sweet decomposition smell becomes noticeable.

Disposal
The black substance, which is laden with concentrated sulfuric acid, is carefully wrapped in paper and disposed of in the container for solid waste. The beaker is cleaned with paper and rinsed with plenty of water.

V3.9 Electricity from a Lemon

Problem
To motivate in a subject-related way for electrochemistry or with regard to the voltage series of metals, two different metal strips can be immersed in a salt solution and electrical voltages up to 2 V between them can be detected with a voltmeter. It is also confirmed that no voltage can be measured with the same metals. It is even more motivating to use a lemon: The juice of the lemon is suitable as an electrolyte solution to produce the same phenomena as a salt solution.

3.6 Experiments

Materials
Beaker, voltmeter, cable cords and crocodile clips, knife; sodium chloride solution, metal strips of copper, zinc and magnesium, lemon

Procedure
a. The beaker is half filled with the salt solution. Two different metal strips are provided with crocodile clips and cable cords and connected to the voltmeter via the cables. The metal strips are immersed in the solution, voltages measured. The experiment is repeated with other metal combinations and also with the same metals.
b. Two cuts are made in the lemon with the knife and the partitions within the lemon are possibly destroyed with the knife. Two metal strips are inserted into the lemon and the voltages between the metals are measured.

Observation
Voltages of about 1.5–1.8 V can be detected between copper and zinc or copper and magnesium, much smaller voltage values between zinc and magnesium, no electrical voltages between the same metals.

V3.10 Brass Nameplate

Problem
Students are always highly motivated when they can make something themselves and take it home. For example, to introduce the topic "Acids dissolve metals", they can make a metal sign with their own name by scratching their name onto a wax-prepared metal sheet and etching these areas with a suitable acid. This sign can not only be used at home as a door sign, but the students can also proudly report to family and friends how they made it themselves.

Materials
Crystallizing dish, beaker, plastic pipette; brass plate or copper sheet, tealight, iron nail, concentrated nitric acid, boiling stones, gasoline, filter paper

Procedure
One side of the metal plate is thinly coated with the liquid paraffin of a burning tealight, a name or the desired figure is strongly scratched into this wax layer with the nail. The plate is placed with the wax layer facing up on some boiling stones in the crystallizing dish and placed in the fume hood. In the beaker, mix 5 ml of water with 10 ml of nitric acid and drop the solution onto the damaged areas of the wax layer with the pipette. After a few minutes, rinse with plenty of water and remove the wax layer with a spatula or with the help of paper and gasoline.

Observation
The acid solution reacts with the scratched areas of the metal with gas development: A blue solution of copper nitrate and a brown gas, nitrogen dioxide, are formed. After removing the remaining wax, the name or the marking is clearly visible in the metal sheet.

V3.11 Sphere packing model of a salt crystal

Problem
The self-assembly of structural models is very motivating for many students—especially when they can take these models home. For example, it is possible to illustrate the composition of salts using the example of a sodium chloride crystal, showing its structure made up of sodium and chloride ions in a 1:1 ratio (see also detailed construction instructions for chapter 7 "Models and model conceptions").

Material
Sodium chloride or rock salt crystals, cellulose spheres (about 18 white spheres Ø 30 mm and 18 red spheres Ø 12 mm), adhesive

Procedure
The spheres are glued together in the form of two layers twice (see picture) and stacked alternately to form a square column. It is determined how many small spheres touch a large sphere and how many large spheres touch a small sphere. The numerical ratio of large and small spheres is determined. The characteristics and limitations of the model are discussed in comparison with a sodium chloride crystal as the original.

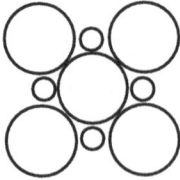 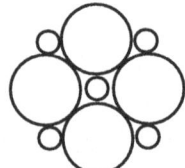

Observation

The layers fit tightly into each other and form a square column. A large sphere is touched by six small spheres inside the sphere packing, a small sphere by six large spheres. The numerical ratio of the types of spheres in the packing is 1:1.

References

1. Prenzel M, Krapp A (1992) Interessen, Lernen und Leistung. Aschendorff, Münster
2. Schiefele U (1990) Einstellung, Selbstkonsistenz und Verhalten. Hogrefe, Göttingen
3. Gräber W, Stork H (1984) Die Entwicklungspsychologie Jean Piagets als Mahnerin und Helferin im naturwissenschaftlichen Unterricht. MNU 37:257
4. Jäckel M, Risch KT (1988) Chemie heute. Sekundarbereich II. Schroedel, Hannover
5. Heilbronner E, Wyss E (1983) Bild einer Wissenschaft: Chemie. ChiuZ 17:69
6. Barke H-D, Hilbing CH (2000) Image von Chemie und Chemieunterrichts. ChiuZ 34:16
7. Harsch G, Bussemas HH (1985) Bilder, die sich selber malen. DuMont, Köln
8. Harsch G, Heimann R (1998) Didaktik der Organischen Chemie nach dem PIN-Konzept. Vom Ordnen der Phänomene zum vernetzten Denken. Vieweg, Braunschweig
9. Benmokhtar S, Harsch G, Heimann R, Wagner A (2014) Das START-Konzept im Anfangs-Unterricht. Aulis, Köln
10. Haupt P, Moritz P (2008) Modelle chemischer Substanzen für den Anfangsunterricht. Aulis, Köln
11. Wagenschein M (1971) Die Pädagogische Dimension der Physik. Westermann, Braunschweig
12. Ministerium für Schule und Weiterbildung des Landes NRW (2011) Kernlehrplan Chemie (G8). Ministerium für Schule und Weiterbildung des Landes NRW, Düsseldorf
13. Asselborn W, Jäckel M, Risch KH (2012) Chemie heute SI NRW. Schroedel, Braunschweig
14. Eisner W et al (2009) elemente chemie 1. Klett, Stuttgart
15. Tausch M, Wachtendonk M (2010) Chemie 2000 + Sekundarstufe I. Buchner, Bamberg
16. Schwedt G (2004) Experimente mit Supermarktprodukten. Wiley, Weinheim
17. Wanjek J, Barke H-D (1998) Einfluß eines alltagsorientierten Chemieunterrichts auf die Entwicklung von Interessen und Einstellungen. In: Behrendt H (Eds) Zur Didaktik der Chemie und Physik. Leuchtturm, Kiel
18. Lind G (1975) Sachbezogene Motivation. Beltz, Weinheim
19. Historische chemische Versuche. Köln 1997 (Aulis)
20. Roesky HW, Möckel K (1994) Chemische Kabinettstücke. VCH, Weinheim
21. Haupt P (1997) Die Chemie im Spiegel einer Tageszeitung. Bände 1–4. Universität Oldenburg, Oldenburg
22. Haupt P (2000) Da schmunzelt der Chemiker! NiU-Chemie 11:92
23. Werbung im TV-Programm „Prisma" Nr. 52/2017 für die Woche vom 30.12.2017–5. Jan. 2018
24. Ducci M, Rubner I, Friedrich J, Oetken M (2009) Chemistry und Cinema – das Projekt ChemCi: Eine Unterrichtseinheit zum Themenfeld Diamant und Graphit – inszeniert und illustriert mit Szenen aus dem Spielfilm „James Bond – Diamantenfieber". PdN-CidS 58:H1
25. Ducci M, Rubner I, Friedrich J, Oetken M (2010) Eine Unterrichtseinheit zum Themenfeld Atmung – inszeniert und illustriert mit Szenen aus dem Spielfilm „Das Boot" und „Apollo 13". PdN-CidS 59:H4
26. Ducci M, Rubner I, Friedrich J, Oetken M (2011) Eine Unterrichtseinheit zum Themenfeld Schwefel und seine Verbrennungsprodukte – inszeniert und illustriert mit Szenen aus dem Spielfilm „Dantes Peak". PdN-CidS 60:H1
27. Ducci M, Rubner I, Friedrich J, Oetken M (2012) Eine Unterrichtseinheit zum Themenfeld Redoxchemie und Toxikologie von Chromverbindungen – inszeniert und illustriert mit Szenen aus dem Spielfilm „Erin Brockovich". PdN-CidS 61:H8

Teaching Objectives

4

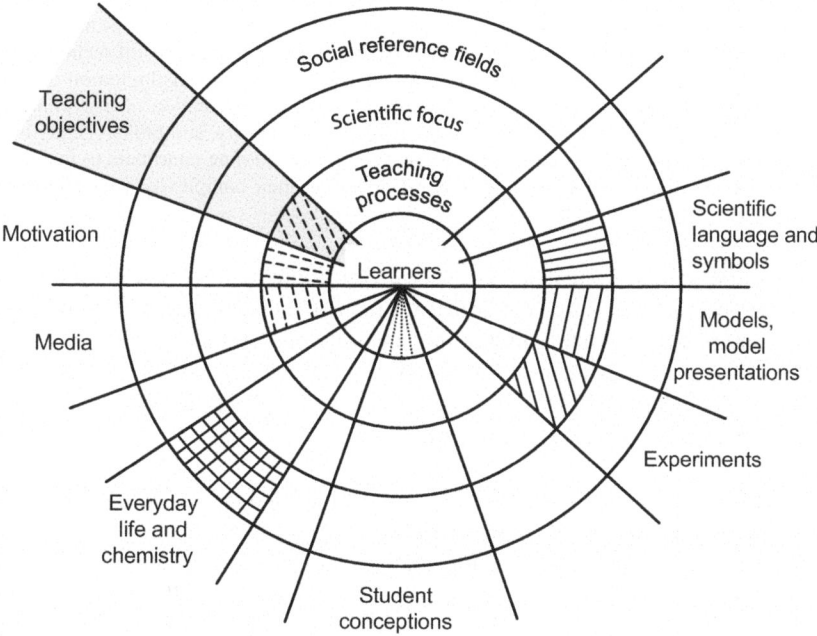

When we ask adults what they remember from their school chemistry lessons, we recognize a difficulty that often plagues our students as well. The answer is usually: "Oh, there were formulas", and one is proud if one still knows what H_2SO_4 is—not what it means—because usually little chemical knowledge has remained. A high-ranking school board official, who even was a scientist, said "Keep me away from the educational value of chemistry, it's just formula stuff". [1]

4.1 General Didactic Introduction

In our curricula, "formulas" are often very much in the foreground. As long as the learners do not understand the information content of these symbols, they also do not understand the content of the subject chemistry. Often they also flirt with their misunderstanding of the subject:

> How often have I heard from adults: "I never understood it". Most then acted as if they were even proud of it. After all, they have "become something anyway" [2].

Many teachers have obviously failed to convey the goals of scientific education from the perspective of the subject chemistry as experts of important professional associations see it:

> Mathematical and scientific education is an essential part of general education; it serves personality development by imparting methodological competence, factual knowledge and attitudes, and it enables a basic technical understanding for questions of technology and thus provides the basis for responsible participation in the social discussion about possibilities and limits of technical development. A specific contribution of disciplinary knowledge is provided by each of mathematics, physics, chemistry, and biology. Only on the reliable basis of subject teaching does interdisciplinary learning contribute to making problems from nature and technology understandable in their complexity and entanglement [3].

In order to be able to contribute to general education in this sense with the content of the subject chemistry, the goals of chemistry teaching must be constantly reflected and formulated according to the latest findings so that a good understanding of chemistry can result in learners. For a meaningful discussion of this basic question of chemistry didactics, general educational and school pedagogical areas must be reflected beforehand. These include:

- Goals of education and teaching, their dimensions and taxonomies
- Goals based on various "didactic models"
- Educational objectives and teaching analysis based on these models

4.1.1 Educational Objectives and Their Dimensions

In the natural sciences, the body of knowledge doubles approximately every 10 years. Accordingly, there is a need to discuss what attitudes, behaviors, abilities, skills, and knowledge should be demanded from learners today. The German Education Council [4] therefore asks:

> With what subjects and content should the student be confronted? What should he learn? In what learning steps, in what way and with what materials should he learn? How should the achievement of learning objectives be determined?

4.1 General Didactic Introduction

With regard to learning objectives, different levels of objectives must first be distinguished. General *guiding objectives* or *educational objectives* cover all areas of education in schools or for school education as a whole. *Broad objectives* concern specific teaching intentions of the subjects, *detailed objectives* refer to individual learning steps in a specific teaching unit. *Operationalized learning objectives* specify in detail the operations of the learners that are to be acquired in the classroom. They provide precise *behavioral dispositions* that are expected, for example: "The students should select three indicators from five given acid-base indicators and name their colors in the acidic and alkaline range." Since such concrete operationalizations require constant learning objective controls to collect the expected behavior of the learners, this type of objectives has not been established.

In learning objectives, aspects of the teaching content should already be linked with aspects of behavior. Behavioral changes can occur in three specific dimensions [5]:

- In the perception, memory, and thinking area: cognitive dimension
- In the interest, attitude, and value area: affective dimension
- In the area of manual and motor skills: psychomotor dimension

Learning objectives of these three dimensions can be hierarchized or transferred into a learning taxonomy. Especially for the learning objectives of the cognitive dimension, *learning objective hierarchies* have been proposed according to increasing complexity, for example by Bloom [6]:

1. *Knowledge*: Knowledge of concrete facts, rules, laws, and symbols
2. *Comprehension*: Linking facts, interpreting, extrapolating, deriving, inferring
3. *Application*: Being able to apply knowledge in new situations, transfer
4. *Analysis*: Breaking down complex information, recognizing existing relationships
5. *Synthesis*: Being able to assemble individual information into a complex
6. *Evaluation*: Evaluating complex facts.

The German Education Council [4] has reduced this hierarchy to the following points:

1. *Reproduction*: Reproduction of knowledge from memory
2. *Reorganization*: Independent reorganization of known facts
3. *Transfer*: Transferring known fact structures to new facts
4. *Problem Solving*: Solving novel tasks, finding new explanations.

The different levels of learning objectives can only be assessed against the background of previous learning and the knowledge level of the respective student—it is not possible to assign isolated educational objectives to these levels independently of this.

4.1.2 Didactic Models

Essentially, five didactic models have been developed:

- Educational Theory Didactics by Klafki [7]
- Learning Theory Didactics by Schulz [8]
- Curricular Didactics by Möller [9]
- Critical-Communicative Didactics by Winkel [10]
- Information Theoretical-Cybernetic Didactics by Cube [11]

In contrast to the *normative didactics* of past centuries, which formulated action instructions with goals, what should come out of education, current didactic models are oriented towards the **is** of educational reality. Different models for didactics [12] have therefore been developed because the teaching process is so complex that the essentials cannot be captured by one model alone. The models are therefore not in contradiction to each other, but complement each other:

> In didactics, we live in a pluralistic perspective world. The individual perspectives are represented by the different models of didactics [13].

The following will compare the first two models mentioned as examples.

The *educational theory didactics* starts from the "didactic analysis as the core of lesson preparation" [14] and develops a "thematic structure" through justifications for present and future significance. Only on this basis does the reflection of the "representability by media" and the "methodical structuring" take place (Fig. 4.1).

While the aforementioned model assumes the *primacy* of content, the founders of the *learning theory didactics* favor the thesis of the *interdependence* of decision fields: intentions, content, methods, and media belong together and also condition each other: The didactic analysis can start from any of these fields [15]. In his contribution, Schulz [8] later changed the decision fields, but continued to sketch the interdependence with double arrows (Fig. 4.2).

4.1.3 Lesson Planning and Analysis

Based on the learning theory didactics and the interdependence of prerequisites, goals, content, methods, and media, Bönsch [16] compiled an overview of lesson planning (Fig. 4.3). For the specific needs of chemistry teaching, a block "*Medium: Experiment/Model*" may be added: What experiments/models are planned, what alternatives are possible? How is the selection of experiments/models justified? Are the experiments used appropriately according to their specific function (introduction, hypothesis testing, data collection, repetition, etc.)? Is the experiment planned as a teacher, student, or group experiment (same work or division of labor)? Are necessary safety measures provided?

4.1 General Didactic Introduction

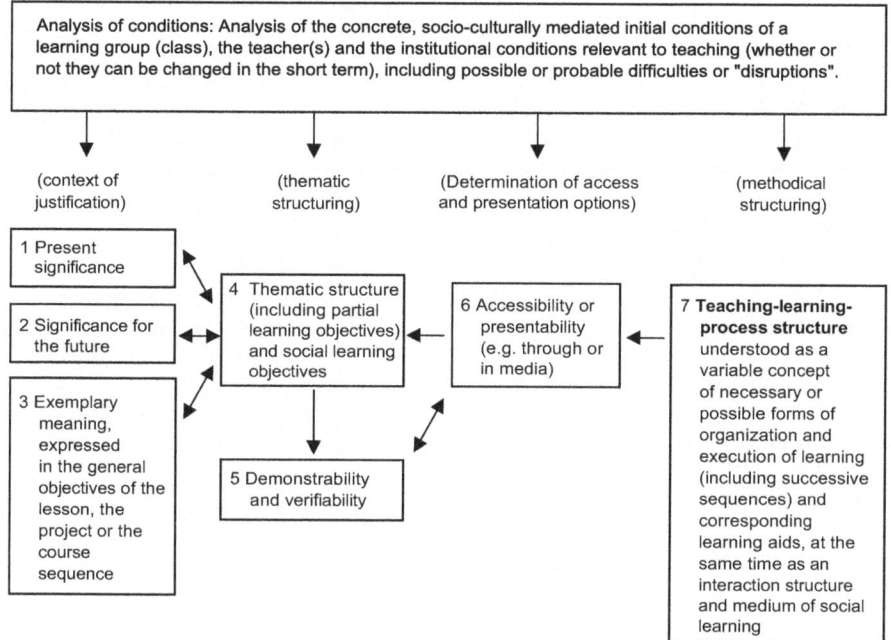

Fig. 4.1 Perspective scheme for educational theory didactics according to Klafki

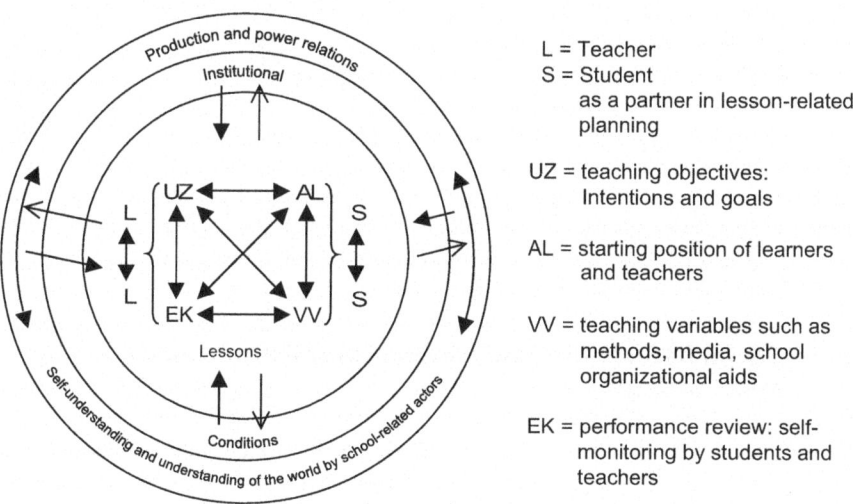

Fig. 4.2 Action moments of learning theory didactics according to Schulz [8]

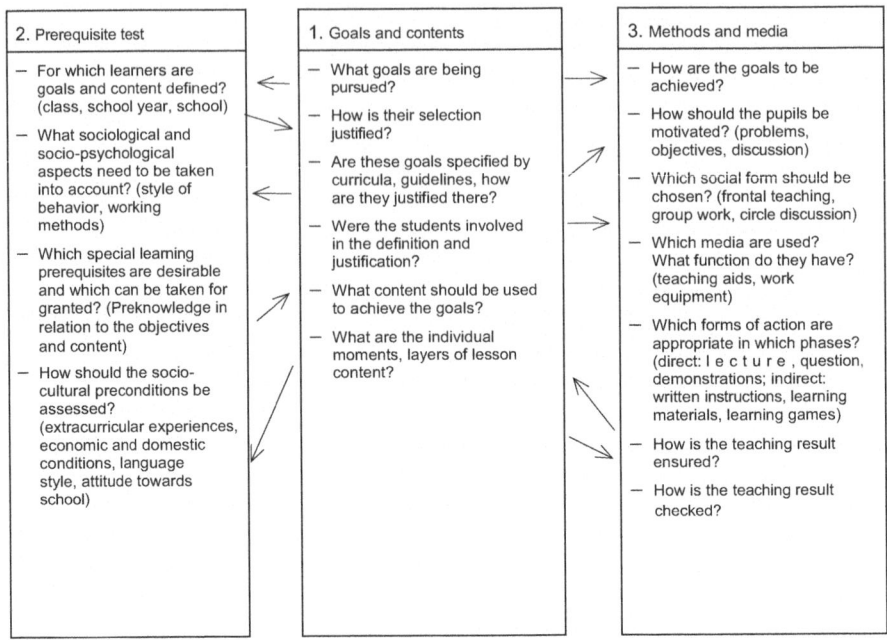

Fig. 4.3 Planning analysis based on the learning theory didactics according to Bönsch [16]

The planning analysis forms the basis for the **process analysis** of teaching—in this case too, a fourth question category for experimental teaching and a fifth question category for the use of models have been added (Table 4.1).

For lesson planning, Meyer [17] distinguishes in his "grid for student-oriented lesson preparation" between the teacher's teaching objectives and the **students' action objectives**. Based on this, the "student factor" is taken into account more than in other overviews: Examples are "socialization problems in the course system, everyday consciousness of the students, interest in the course (choices, compulsory assignment), performance and point orientation, temporal resilience".

4.2 Social Reference Fields: Educational Standards and Curricula

The "PISA shock" due to the mediocre performance of German students led to a heated education debate. In an effort to improve performance, the Conference of Ministers of Education agreed to standardize school education by prescribing uniform educational standards for all federal states [18]. The goal was not to list mandatory subject content, but rather to formulate *competencies* that are to be achieved in the sense of increased output orientation at the end of secondary level I. According to Weinert, competencies describe "the cognitive abilities and skills

4.2 Societal Reference Fields: Educational Standards and Curricula

Table 4.1 Process analysis based on planning analysis according to Bönsch [16]

1. General:	*Comparison of the observed teaching process with the teaching plan* Was the lesson able to be implemented according to the plan? If not, why not? What unexpected events were observed? Were the goals set too high, too low? Were the contents possibly already known? Were the methods, media appropriate for the goals/contents and the students? What should have been done differently, if necessary? Were the goals of the lesson achieved? Has this been checked, or can only assumptions be made?
2. Specific:	*Teaching in classes is always also a complex contact event* How strong (and disturbing) was the dominance of the teacher (language shares, leadership style)? To what extent were all students involved in the conversation, in the dispute? To what extent were conversations possible? Did the dominance of some students become noticeable in a disturbing or stimulating way? What kind of spontaneous methodological aids such as encouragements from the teacher were needed? To what extent were the students involved in decisions on goal determination, achievement? To what extent were independence and co-determination possible?
3. Specific:	*Teaching always aims to initiate, control, and successfully design learning processes in learners* How were the students motivated? Which problem situations were chosen? What could be observed from the learning processes of individuals? (Teaching events are not the same as individual learning events.) To what extent was individual learning possible? Were the starting position and final behavior recorded, and can the result, the learning gain of the individual, be stated with this? Is the learner essentially an object of organized learning, or is he a subject in the sense of having a say in learning goals, contents, methods, media, success controls? What type of learning processes were intended, which were realized? Which media were used, for what purpose, with what effect?
4. Specific:	*Chemistry teaching is experimental teaching* (see Chap. 6 "Experiments") Did the students understand the problem statement for the experiment? Was the execution of the experiment successful? Was the experimental apparatus drawn, was it understood by the students? Did the experiment actually contribute to problem solving? What questions remained open? Were the observations visible to all students—possibly through projection of the experiment? Were the instructions for group and student experiments understandable and complete? Was the evaluation of the experiment based on the observed phenomena or measured values? Did the students have the opportunity to record the execution, observations and evaluation of the experiment in writing? Were measurement errors discussed?
5. Specific:	*Chemistry teaching and the use of models* (see Chap. 7 "Models") Were the students able to recognize and understand the reasons for using models? Did they understand all the features of the model? Did they become aware of irrelevant ingredients of the model—perhaps through the comparative use of two or three structural models for the same fact? Did the students have the opportunity to grasp the submicroscopic structure of matter through the peculiar construction of structural models—such as sphere packing, space lattices or molecule models? Were structure and bonding models only verbally conveyed? Or did they become vivid for learners through drawings, spatial models or other media? Did the students discuss these media in terms of their imaging and shortening features? Did the differences between concrete visual models and thought models become clear? Were experiments and models related to each other?

available to individuals or learnable by them to solve certain problems, as well as the associated motivational, volitional and social readiness and abilities to use the problem solutions successfully and responsibly in variable situations" [19].

The goal of chemistry teaching is a comprehensive *scientific education*, which goes far beyond the acquisition of subject knowledge:

> Scientific education enables the individual to actively participate in societal communication and opinion formation about technical developments and scientific research and is therefore an essential part of general education. The goal of basic scientific education is to make phenomena tangible, to understand the language and history of the sciences, to communicate their results, and to engage with their specific methods of knowledge acquisition and their limits. This includes theory- and hypothesis-driven scientific work, which enables an analytical and rational view of the world [18].

The *educational standards* in chemistry for the intermediate school leaving certificate distinguish four areas of competence: subject knowledge, knowledge acquisition, communication and evaluation (Table 4.2).

In the area of subject knowledge, the school-relevant chemical subject content is reduced to four *basic concepts*, which on the one hand are intended to structure subject content, and on the other hand to network this with the knowledge of other scientific subjects.

In the area of the basic concept *"Substance-Particle-Relationships"*, students should be enabled to name substances with their typical properties, describe the submicroscopic structure using suitable atomic models, apply bonding models and explain the variety of substances on the basis of different combinations and arrangements of particles.

Within the basic concept *"Structure-Property-Relationships"*, students learn important ordering principles for substances, use models to interpret substance properties at the particle level, and infer possible uses from the properties of substances.

The basic concept of *"Chemical Reaction"* includes, among other things, the ability to describe phenomena of substance and energy transformation in chemical reactions, to determine types of reactions, to create reaction schemes, to recognize the reversibility of chemical reactions, and to describe possibilities for controlling chemical reactions by varying reaction conditions.

Table 4.2 Areas of competence of the educational standards in chemistry [18]

Areas of competence in chemistry	
Subject knowledge	Know chemical phenomena, terms, laws and assign to basic concepts
Knowledge acquisition	Use experimental and other research methods as well as models
Communication	Access, discuss and exchange information in a subject- and fact-related manner
Evaluation	Recognize and evaluate chemical facts in different contexts

4.2 Societal Reference Fields: Educational Standards and Curricula

In the basic concept of *"Energetic Consideration in Substance Transformations"*, students should recognize that the energy content of the reaction system also changes during chemical reactions through exchange with the environment, that energetic phenomena can be traced back to the conversion of a part of the chemical energy stored in substances into other forms of energy, and that chemical reactions can be influenced by the use of catalysts.

The following standards are set within the remaining three competence areas:

Competence Area Knowledge Acquisition
The students …

- recognize and develop questions that can be answered with the help of chemical knowledge and investigations, especially through chemical experiments,
- plan suitable investigations to verify assumptions and hypotheses,
- carry out qualitative and simple quantitative experimental and other investigations and document them,
- observe safety and environmental aspects when experimenting,
- collect relevant data in investigations, especially in chemical experiments, or research them,
- find trends, structures, and relationships in collected or researched data, explain these and draw appropriate conclusions,
- use suitable models (e.g., atomic models, periodic table of elements) to work on chemical questions,
- demonstrate exemplary connections between societal developments and findings in chemistry.

Competence Area Communication
The students …

- research a chemical issue in different sources,
- select topic-related and meaningful information,
- check representations in media for their technical accuracy,
- describe, illustrate or explain chemical facts using technical language and/or with the help of models and representations,
- establish connections between chemical facts and everyday phenomena and consciously translate technical language into everyday language and vice versa,
- document the course and results of investigations and discussions in an appropriate form,
- document and present the course and results of their work in a situation-appropriate and recipient-related manner, argue technically correct and logically,
- represent their positions on chemical facts and reflect objections self-critically,
- plan, structure, reflect and present their work as a team.

Competence Area Evaluation
The students ...

- present application areas and professional fields where chemical knowledge is significant,
- recognize questions that have a close relation to other subjects and show these connections,
- use typical and networked knowledge and skills to open up life-practical contexts,
- develop current, life-world-related questions that can be answered using scientific findings in chemistry,
- discuss and evaluate socially relevant statements from different perspectives,
- integrate chemical facts into problem contexts,
- develop solution strategies and apply them.

For the upper secondary school level, there are currently no specifications in the form of educational standards in the subject of chemistry. To graduate from German grammar schools, some state requirement, namely the "Einheitliche Prüfungsanforderungen in der Abiturprüfung Chemie" [20] apply. In addition to the basic concepts mentioned above for the middle school leaving certificate (substance-particle, structure-property, energy), there are two more: the "Donor-Acceptor Concept" and the "Equilibrium Concept". The uniform examination requirements are to be replaced in the near future—as has already happened in the subjects of German and mathematics—by educational standards.

At the state level, the educational standards for the middle school leaving certificate have been implemented in the form of *core curricula*, which precisely specify the individual competences, integrate them into contexts or illustrate them with examples. The final stage of design is the so-called *school-internal curricula*, which are designed by the chemistry conferences of the schools with the participation of chemistry teachers, parents, and student representatives.

Table 4.3 exemplarily shows a section from the core curriculum for chemistry for the introduction to secondary education at grammar schools in North Rhine-Westphalia (NRW), Germany [21]. Here, professional contexts are assigned to individual content areas, which can be replaced by equivalent contexts at the schools through a decision of the subject conference. Such an obligation to contextualize subject content is not found in all federal states; the individual core curricula vary significantly in their design.

Whether the educational standards have actually led to a stronger standardization of chemistry teaching is open. In 2012, the Institute for Quality Development in Education conducted a comparison of federal states in the field of natural sciences for the first time, focusing on the competence areas "subject knowledge" and "knowledge acquisition" [22]. About 44,500 students from 1300 schools participated in this. Figure 4.4 shows a sample task for the subject of chemistry in the

4.2 Societal Reference Fields: Educational Standards and Curricula

Table 4.3 Excerpt from the core curriculum for chemistry at secondary level I for grammar schools in NRW [21]

Content areas	Professional contexts
Substances and substance changes	*Food and drinks—all chemistry?*
Mixtures and pure substances Material properties Substance separation methods Simple particle representation Characteristics of chemical reactions	What's inside? • We examine food for its components • We extract substances from food • We change food by cooking or baking
Substance and energy changes in chemical reactions	*Fires and firefighting*
Oxidations Elements and compounds Analysis and synthesis Exothermic and endothermic reactions Activation energy Law of conservation of mass Reaction schemes (in words)	Fire and flame Fires and flammability The art of extinguishing fires Burned is not destroyed

---

Sample task for the comparison between federal states 2012 Competence area: Knowledge acquisition

Knowledge acquisition

Task: "Temperature and reaction rate"

In chemistry lessons, hydrochloric acid is added to zinc in an experiment.
The question of whether the temperature of the starting materials has an influence on the reaction rate is to be clarified.
Which of the following combinations V1-V4 is suitable to answer this question?
Mark the correct combination.

	Attempt	Mass of the zinc piece	Volume of hydrochloric acid solution	Temperature of the hydrochloric acid solution
☐	V1	5 g	10 ml	30 °C
		5 g	10 ml	50 °C
☐	V2	5 g	10 ml	30 °C
		10 g	5 ml	50 °C
☐	V3	10 g	10 ml	30 °C
		10 g	10 ml	30 °C
☐	V4	10 g	10 ml	30 °C
		5 g	10 ml	50 °C

Fig. 4.4 Sample task Chemistry, federal state comparison 2012 [22]

competence area of knowledge acquisition. Testing of the areas "communication" and "evaluation" is planned for the federal state comparison in 2018.

The teaching of chemical content does not start only during secondary school; selected aspects are already addressed in primary school science classes. For example, the NRW curriculum includes the focus areas "substances and their transformation", "heat, light, fire, water, air, sound", "resources and energy" or "environmental protection and sustainability". There are numerous experiments suitable for primary school on the various focus areas (see [23] and [24]).

The touching of the listed focus areas with topics of the initial chemistry teaching speaks for intertwining primary school and secondary school teaching more than has been done so far, in order to build on phenomena and knowledge that the children bring from primary school.

4.3 Learners: Cognitive development, student conceptions, attitudes, interests

The "student-oriented lesson preparation" by Meyer [17] most clearly shows that the learner himself must also be taken into account in the objectives of teaching and in lesson planning. It is primarily the attitudes, interests, and original ideas of young people that have a major influence on the successful design of lessons. However, aspects of developmental psychology should be discussed beforehand.

4.3.1 Learning Objectives and Developmental Psychology

Goals and content of school chemistry differ at various school levels due to developmental psychological conditions. Thus, Piaget [25] distinguishes the following four stages of thinking :

- Stage of sensorimotor intelligence (0–2 years)
- Stage of preconceptual-symbolic thinking (2–7 years)
- Stage of concrete operational thinking (7–13 years)
- Stage of formal operational thinking (from 13 years)

The transition from concrete to formal operational thinking is characterized by the learner increasingly detaching himself from actions on the concrete object and thinking more and more abstractly: considering variables, deriving potential relationships, understanding formal-mathematical descriptions. However, recent studies show that the age of 13 years for this transition is a very arbitrary determination: it can vary considerably. Studies show that often only 25% of 16-year-old high school students in grade 10 reach the final stage [26].

To describe the *development of cognitive structures*, Piaget uses the terms assimilation, accommodation, and equilibration. New perceptions are initially incorporated into the existing cognitive structure by learners or compared to existing prior knowledge. As long as new experiences are classified into the existing cognitive structure without changing it, this is referred to as *assimilation*. If there is a conflict between new information and prior knowledge, the individual must therefore change his cognitive structure and process the new appearance with a changed cognitive structure: this is referred to as *accommodation*. The organism's tendency to constantly establish this balance between structure preservation and readjustment is also called *equilibration* or *equilibrating*.

Current theories of learning adopt a *constructivist perspective*: each learner must individually build his own cognitive structure; changes in knowledge are constantly reconstructed individually. Depending on the type of knowledge change that occurs, the concept of constructivism speaks of *conceptual growth* and *conceptual change* [27]. *Conceptual growth* is comparable to assimilation in Piaget and takes place on a continuous learning path. *Conceptual change* is based on discontinuous learning paths, on which revisions of existing thought structures take place: often entire conceptual concepts have to be changed and radical restructurings have to be carried out. *Conceptual growth* and *conceptual change* are in a reciprocal relationship like assimilation and accommodation.

4.3.2 Student Conceptions

As the terms student and everyday conceptions or **misconceptions** suggest, at the beginning of chemistry lessons, learners have thought structures that have been constructed by the learners over many years based on their experiences. These are deeply rooted student conceptions, such as those that exist for the combustion process (Chap. 2).

Special attention must be paid to student conceptions, which are firmly anchored in the cognitive structure of students, for the determination of teaching objectives and content: on the combustion process, on the conservation law, on properties of gases, on the structure of matter, etc. (Chap. 2). The corresponding student conceptions should be carefully discussed in class before the scientific concept is conveyed in a comprehensible way through convincing experiments and models. So-called "common-sense explanations" or "everyday ways of speaking" also belong to the type of deeply rooted student conceptions: they are often also "socially accepted misconceptions" [26].

4.3.3 Attitudes and Interests

When formulating educational objectives, it is important to take into account attitudes and interests. If students have negative attitudes towards a subject, it is much more difficult to achieve the set goals in this subject than in another subject where the attitudes are positive. This also applies to all questions of motivation of young people (Chap. 3).

Heilbronner and Wyss [28] were able to show in 1983 through pictures of young people from Switzerland that the attitudes towards chemistry and chemistry lessons were very negative: "The vast majority of children see themselves and their environment threatened by chemistry." About 40% of the pictures showed images of a destroyed environment, 15% the direct threat of the individual by chemistry, 10% of the pictures were against animal testing. Barke and Hilbing [29] picked up the task for the students "Draw your picture of chemistry" and repeated the study in 1998: The proportion of negative motives, which was still 65% at Heilbronner, was reduced to 40% (see also Chap. 9).

Müller-Harbich, Bader and Wenck [30] examined the attitudes of secondary school students. They found a neutral to rejecting attitude among young people: A significant difference between the genders could not be determined. The place of residence had a significant influence on the attitudes: Students from places of residence in the Ruhr area with a relatively massive presence of chemical industry show a more negative attitude than young people who do not come from industrial locations.

If there are positive attitudes towards a subject, it is easier for young people to develop an **interest** in the content of this subject than with a negative attitude. If specific interests of the young people are known, it is advantageous for successful teaching to take these interests into account when defining educational objectives.

Gräber [31] examined the interests of young people and was able to prove that about half of all students find chemistry lessons interesting:

> What is remarkable is the course insofar as there is an increase with the beginning of chemistry lessons in the 8th grade and after a "hole" in the 9th grade again an increase in the 10th grade.

When asked about interest in contexts in chemistry and various activities, differences emerged between these items and also between boys and girls (Fig. 4.5). For the activities, the approval was mainly for the item "Perform experiment", also the items "Plan and observe experiments", "Watch films" and "Build chemical models" were rated as interesting.

Barke and Wanjek [32] were able to confirm this positive interest in student experiments. For the topic "Acids and Bases", substances known to the students from the kitchen and bathroom were used instead of the laboratory chemicals and used in the form of student experiments. This teaching unit not only received high approval from all students, but especially the girls expressed a higher interest in this teaching with everyday chemicals than before in the usual chemistry lessons.

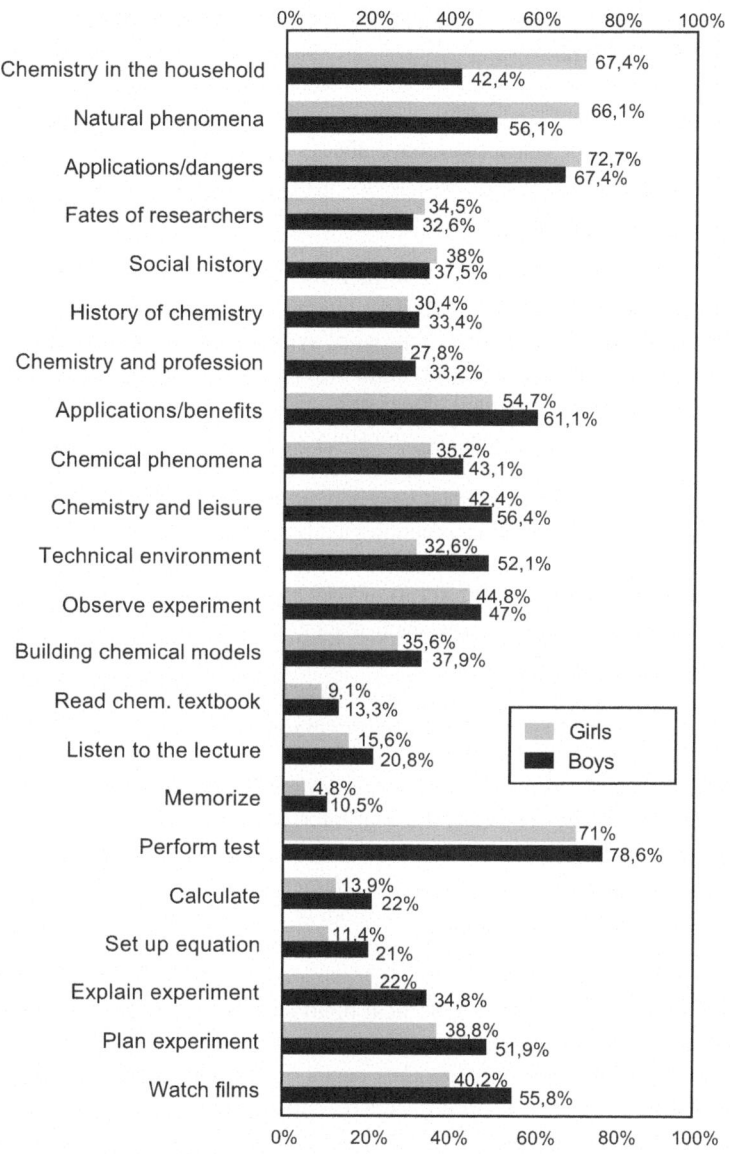

Fig. 4.5 Interest in contexts in chemistry and in various activities [31]

4.4 Scientific Focus: Chemistry Lessons as a Spiral Curriculum

Regarding the question of which phenomena and how much theory students should work on in chemistry lessons, there have always been heated discussions. "Save the phenomena", some say, "Model ideas are the actual domain of chemistry" say others. As so often, the truth lies in the middle: The phenomena naturally predominate at the beginning.

Guidelines and curricula (Sect. 4.2) therefore always aim for the *importance of phenomena* at the beginning of chemistry lessons: A description of first substances and chemical reactions takes place with word symbols before chemical symbols are added later. Harsch and Heimann [33] build the approach to organic chemistry strictly phenomenon- and action-oriented. Their curriculum is designed in such a way that students can first discover many relationships themselves by ordering phenomena before structure ideas and formulas play a role. New substances, phenomena, molecular structures and formulas are added later and are integrated into the growing system and thus easily assimilated: "**P**henomenologically **I**ntegrative **N**etwork" [33].

To what extent a simple *particle model* is introduced initially to interpret changes in states of matter, dissolution processes, diffusion, or simple chemical reactions, must be decided by each teacher for the respective student group. As long as particle associations—such as through sphere packing models for the structure of metal crystals—are conveyed clearly and students even build simple structural models themselves, this preliminary model is very useful for understanding (see also Chap. 7). Studies in grades 3 and 4 of primary school have shown that the usual drawings for particle arrangements are well understood and can still be successfully reproduced a year later.

As the adolescents progress into the stage of formal thought operations, the curriculum can include *Dalton's atomic model*. The terms element, compound, atom, io, and molecule introduce first models for the structure of elements and compounds, and formulate atom, ion, and molecule symbols or reaction symbols. Finally, the topics *atomic structure and chemical bonding* should be addressed: Corresponding symbols for the chemical bond and the structure of individual particles become possible (Chap. 8).

Models, especially *structural models* based on Dalton's atomic model, are very helpful for understanding formulas and reaction symbols—an important goal of chemistry teaching. Structural models—such as sphere packings for the structure of crystals or molecule models for molecule structures—are very suitable for conveying the structure of the substances involved at a medium level of abstraction before the chemical symbols and their different information content are introduced at the abstract level (Fig. 4.6, also Chap. 7 and 8).

The structuring of lessons based on models and symbols has two essential advantages. On the one hand, they can be arranged in a *curriculum spiral* and open up an overarching teaching concept that begins at the concrete-operational level of thinking in primary school and ends at an abstract level at a higher educational level (Fig. 4.7).

4.4 Subject Focus: Chemistry Lessons as a Spiral Curriculum

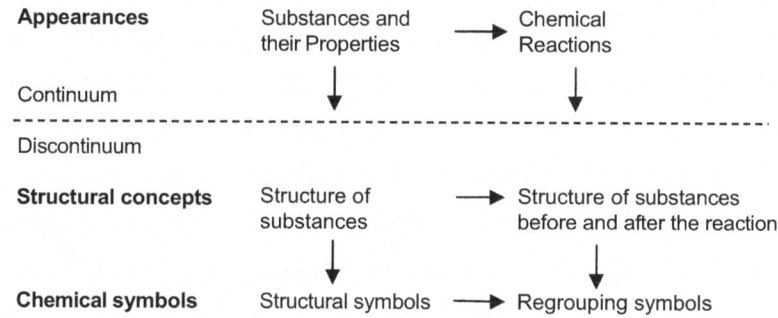

Fig. 4.6 Structural concepts as mediators between phenomena and symbols

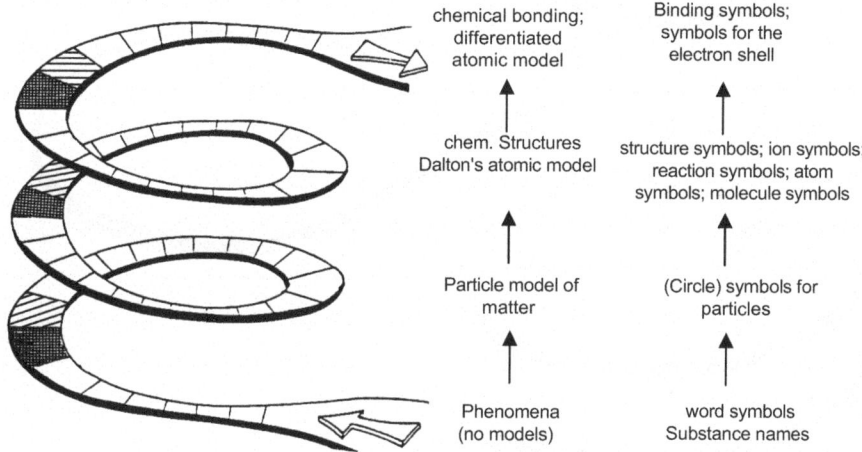

Fig. 4.7 Models and chemical symbols in the curriculum spiral

In Fig. 8.4 (Chap. 8), two further examples of spiral curricular approaches can be found for the terms solubility and acids, which start from everyday knowledge or student ideas derived from everyday knowledge and are increasingly sharpened over various learning levels until the descriptions reach the initially highest level of abstraction in school. Schmidkunz and Büttner [34] have tried to represent the entire teaching that concerns the subject of chemistry using a large curriculum spiral (Fig. 4.8).

On the other hand, the different models and symbols are very well suited to determine the times or grade levels at which these models and symbols should be used or which phenomena and reactions should be interpreted based on which model ideas when roughly fixing teaching objectives. These decisions need to be made depending on the type of school, grade level, prior knowledge, and quality of the student group and need to be made anew time and again.

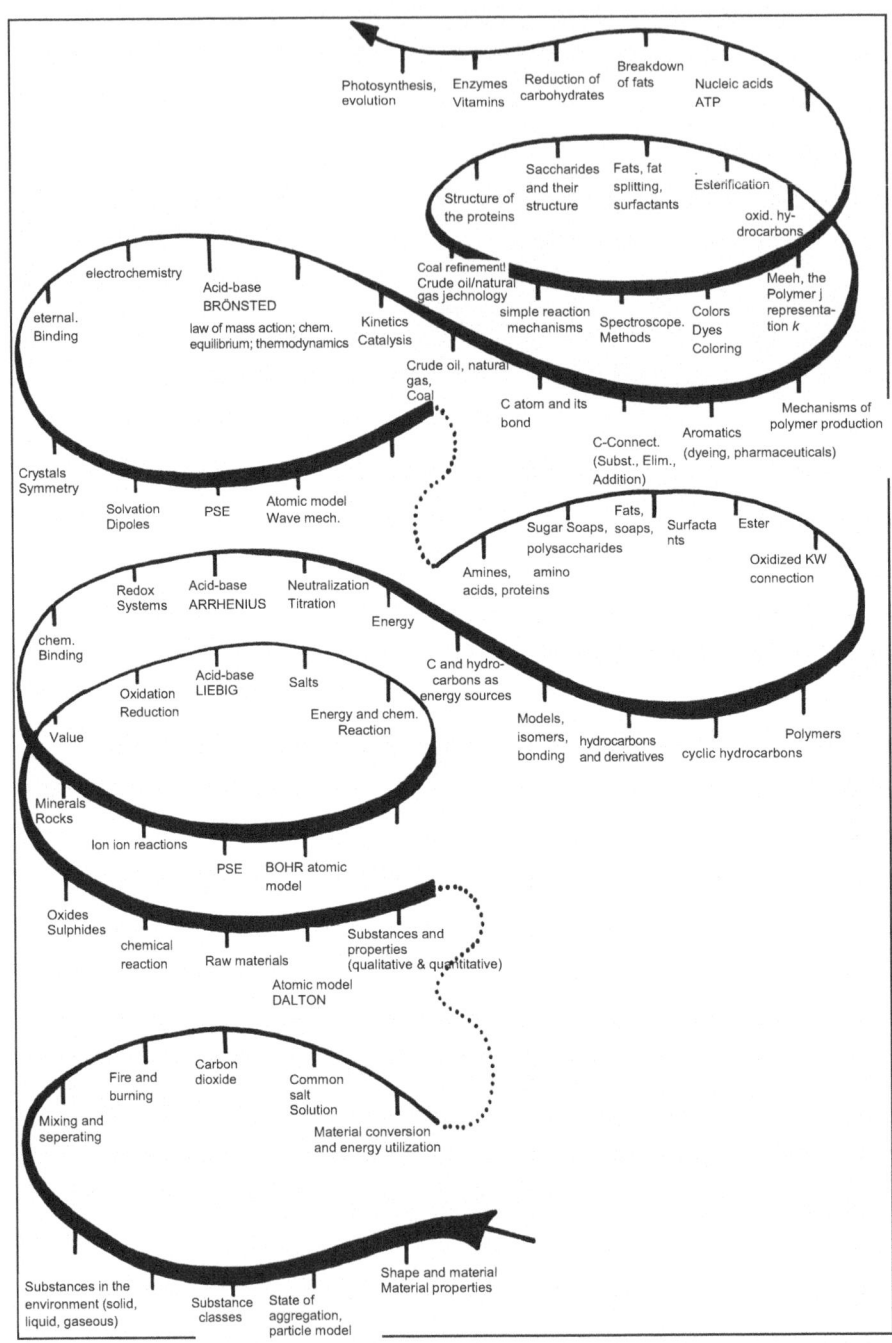

Fig. 4.8 Spiral curriculum for the subject of chemistry [34]

Finally, in addition to all concepts and their spiral curriculum arrangement, an important goal of chemistry education is to clarify the scientific method of knowledge acquisition to learners. In chemistry, knowledge is primarily gained using the empirical method. Understanding this approach is made easier for students if empirical work is carried out as often as possible in class. This *inductive-empirical method* can be briefly described as follows (see also Chap. 6):

- Observing a phenomenon
- Collecting further empirical findings, finding a regularity
- Formulating a *hypothesis* to explain the findings or regularity
- Seeking and experimentally verifying conclusions from the hypothesis
- Developing theories and models by linking supported hypotheses

In class, it is certainly not always possible to trace all the steps mentioned in every example. However, it makes little sense to arbitrarily omit steps and too quickly draw a universally valid connection from a single phenomenon. *Inductive approach* refers to deriving a regularity from individual phenomena, while *deductive* approach refers to the path from theory to the derived individual phenomena.

4.5 Teaching Processes—Teaching Methods for Achieving Educational Goals

> Teaching must be experienced as meaningful, otherwise it is meaningless. Learning must build on interest, otherwise it passes students by. "To be boring," wrote Johann Friedrich Herbart as early as 1806, "is the worst sin of teaching." Therefore, we teachers must overcome rigid frontal teaching (but also misforms of "open teaching") in favor of a teaching that responds to diverse learning opportunities with a rich choreography of teaching [35].

The demands on chemistry teaching appear high: It should not only impart specialist knowledge, but also introduce methods of scientific knowledge acquisition, promote evaluation and communication skills, relate to the real world, be exciting and ideally be experienced as meaningful by the students. It is undisputed: These goals can hardly be achieved in rigid frontal teaching. Rather, student-oriented teaching methods are needed that actively involve learners in all phases of scientific work.

Teaching methods that are specifically tailored to the requirements of a scientific lesson can help teachers in planning and designing these learning processes. In contrast to *methods* (such as group puzzles, station learning, etc.), which can be used equally in many subjects, teaching *procedures* or *concepts* not only structure the learning processes, but always also pursue a didactic idea. The methods presented below represent different orientations: problem orientation, orientation towards the history of chemistry, context orientation, orientation of teaching towards student ideas, and the promotion of evaluation competence. Another teaching concept provides suggestions for designing learning processes in inclusive learning groups. An overview of typical teaching concepts for chemistry teaching can be found under [36].

4.5.1 The Research-Developing Teaching:

The research-developing method according to Schmidkunz and Lindemann is one of the problem-oriented teaching approaches [37]. The teacher gradually introduces the students to a problem that they should ideally solve independently ("research") during the unit. False paths and repeated attempts at solutions are not viewed negatively, but are understood as an expression of a realistic research approach. The teacher's task is to accompany (to "develop") the students' learning processes, but not to determine them.

The method divides the lesson into five stages of thought, which in turn can be divided into three phases of thought (Table 4.4).

Especially in the early days, this phasing was misunderstood by many teachers and subject leaders: They believed that a teaching unit must always include all phases—and preferably in a single lesson. In contrast, the authors emphasize that depending on the topic, individual phases of thought or even stages can be omitted. Moreover, independent planning, implementation, and reflection of solution proposals by the students can hardly be realized in 45 minutes.

Depending on the students' prior knowledge, the authors distinguish two "branches" of the method: In the "deductive branch", the students' prior knowledge is so great that they can largely predict the solution to the problem; it only needs to be secured by a *confirmation experiment*. The phases of thought 3c to 4c can be omitted here. In the "inductive branch", on the other hand, the prerequisites for safely predicting the solution to the problem are lacking. Possibly, the students formulate various hypotheses that need to be tested and interpreted by *further experiments*. In this case, the second and third stages of thought can be run through several times.

Table 4.4 Stages and phases of thought in the research-developing teaching method [37]

Stage of thought		Phase of thought
1	Problem acquisition	Problem basis Problem understanding Problem formulation
2	Considerations for problem solving	Analysis of the problem Solution proposals Decision for a solution proposal
3	Implementation of problem-solving proposals	Planning of the experimental solution project Practical implementation of the solution project Discussion of the results
4	Abstraction of the gained insights	Iconic abstraction Verbal abstraction Symbolic abstraction
5	Securing of knowledge	Application examples Repetition Learning objective control

Examples

The authors describe various examples that illustrate that the method can be applied from primary school teaching to upper secondary level [37].

A possible problem in primary school social studies is the question of how to prevent a paper boat brought by the teacher from soaking up water and sinking after a short time. Various solutions are conceivable (e.g., choosing other types of paper or multiple layers of paper; painting with color or varnish). Painting the boat with wax crayons ultimately leads to success. In the abstraction phase, the water-repellent effect of the wax is figured out and secured in the last stage of thought using application examples such as waxing shoes.

An example for middle school chemistry addresses the question of whether specific cleaners are necessary for the removal of certain types of dirt. For this purpose, the teacher brings various household cleaners to class. The list of ingredients on the bottles shows that agents for removing lime deposits always contain acids. Whether acids are indeed a necessary prerequisite for removing lime stains is experimentally tested by the students by treating lime deposits with different types of cleaners. The assumption is confirmed: Only acidic cleaners prove to be suitable.

A research-developing teaching unit for the upper level by K. Himmerich and M. Weiß works out the principle of redox titration using the example of determining the oxalic acid content in rhubarb [38].

For Reflection

An investigative-developmental teaching approach familiarizes students with essential aspects of scientific knowledge acquisition, such as hypothesis formation and experimental hypothesis testing. The authors also refer to a long-term study that proves that insights from an investigative-developmental teaching approach are significantly better retained in memory than knowledge acquired through "conventional" methods [37].

A possible danger lies in conveying an idealized image of science in the sense that "every scientific problem also finds a solution". This should be reflected upon in the classroom.

A challenge for teachers is to anticipate students' solution proposals when preparing the lesson and to have suitable materials available for their implementation. Here, a temporal separation of thinking stage 2 (considerations for problem solving) and 3 (implementation of problem-solving proposals) helps by dividing them into two lessons. The use of *interaction boxes* also facilitates preparation for the teacher. Interaction boxes provide students with a selection of devices and chemicals with which they can solve the problem. The provided materials should allow for different solution proposals. Interaction boxes thus offer a balance between the desire for good lesson planning for the teacher and the need to give students sufficient degrees of freedom in developing solutions (see the special issue *Interaction Boxes* [39]). ◄

4.5.2 The Historically Problem-Oriented Teaching

The idea of orienting chemistry lessons on historical developments was already developed by Rudolf Winderlich at the beginning of the 20th century. The current understanding of a historically problem-oriented teaching approach goes back to W. Jansen. The aim is to pick up historical problem situations in the classroom, which are reconstructed, understood, and—as far as possible—solved using historical sources and experiments. The students should be enabled to work out parts of the solution path as independently as possible.

According to Jansen, the main aim is "… to give students fundamental insights into a scientific understanding of the world. The teacher must make it clear that theories often cannot be derived empirically-inductively in simple causal chains, but must be invented and thus must bear speculative features. Older theories often touch on latent ideas of the students" [40].

According to Jansen, a historically oriented teaching can also show the dependence of research on the demands of society and clarify the human dimension of science. Especially the human component can help to reduce the emotional distance to the science of chemistry.

Examples
Examples of historically oriented teaching units can be found under [41, 42] and [43]. Matuschek et al. developed a teaching sequence that focuses on van't Hoff's considerations for the structural elucidation of the methane molecule in 1834 [41]. The students receive ball models made of cellulose or styrofoam (alternatively, clay or wooden beads would be conceivable) and connecting rods made of wood or wire (alternatively: matches). The task is to develop possible spatial models for the methane molecule. A rectangular-planar, a pyramidal, or a tetrahedral arrangement are conceivable, for example. As a stimulus for hypothesis formation, the authors suggest a historical source:

> Since the starting point of the following considerations was found in the chemistry of carbon compounds, I will initially only share the part that relates to this. It is becoming more and more apparent that today's constitutional formulas are unable to explain certain cases of isomerism; perhaps this is due to the lack of a more specific conception of the actual arrangement of the atoms. … If one assumes that these are spread out in a plane …, where the four affinities of each carbon atom are represented by four perpendicular directions in the plane, one arrives at the following number of isomers (the hydrogen atoms are replaced in turn by univalent groups R_1, R_2 etc.) when applying to the derivatives of methane CH_4 (if we choose the simplest case as a starting point): …
> A second assumption brings the theory into agreement with the facts, namely that the affinities of the carbon atom are thought to be directed towards the corners of a tetrahedron, the center of which is formed by this atom itself. The number of isomers then simply becomes the following … [44]

Following the discussion of the text passage, the students consider how many isomers there would be in the case of different spatial arrangements if one or more hydrogen atoms were replaced by other substituents. In the first step, the chlorination products are considered, and in the second step, further halogenation products of methane are examined. The number of isomers is then compared with the actual number of isomers. In this way, the students can recognize that only the tetrahedral model can explain the actual number of isomers found (Table 4.5).

The fact that science always has a human dimension is made clear to the students in an article by Hermann Kolbe in the "Journal for Practical Chemistry". In it, Kolbe reacts in a highly polemical form to van't Hoff's article "On the Arrangement of Atoms in Space":

> In a recently published essay with the same title, I identified one of the causes of the current decline in chemical research in Germany as the lack of general and at the same time thorough chemical education, which a not insignificant number of our chemical professors suffer from to the great detriment of science. The result of this is the rampant growth of the weed of scholarly and seemingly witty, in reality trivial, mindless natural philosophy, which, eliminated by exact natural research 50 years ago, is currently being brought out again by pseudo-natural researchers from the junk room that harbors the aberrations of the human mind and, like a modernly dressed up and made-up whore, is trying to smuggle into good society where she does not belong.
>
> Anyone who considers this concern exaggerated should read, if they can, the recently published, fantasy-filled treatise by Messrs. van't Hoff and Hermann on "The Arrangement of Atoms in Space". I would ignore it like many others if a notable chemist had not taken it under his protection and warmly recommended it as a meritorious achievement. A Dr. J. H. van't Hoff, employed at the veterinary school in Utrecht, seems to find no taste in exact chemical research. He found it more convenient to mount Pegasus (obviously borrowed from the veterinary school) and proclaim in his "La chimie dans l'escape" how the atoms appeared to him in the universe on the bold flight to Parnassus. … It is characteristic of today's uncritical and criticism-hating time that two virtually unknown chemists [van't Hoff and le Bell], one from a veterinary school, the other from an agricultural institute, judge the highest problems of chemistry, which will probably never be solved, especially the question of the spatial arrangement of atoms, with a certainty and undertake their answer with a boldness that astonishes the real natural researcher [45].

Another historically-problem-oriented teaching example by Berger et al. [42] addresses the discovery of the alkali metals by retracing the historical experiments of Humphry Davy (1778–1829). He found that the electrolysis of an aqueous

Table 4.5 Number of isomers of halogenation products of methane [41]

Number of isomers			
Compound	Square-planar arrangement	Tetrahedral arrangement	Actually found isomers
CH_3Cl	1	1	1
CH_2Cl_2	2	1	1
$CHCl_3$	1	1	1
CCl_4	1	1	1
CH_2BrCl	2	1	1
$CHBrClF$	3	2	2

solution of zinc, copper, or lead salt led to the deposition of the respective metal at the negative pole; however, the electrolysis of "sulphuric natron or potash" (sodium or potassium sulfate) resulted in no metal deposition, but rather the formation of hydrogen. The assumption that these compounds, too, must contain—hitherto unknown—metals led to the consideration of attempting electrolysis in a non-aqueous state. Indeed, the electrolysis of a *melt* of caustic soda (sodium hydroxide) or caustic potash (potassium hydroxide) led to the deposition of metallic globules:

> Immediately a lively reaction was observed. The potash began to melt at the two points where it was electrified. A violent effervescence was seen on the upper surface; on the lower, or negative, no release of an elastic fluid (gas) was noticeable, but I discovered small globules that had a very lively metallic luster and looked exactly like mercury. Some burned the moment they were formed, with explosion and bright flame; others remained, but tarnished, and finally covered themselves with a white crust that formed on their surface [46].

According to a nephew, this discovery made Davy dance around the laboratory table in excitement. During the course of the teaching unit, Davy's discovery is retraced using student and teacher experiments as well as various historical sources.

For Reflection

The teaching concept can help counteract idealized notions students may have about processes of scientific knowledge acquisition. They can realize that scientific progress is neither linear nor exclusively follows logical criteria. Above all, the human component, such as the influence of personal reputation or envy on the enforcement of a theory, offers students a new perspective on science and can increase interest in a subject matter.

Of course, not all historical questions are suitable for problem-oriented processing. In some cases, the historical component of chemistry may also be limited to the inclusion of historical sources, experiments, or anecdotes in the classroom. The procedure also requires careful selection of sources, as comprehension can be hindered by outdated technical terms or linguistic expressions. ◄

4.5.3 ChiK – Chemistry in Context

The concept of "Chemistry in Context" (orig.: "Chemie im Kontext" [47], developed by Demuth, Parchmann and Ralle, is based on insights and experiences from the "salters advanced chemistry project" [48], which was developed at the University of York. Both projects start from the premise that *situated* knowledge can be better applied and transferred. Therefore, the starting point for learning processes in chemistry teaching should be a real-life context that illustrates the relevance of chemistry for daily life. According to ChiK, chemical subject matter is developed from this context and traced back to basic chemical concepts. In addition to context orientation and networking with basic concepts, methodological diversity is listed as a third basic principle.

The teaching concept proceeds in four phases: The *encounter phase*, in which the students familiarize themselves with the context, is followed by the *curiosity and planning phase*, in which the students participate in further planning and structuring of the teaching. They can ask questions about the context and plan strategies by which experiments, research or other measures they can clarify these questions. However, questions and impulses are also introduced by the teacher. The *elaboration phase* serves to pursue the previously developed questions. This phase is again characterized by a high level of student activity, which is supported and moderated by the teacher. Different methods are used. In the final *networking phase*, the chemical subject matter is extracted from the context, linked to basic concepts and applied in new contexts.

Examples

Examples of teaching sequences according to ChiK can be found both on the internet and in journal articles as well as the published textbook series. An overview of developed teaching units can be found at [49]. A teaching sequence on the topic "Acids in the Pantry" addresses various aspects of acids in food [50]. Possible questions could be: What are acids? Why do some foods taste more acidic than others? Do acidic foods harm the teeth? What is the purpose of stomach acid? How does Helicobacter survive in stomach juice? How does milk turn sour?

Some questions are pursued through experiments (e.g., taste tests, pH measurements, yogurt production, etc.) which are partly planned independently by the students (effects of acids on teeth), and partly prescribed by worksheets. In addition, the students receive various information sheets; these contain, for example, expert knowledge about the functioning of stomach acid, the bacterium *Helicobacter pylori*, the discovery of ascorbic acid, or the "donor-acceptor concept". Furthermore, the students are given a glossary of basic terms of the teaching unit, which explains terms such as *protolysis*, *pH value*, or *indicator*.

In the final networking phase, technical content from the development phase should be taken up in a new context; for this, the context "acid rain" is suggested.

Table 4.6 provides an overview of another teaching series according to ChiK, which was developed under the title "Coca Cola—More than a Refreshing Drink" for introductory chemistry lessons [51].

For Reflection

Context-oriented teaching proves to be well suited for establishing connections between chemical subject matter and everyday phenomena. It rightly puts the function of natural science in the foreground, to understand and shape the world. Context-oriented teaching can therefore be meaningful and motivating for students. However, the elaboration of the subject matter and networking

Table 4.6 Context-oriented teaching unit on Coca Cola® according to ChiK [51]

	Contextual Content	Chemical Content
Encounter and Curiosity Phase	*OHP template*: A taster in distress Taste samples with different types of cola	Material property Taste?
Planning Phase	*OHP template*: Chemistry replaces the taster	
	Collection and structuring of suggestions leading to the investigation of Coca-Cola, e.g. • How much sugar is in Coca-Cola? • What "ingredients" does Coca-Cola contain? • What effect does Coca-Cola have, e.g. on meat? • What happens when Coca-Cola is heated? • ...	
Elaboration I	Why does something taste sweet? How much sugar is in Coca-Cola?	Particle model Key-lock principle Density Solubility, concentration
Elaboration II		
Group Work	**Experiments with Coca-Cola:** Evaporation of different types Boiling curves Distillation Decolorization of Coca-Cola ...	
Division of labor group work with posters/presentations as final products	**Ingredients of Coca-Cola:** A: Sugar B: Carbonic acid C: Phosphoric acid D: Caffeine/Sweeteners E: Sugar color F: History G: Recipe H: Health	States of aggregation and transitions, explanation with the particle model Melting and boiling points of substances Separation methods Property combinations/"profiles" Detection reactions Basic concepts: substance-particle concept
Deepening and networking phase—textbook work	Decontextualization	Characterization of chemical reactions Basic terms of chemical reactions

with the basic concepts at the end of the series poses a challenge for teachers and students, as a wide variety of different aspects are usually worked out in the course of a context-oriented unit. Bundling these and applying them in a further context is demanding and requires good structuring by the teacher. ◀

4.5.4 choice²learn

Students develop numerous ideas that deviate from our professional concepts in chemistry (Chap. 2). Reasons for this include everyday experiences and language, but also misleading models, illustrations, and definitions that we use in class [52, 53]. The teaching concept choice²learn developed by A. Marohn [54] is based on the insight from teaching-learning research that teaching must be oriented towards the students' ideas, as these significantly influence the success of learning processes.

However, according to the constructivist approach, students' ideas cannot be replaced by scientific concepts in a simple "exchange process". Often, measures such as *explaining* the scientific concept or *refuting* the student's idea (e.g., through an experiment) are not sufficient—especially when ideas have been imprinted and consolidated over years due to everyday experiences. For this reason, choice²learn initiates a conscious and independent examination of the ideas of the learning group.

The core idea of the concept is based on the approach of (socio-)cognitive conflict [55, 56] and collaborative argumentation [57, 58]: Students with different ideas are confronted in small groups with phenomena and information that they can use to check their ideas.

The design takes into account various criteria of *conceptual-change* research: that students should negotiate different viewpoints among themselves and act independently and without teacher intervention [59]; the integration of the learning process into a context close to the students [60]; a balance between independent learning and impulses from the teacher [61].

The starting point of the learning process is a multiple-choice task on a scientific fact (e.g., the evaporation process), which is embedded in a context close to everyday life (Fig. 4.9). The multiple-choice answers represent typical student ideas that have been empirically determined in studies [52, 53, 62]. After each student has individually chosen a multiple-choice answer, the students work on various learning impulses in small groups, e.g., experiments, thought experiments, models, or questions. From these, limits and contradictions to the various answers can be derived. In this way, the students gradually develop the technically "correct" idea in an argumentative process.

The multiple-choice task thus serves not only for the *diagnosis* of student ideas in the context of the teaching method, but is also used as a *learning task*—this also results in the name of the teaching concept: choice²learn.

The teaching phases (Fig. 4.10) are designed to last a double lesson, which can be used flexibly over the course of a school year. The developed learning materials refer to "elementary", i.e., significant and stable ideas, which students often hold on to until the upper grades, even though the scientific concept has already been addressed several times in chemistry lessons [52]. Choice²learn can thereforenot only be used for the development of a new subject content, but also as a review and repetition of previously covered content.

TASK

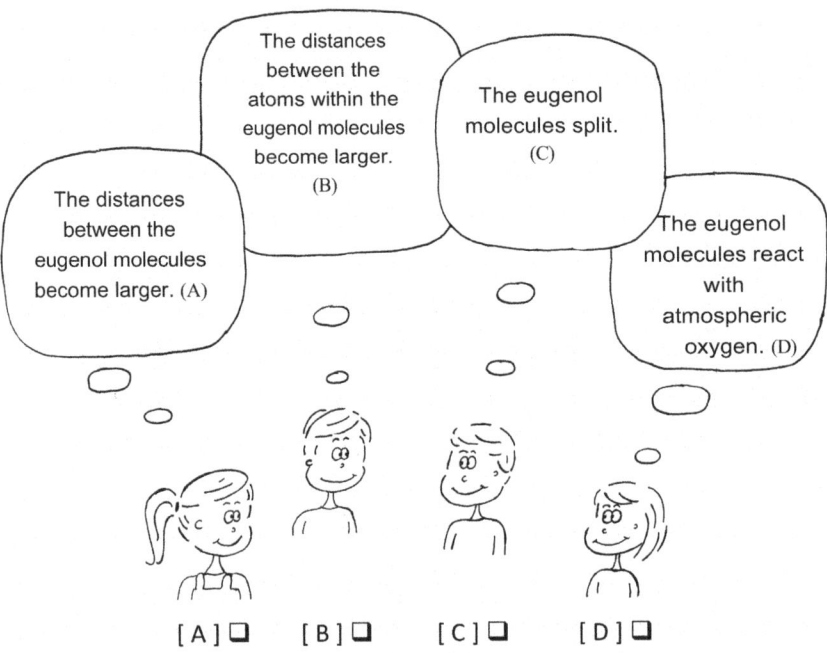

Fig. 4.9 choice²learn: Learning task

Examples
Examples of lesson planning can be found, for example, under [63–67]. The lesson unit "Evaporation of Eugenol" [63] begins with the teacher placing a fragrance lamp with clove aroma on the teacher's desk (**contextualization**). The spread of the scent in the classroom leads to the question of how this spreading process can be explained at the particle level. Alternatively, the teacher can use the text in Fig. 4.11 as a written context for the subsequent multiple-choice task.

4.5 Teaching Processes—Teaching Methods for Achieving Educational …

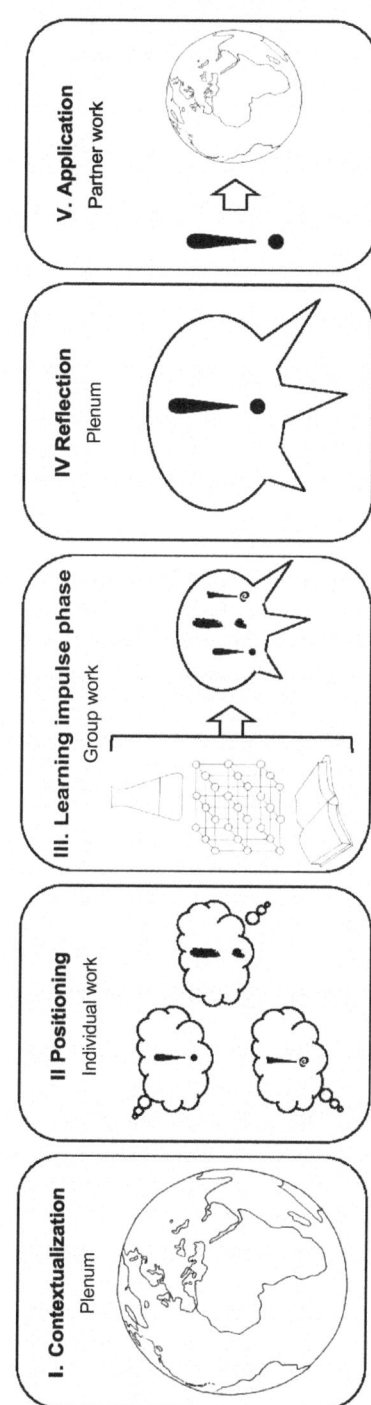

Fig. 4.10 choice²learn: Teaching phases

It's Christmas time and Lisa is happy that she can use her scented lamp with wintery fragrances again. She pours some clove oil into the small bowl and places a lighted tea light underneath. After just a short time, Lisa can smell the scent of cloves.

One component of clove oil is eugenol. This substance is responsible for the clove fragrance.

How can you explain that you can smell the scent of cloves throughout the room when the scented oil is heated?

Fig. 4.11 choice^{2}learn: Context

In the second step, each student completes the multiple-choice task in Fig. 4.9 in individual work, coding his or her task sheet with a personal abbreviation (**positioning**). The task, using the example of eugenol, a component of clove aroma, reflects the most common conceptions of high school students about the evaporation or condensation process. Each student must now consider whether the evaporation process is associated with an increase in the molecular distance, an increase in the distance of the atoms, a molecular splitting, or a reaction with oxygen.

Subsequently, students who have chosen different selection answers are grouped into small groups, with the aim of clarifying the correct answer. To this end, the teacher or a student collects the task sheets, sorts them according to the ticked answer on four stacks (A, B, C, D) and then sorts the sheets into small groups, so that advocates of different answers come together in one group. The members of a group are called up based on the coding of the task sheets and sit together. Due to the coding, it remains anonymous to the teacher which student has chosen a wrong answer.

At the beginning of group work (**learning impulse phase**), each student first presents his or her own selection and tries to justify it. Arguments for or against a selection answer are entered into the argumentation sheet (Fig. 4.12). The basic principle of the group work phase is that the students try to respond to the considerations of the other group members by further thinking them through or perhaps even refuting them.

Since the groups are usually not able to definitively clarify the question without assistance, they can gradually draw on learning impulses, which are available in (multiple) copies on the group tables. The order of the impulses is predetermined. The aim is to derive supporting or contradicting statements from the learning impulses to individual selection answers; these are again entered into the argumentation sheet.

Learning Impulse 1 (Fig. 4.13) is intended to help students illustrate and justify their own position to the other group members. The drawings are particularly helpful for those students who have difficulty distinguishing between the terms "atom" and "molecule". They also provide a basis for discussion for the further course of the conversation.

4.5 Teaching Processes—Teaching Methods for Achieving Educational ...

Learning impulse	Argument	Speaks for the selection answer(s)	Speaks against the selection answer(s)	Evaluation of the argument Strong: + Medium: o Weak: -
	The following arguments were entered before the learning impulses were added:			
	It remains the same substance; it has only become gaseous.	A	B, C, D	+
	The heat causes the atoms to vibrate more strongly and therefore they need more space.	B	A, C, D	o
	Eugenol is broken down by heat, spreading the fragrance throughout the room.	C	A, B, D	o
	Eugenol reacts with acid and can be released into the air be worn.	D	A, B, C	+
	The following arguments were entered during the learning stimulus phase:			
2	There is no oxygen in the balloon; Eugenol cannot react with oxygen. Balloon nevertheless expands.		D	+
3	The distance between the atoms to each other would be so great that there would no longer be a bond, but individual atoms.		B	+
4	Carbon dioxide and wate cannot be smelled; the eugenol must be preserved.	A, B	C, D	+
5	Oxygend and hydrogend would be produced and a flame would trigger an oxyhydrogen reaction		C	+

Fig. 4.12 choice²learn: Argumentation sheet, filled out by a small group of 10th grade students

> **Learning impulse 1**
>
> Draw how you imagine the processes on the particle level!
>
> ***Explain*** *your drawing to the other group members and **give reasons for** it.*
> *the answer you have chosen!*
> *give reasons for your chosen answers!*

Fig. 4.13 choice²learn: Learning Impulse 1

Learning Impulse 2 (Fig. 4.14) serves to exclude selection answer [D]. Among the students who choose answer [D], two ideas dominate: on the one hand, the assumption that eugenol reacts with oxygen from the air to form CO_2 and water during evaporation or condensation; on the other hand, the idea that the eugenol molecules attach to the oxygen molecules of the air and are only "carried into the gas phase" by these. However, Learning Impulse 2 clarifies through a model experiment that an evaporation process and an increase in volume occur even without the involvement of oxygen. The fact that the substance has actually transitioned into the gas phase is supported by the "shake test", which shows that after heating, no liquid remains in the balloon.

Learning Impulse 3 (Fig. 4.15) contradicts answer [B], which assumes that the atoms within the eugenol molecules gain greater distances from each other when evaporating. If one follows this assumption, the molecules would have to expand over many meters; otherwise, it would not be explainable that the scent of cloves can still be smelled at the end of the room. This seems very unlikely, especially since these "giant molecules" would have to dock at the receptors of the human nose to be perceived by smell. Some students argue at this point that with such a large expansion, there would no longer be any cohesion between the atoms within the molecules, so the molecules would actually no longer exist.

Learning Impulse 4 (Fig. 4.16) serves—in addition to Learning Impulse 2—to weaken answer choice [D]. If eugenol actually reacted with oxygen from the air to form CO_2 and water when heated, two odorless substances would be produced. This would not be compatible with the strong aromaticity of eugenol. The considerations can also lead students to exclude answer [C], as a *split* of the eugenol molecules would also lead to new substances with different properties.

Answer [C] can also be excluded based on another consideration (Learning Impulse 5, Fig. 4.17): If one assumes a molecule split during the evaporation process, hydrogen and oxygen, and thus a dangerous gas mixture, would have to be produced when simply boiling water in the kitchen. The fact that Learning

Learning impulse 2

Thought experiment
Put a few drops of eugenol into a balloon, squeeze out the air and seal the balloon. Make well-founded hypotheses about the following question:

Based on the individual answers, how would the volume change if the eugenol was heated in the absence of air?

Assumption	Hypotheses	
	Volume increases	Volume remains the same
A		
B		
C		
D		

Review
Carry out the experiment to test the hypothesis. Instead of Eugenol, acetone is used as the model substance for this test experiment.

Hold the balloon in hot water for approx. 30 seconds using the crucible tongs. Then take it out and shake it to check whether there is still liquid inside.

Observation: The volume...
Note the result in the argumentation sheet!

Fig. 4.14 choice²learn: Learning Impulse 2

Impulse 5 is very impressive is shown by the following conversation excerpt from a small group:

> S1: Hydrogen and oxygen, yes, oxygen would be produced. S2: I have never had hydrogen in the kitchen. Because, then cooking would be too dangerous. [...] S3: Do they want to suggest that it explodes? S2: Yes, it blows up with hydrogen. And with oxygen, the reaction gets even worse. S4: And since hydrogen is highly explosive—or what should I write there? S3: highly flammable—Let's put it this way: It leads to an explosion [...] but since that doesn't happen and we don't know it from the kitchen, answer C can't be correct. You would notice if something explodes. S2: Then suddenly the house flies into the air. S4: Then every housewife S1: would die! S4: Then every housewife S1: the pot would fly around their ears [58]!

Learning impulse 3

How can it be explained that Lisa can smell the scent even at the end of the room?

Is answer [B] strengthened or weakened by your considerations?

Fig. 4.15 choice²learn: Learning Impulse 3

Learning impulse 4

Eugenol is a hydrocarbon. When hydrocarbons react completely with oxygen, the reaction products are **water** and **carbon dioxide**.

What properties (appearance, smell, ...) do water, carbon dioxide and eugenol have that we can perceive?

Which answers are strengthened and which are weakened by this information?

Fig. 4.16 choice²learn: Learning Impulse 4

4.5 Teaching Processes—Teaching Methods for Achieving Educational ...   119

Learning impulse 5

Which substances could be formed when water evaporates if the water molecules split as described in answer [C]?

What would be the consequences of boiling water over an open flame - for example on a gas stove?

Which answers are strengthened and which are weakened by this information?

Fig. 4.17 choice²learn: Learning Impulse 5

At the end of the group work, the groups should make a statement based on their argumentation sheet, which answer choice is correct. Afterwards, the groups present their considerations and their result in the plenum, where open questions can be clarified together—now also with the participation of the teacher (**Reflection**).

In the last phase of the **Application**, the students work on a task in pairs, which serves to apply and thus secure the professional concept developed (here: the change of state) in another context (Fig. 4.18).

While the first choice²learn teaching units were aimed at grades 9 to 12, **units for initial instruction** were developed as part of R. Schillmüller's doctoral project. For this purpose, the materials had to be adapted to the starting situation of the younger students (less prior knowledge, less experimental experience, less practice in scientific reasoning, etc.).

It is hard to imagine life today without butter in cooking and baking or as a spread on bread. Butter consists partly of palmitic acid:

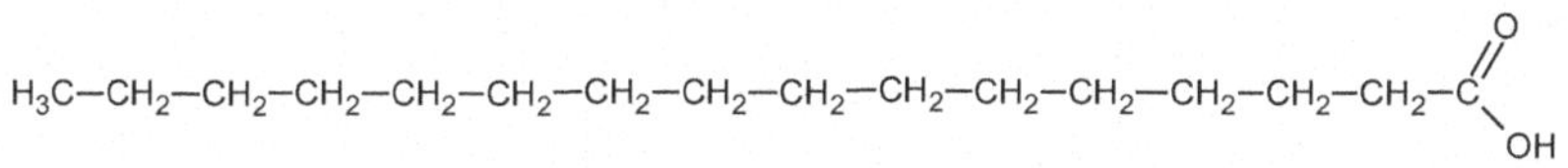

$H_3C-CH_2-CH_2-CH_2-CH_2-CH_2-CH_2-CH_2-CH_2-CH_2-CH_2-CH_2-CH_2-CH_2-CH_2-C\begin{smallmatrix}O\\ \\OH\end{smallmatrix}$

What happens when solid palmitic acid melts?

Draw and describe the processes taking place at particle level!

Fig. 4.18 Application task

An example of this is the unit "Why does it fizz in the soda?" [64]. Figure 4.19 shows the typical ideas of seventh and eighth graders on this question.

The learning material (Fig. 4.20) uses as little text as possible and employs simple language with short sentences and simple sentence structure. Recurring symbols facilitate the capture of learning impulses: An Erlenmeyer flask, for example, serves as a symbol for an experiment to be conducted; a pen serves as a hint that something should be entered into the argumentation sheet.

The proportion of experiments compared to purely mental learning impulses was deliberately increased to strengthen basic experimental skills in initial instruction. A color marking of the learning impulses and the respective experimental material facilitates the assignment of both. Experiment instructions are divided into individual steps and are supported by illustrations (Fig. 4.21). Conclusions from the learning impulses are guided by step-by-step questions.

The argumentation sheet was also simplified: The crossed-out fields show the students which multiple-choice answers they do not need to consider for the respective learning impulse (Fig. 4.22). Each learning impulse is also accompanied by a picture, which makes it easier for the students to remember the experiments they have previously conducted.

> **For Reflection**
>
> A detailed analysis of the learning processes by M. Egbers [58] shows that ideas can be changed—even in the long term—through the group work phase; the learning impulses lead to argumentatively dense and long-lasting discussions even in weaker learning groups. The students find the "detective" character of the

2) What happens when you put an effervescent tablet in water?

Tick one answer.

Justify your answer in writing.

The aggregate state changes.	☐	[A] A substance from the effervescent tablet changes from a solid to a gas.
	☐	[B] The whole brewing seta blette changes from a solid to a gas.
A new substance is formed.	☐	[C] A substance from the effervescent tablet forms a new gas with water.
	☐	[d] Several substances from the effervescent tablet form a new gas with water.
An existing gas rises.	☐	[E] A gas rises from cavities in the effervescent tablet.
	☐	[F] A gas rises from the water through the effervescent tablet.

Fig. 4.19 Learning task "Why does it fizz in the soda?"

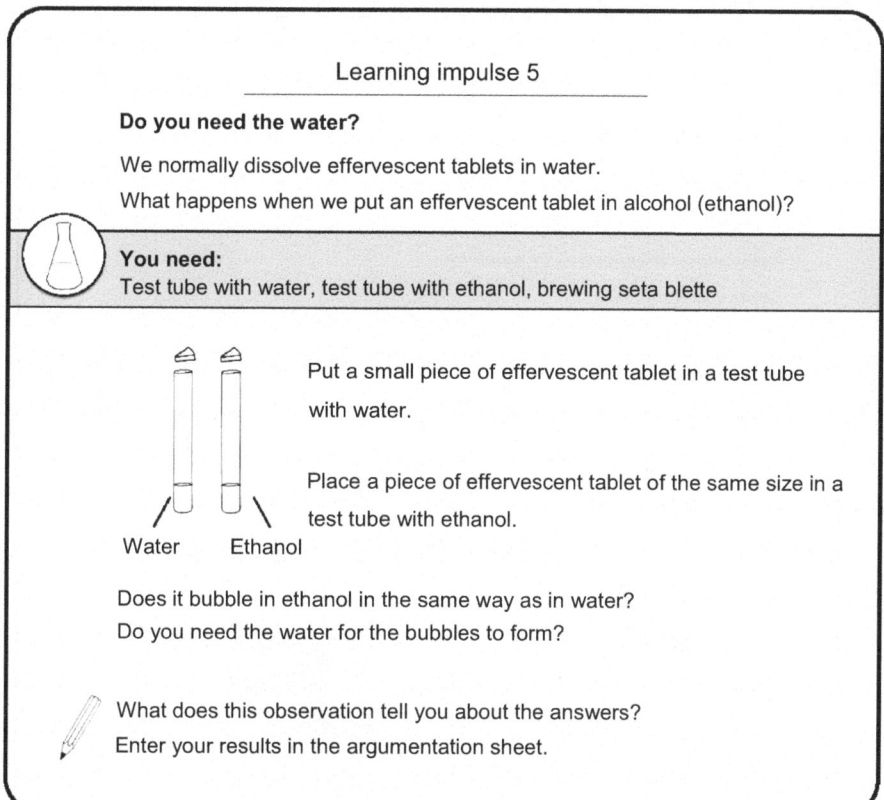

Fig. 4.20 choice2learn: Design of learning impulses in initial instruction (Low text content, simple sentences, recurring symbols)

concept motivating, among other things: The students must jointly clarify a question, the solution of which remains open until the presentation phase.

The study also shows that the success in the learning groups does not depend on which ideas were represented in the group at the beginning—even groups in which all students had chosen the same wrong answer arrived at the correct result.

Groups that work co-constructively, i.e., interpret impulses together and draw conclusions, prove to be particularly successful. This type of collaboration can compensate for unfavorable initial factors such as lower performance of the group members and low motivation.

The teaching concept also proves to be suitable for providing insights into scientific thinking and working methods [58, 66]. By contrasting different ideas in the group phase, the students learn that different model ideas can exist in science for the same fact; it must be checked which is best suited to explain a

Learning impulse 4b

Does a solid become a gas?

It is well known:

Some substances change from a solid to a gas.

These substances also change from a gas to a solid again.

Caffeine is one such substance.

Is sherbet also such a substance?

You need:
Aluminum bowls, wire frame, 2 watch glasses, tea light, matches, dark paper, spatula, crucible tongs, caffeine, water, effervescent tablet

Note: In the case of caffeine, the gas is produced by the addition of heat.	Note: With the effervescent tablet, the gas is produced by adding water.
Experiment 1	**Experiment 2**
Place the aluminum bowl in the wire rack. Add some caffeine to the aluminum bowl.	Place the aluminum bowl on the table. Put some effervescent tablet in the aluminum bowl.
Light the tea light. Place the wire rack over the tea light.	Add a splash of water.
Immediately place a watch glass on the aluminum bowl. *Caution: it gets hot!*	Immediately place a watch glass on the aluminum bowl.
Wait 3 minutes. Blow out the tea light. Remove the watch glass from the aluminum bowl using a pair of tongs.	Wait 3 minutes. Remove the watch glass from the aluminum bowl.
Place the watch glass on the dark paper. Take a close look at the watch glass. Has solid caffeine settled?	Place the watch glass on the dark paper. Take a close look at the watch glass. Has solid effervescence settled?
Compare the watch glasses from the two experiments. Has a solid formed again in the effervescence (as with caffeine) or not?	

Does the result speak for (+) or against (−) the answers [A] and [B]?

Enter this in the argumentation sheet.

Fig. 4.21 choice²learn: Design of learning impulses in initial instruction (Division of experiments into individual steps)

Argumentation sheet

What happens when an effervescent tablet is added to water?

[A] A substance from the effervescent tablet changes from a solid to a gas.
[B] The whole effervescent tablet changes from a solid to a gas.
[C] A substance from the effervescent tablet forms a new gas with water.
[D] Several substances from the effervescent tablet form a new gas with water.
[E] A gas rises from cavities in the effervescent tablet.
[F] A gas rises from the water through the effervescent tablet.

	Argument	Speaks for (+) or against (−) answer...					
		[A]	[B]	[C]	[D]	[E]	[F]
2							
3 a+b+c							
4 a+b							
5							
6a + b							
6c							

We have decided on answers !

Fig. 4.22 choice²learn: Design of the argumentation sheet in initial instruction

chemical phenomenon (such as the evaporation process). Since they gradually exclude model ideas with the help of the learning impulses, the students experience falsification as an essential element of scientific knowledge acquisition.

For teachers, it is recommended to test the concept first using the elaborated learning units [e.g., 63, 64, 65, 66, 67]. However, feedback from teachers and department heads shows that with increasing experience, own tasks and learning impulses can also be developed. In doing so, the teachers can incorporate answers from students that they have observed in class when designing the learning task. ◄

4.5.5 The teaching method oriented towards student conceptions

A further teaching method, which is oriented towards student conceptions, can be traced back to K. Petermann, J. Friedrich and M. Oetken [68]. This method pursues a different approach compared to choice^{2}learn: While students in choice^{2}learn develop the technically "correct" thinking concept based on their own ideas with the help of learning impulses, this is already conveyed in advance in the teaching concept of Petermann et al. within the framework of a *Technical Clarification*.

In the presented teaching example on the law of conservation of mass, the Boyle experiment is initially discussed in the first *Phase of Hypothesis Formation and Problem Acquisition*. In this experiment, charcoal is burned in a closed system to demonstrate the law of conservation of mass. The students express assumptions about the expected result.

In the following *Phase of Technical Clarification of the Subject Matter*, the experiment is demonstrated and the law of conservation of mass is worked out. The learning group is made aware that the experiment is carried out in a closed system and thus nothing can escape or be lost. The aim of the phase is to present the "scientific conception for the students in a logical and plausible way".

Only after this technical clarification does the confrontation with typical student conceptions on the topic, which are known from the literature, take place (*Elaboration Phase and Phase of Consolidation and Knowledge Assurance*). In a newer version of the teaching method [69], students are given a technically incorrect student statement in groups (e.g., "The mass decreases because gaseous substances are lighter than solid substances"); they receive an experimental instruction or a model with which the statement is to be refuted. According to the authors, the students should recognize the "thinking error" behind the statement.

This is followed by the *Phase of Application and Transfer*, in which the concept of mass conservation is applied to other examples (e.g., the combustion of sulfur) as well as a *Phase of Metacognition*, which serves for final reflection.

T. Dörfler also addresses student conceptions known from the literature in his teaching unit on the topic of "Acids and Bases" [70]. Through various worksheets and discussion phases, he achieves changes in conceptions, which are proven in post and follow-up tests.

4.5.6 choice²reflect

From the perspective of many teachers and didacticians, evaluation competence forms the most complex competence area of the educational standards. After all, not only "subject-immanent" evaluations are to be promoted, e.g. that "students use model conceptions for processing, explaining and evaluating chemical questions" [21, p. 19]. The competence area also includes the evaluation of everyday life situations and societal controversies: Students should "discuss socially relevant statements from different perspectives", "critically evaluate information" or also "evaluate measures and behaviors for maintaining their own health" [21, p. 19]. But how can this be realized in chemistry lessons? What do students need in order to scientifically evaluate information and judge societal topics?

The **core idea** of the teaching concept choice²reflect according to Marohn and Jungkamp [71, 72] is to develop scientific "test criteria" together with the students. These include, for example, *objectivity, reproducibility, variable control* or *blinding*. These can help to check statements and studies for their content, their range and their scientific nature. They thus form an important building block for making informed assessments in societal controversies such as nuclear energy, genetic manipulation, vaccination, nutrition or alternative medicine.

The concept thus not only touches on the competence area of evaluation. By illustrating essential criteria that science must follow, it also provides an insight into the *nature of (natural) sciences* (Nature of Science) and thus strengthens the competence area of knowledge acquisition. The method is therefore not specifically designed for chemistry; it can also be used in other natural science subjects, as well as in the field of social sciences or politics.

Fig. 4.23 shows a pool of central scientific criteria. From these, the teacher can selectively choose individual criteria that, in her view, are helpful for evaluating a fact. The action-oriented development of a criterion (Phase III of the teaching concept) always takes place in three steps:

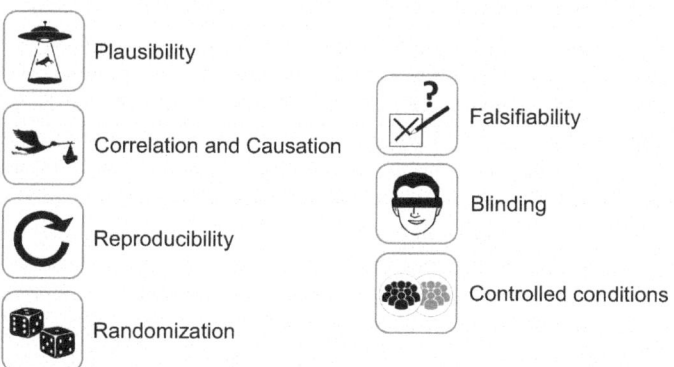

Fig. 4.23 choice²reflect: Scientific test criteria

Firstly, the students should recognize the significance of the criterion. In the case of *blinding*, the students taste natural yogurt samples that have been colored with various food colors. The tasting takes place once with a blindfold, once seeing. While the samples taste the same when tasted blindfolded, some students believe they can taste a strawberry or banana flavor in the case of visible testing—depending on the color. The test thus makes it clear that our evaluations depend on prior experiences and expectations. This must be taken into account in scientific studies, e.g. in the field of medicine.

In the case of *controlled conditions*, the students receive an interaction box with materials for making milk foam and the task of making the most stable foam possible. They consider which factors can play a role here (temperature, fat content, stirring duration, type of vessel, etc.) and plan—supported by various worksheets—test series to check the influence of individual factors. It becomes clear that only the factor to be investigated may be varied; the others must be kept constant, i.e. *controlled* (Fig. 4.24).

In the second step, the students receive a "test card" for the respective scientific criterion (Fig. 4.25), which they can refer to in later evaluation processes. The criterion is described on the front and its significance is explained; on the back, there is information on how the criterion is specifically implemented in practice.

To internalize the criterion more strongly, the students apply it in the last step of the development to a factual situation from science or everyday life. The teacher has various elaborated examples available for this purpose.

The teaching concept choice^{2}reflect takes place in **five phases** (Fig. 4.26). At the center is a topic controversially discussed in society; this should be relevant from the students' perspective and have a reference to science teaching. A possible example of this is *homeopathy*. Although the effectiveness of homeopathic remedies does not go beyond a placebo effect, many people in Germany trust in homeopathic remedies [73]. Students also usually have experience with homeopathic remedies; 40% of the ninth graders surveyed even take them regularly [71].

Homeopathy becomes a relevant topic for chemistry teaching also because it contradicts the most important finding of chemical research, which according to the curriculum is taught in the first weeks of chemistry teaching: that every substance is characterized by typical, unchangeable properties. The homeopathic principle of *potentiation* postulated by Hahnemann in the 18th century, on the other hand, assumes that substance properties can be transferred. If you dissolve a substance in water and hit the vessel ten times on a leather cover, the properties of the substance are transferred to the solvent from the perspective of homeopathy. This property expression becomes stronger the more often the process of dilution and shaking is repeated. Substances like water or ethanol could thus carry millions of different properties—depending on which substances were once dissolved in them.

In the first teaching phase, the *subject-specific problem introduction*, there is initially a neutral introduction to the respective topic. In the case of the topic of homeopathy, the students learn the two principles of action (*principle of similarity* and *potentiation*) using the example of Johanna, who suffers from ADHD. They

The milk foam dispute

Tim and Anna argue about what really matters when making milk foam.

The two of them are planning a series of experiments to test the influence of the **type of milk** and the **Temperature of the Milk**. However, their proposals look slightly different.

Tims Test series	Influencing factors	1st attempt	2nd attempt	3rd attempt
	Type of milk	3,5%	1,5%	0,1%
	Milk temperature	20°C	50°C	80°C

Annas Test series	Influencing factors	1st attempt	2nd attempt	3rd attempt	4th attempt	5th attempt	6th attempt
	Type of milk	3,5%	1,5%	0,1%	3,5%	3,5%	3,5%
	Milk temperature	50°C	50°C	50°C	20°C	50°C	80°C

Task:

Decide: Which of the two series of experiments do you think is better suited to test the influence of these two factors? **Give reasons for your decision.**

Fig. 4.24 choice²reflect: Development of the test criterion "Controlled conditions". (Excerpt from the learning material)

What are controlled conditions?

Experiments help us to scientifically test different claims:

For example: Do plants grow faster when you talk to them?

To answer such a question, an experiment must take place under controlled conditions. This means that two tests are carried out that only differ in one single characteristic (speaking to/not speaking to the plant). All other properties remain the same for both experiments.

How do you create controlled conditions?

This is how experiments are planned under controlled conditions:

Example (milk foam):
*To find out at which **temperature** you get the best milk foam, **all other properties** must remain unchanged.*

Step 1: Identify properties that could have an influence on the outcome of the experiment:

Step 2: Carry out the experiment several times, keeping all properties constant except for the one to be investigated (temperature):

This procedure is absolutely necessary to find out to what extent the temperature influences the quality of the milk foam.

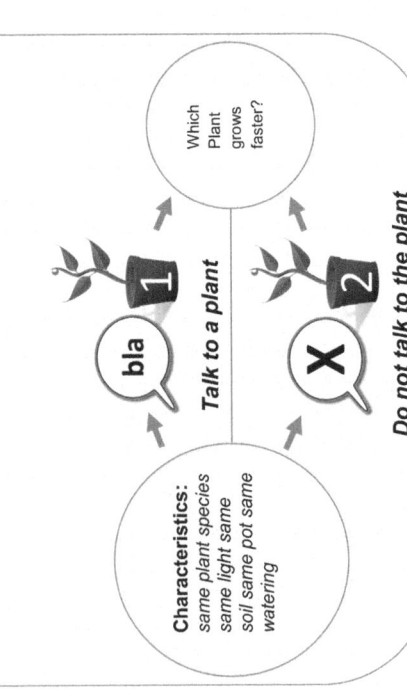

Fig. 4.25 choice²reflect: Test card "Controlled conditions"

4.5 Teaching Processes—Teaching Methods for Achieving Educational ...   129

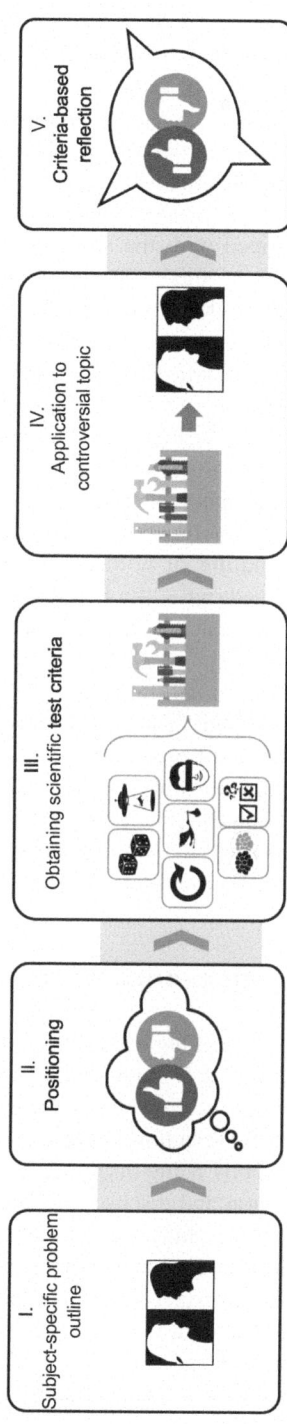

Fig. 4.26 choice²reflect: Teaching phases

choose a suitable homeopathic active ingredient for Johanna according to the principle of similarity (coffee) and make a C30 potentiation remedy from it by dissolving it in water and repeated dilution and beating.

In the following phase, the students *position* themselves individually on the topic using guiding questions. Among other things, they should indicate whether they would take homeopathic remedies themselves.

In the phase of *acquiring scientific test criteria*, as described above, topic-relevant scientific criteria are developed with the students. Relevant criteria for the topic of homeopathy include, among others, the *blinding* and the *controlled conditions*. These can help in the subsequent teaching phase (*application of the test criteria to a controversial topic*) to evaluate studies on homeopathy. Possible test questions are, for example: Were the variables controlled, or were the patients additionally treated with conventional medicine or psychotherapy? Was there a control group? Was it blinded, or was the treatment known to patients and doctors?

In addition, the students receive further topic-specific information: e.g. about the placebo effect and the design of medical studies, about reasons for the importance of homeopathic remedies in society, about typical misconceptions (equating with herbal medicine), or the handling of homeopathy in other countries. They also model the dilution series carried out at the beginning at the particle level and establish connections to their chemical knowledge such as the amount of substance or Avogadro's number.

In the final phase of the *criteria-guided reflection*, the students should re-evaluate the topic based on their knowledge. The controversy is also finally discussed in plenary. As with any controversy, different viewpoints are conceivable, e.g.: "I don't spend money on C30 pellets that only contain sugar. I'd rather buy a tablet with an active ingredient." "I can imagine using the placebo effect of globules when I have a headache." "Remedies that do not work stronger than a piece of sugar should not be approved as medicines."

4.5.7 The socio-critical problem-oriented teaching

The socio-critical problem-oriented teaching according to Eilks [74] also deals with societal controversies. In the tradition of *Socio Scientific Issues (SSI)*-oriented teaching, the concept aims at a general education science teaching that strengthens the ability for self-determination. For the selection of topics, the authors name clear criteria: authenticity (the topic is currently being discussed in society), relevance for the students, an open evaluation situation and discussability, as well as the reference to science and technology. Meanwhile, teaching sequences have been described on various topics, such as bioethanol, climate change, alcopops, diets or tattoos [75–77].

The teaching concept proceeds in five phases: The phase *access and analysis of the controversy* is done through media from the students' everyday life. In the topic of tattoos, for example, the students work on a self-test, which is based on a test from the youth magazine Bravo. This includes questions about aesthetic, societal or medical-scientific aspects. Different views and arguments of the students are collected.

In the *technical clarification*, backgrounds of the topic are worked out, usually involving experiments. The students work on a station learning, in which they examine and compare tattoo colors; health and legal aspects are included.

When *revisiting the controversial problem situation*, the students reflect on where scientific knowledge can contribute to understanding the controversy. Which questions in the field of tattoos can be clarified with the help of science?

As part of the *development and discussion of different perspectives*, the students should respond to a letter from a teenager to the editor of a youth magazine. It is about the question of whether the teenager should get a tattoo without the consent of his parents. In their small groups, the students must weigh up to what extent they want to use medical-scientific aspects for their answer.

In the *meta-reflection*, the students reflect on their answers to the reader's letter. Among other things, it becomes clear that the attitude and scientific understanding of the respective editor has a major influence on the content and tone of the answer.

For Reflection

The success of teaching concepts like choice^{2}reflect or the socio-critical problem-oriented teaching strongly depends on how closely the chosen controversy actually moves to the students' world of life. Topics like *tattoos* or *diets* are likely to trigger a stronger willingness to engage with the content among students than the topic *renewable resources*, which appears much more abstract and removed from their own reality of life. Especially topics that are strongly emotionally charged, because they affect our own behavior or health, such as *nutrition*, *alternative healing methods* or the *vaccination question* can trigger intense discussions and stimulate thought processes. It is therefore worthwhile to integrate such content into chemistry lessons—even at the "risk" of having to continue the topic at the next parents' evening! If we take the goal of scientific literacy seriously, then we must confidently represent scientific findings and standards in teaching and clearly distinguish them from personal belief, wishes or individual experiences. ◀

4.5.8 choice^{2}explore

The teaching concept choice^{2}explore, based on Marohn and Rott, adopts the core idea of choice^{2}learn, but adapts it to the requirements of inclusive learning groups using the example of science-oriented primary school teaching [78–81]. Even though the concept does not refer to chemistry teaching, it provides significant insights for scientific work in inclusive learning groups.

In this teaching method, children test their own ideas about a scientific phenomenon using small experiments. Preliminary studies have shown that the ideas of children with and without support needs hardly differ [82]. Figure 4.27 shows typical ideas of primary school children about "dissolving salt in water".

What do you think?

Name:

What happens to the salt?
Milla and Lutz have lots of ideas.

What do you think? Mark an answer with a cross

- ☐ The salt has turned to water.
- ☐ The salt is distributed in very small particles.
- ☐ The salt has become liquid.
- ☐ The salt is gone.

Give reasons!

© Institute for Didactics of Chemistry - Münster University - Lisa Rott & Annette Marohn

Fig. 4.27 choice^{2}explore: Typical ideas of primary school children about dissolving salt in water

To verify their ideas, the students, for example, evaporate salt water over a tealight and find that salt remains in the bowl—it is therefore not "gone", as many children suspect (Fig. 4.28). In another experiment, they compare salt water with "pure" water by adding a piece of potato to both samples (Fig. 4.29). Since this sinks in the water but floats in the salt water, it becomes clear that the salt cannot have "become water"—otherwise, both samples would have had to behave the same way. Each experiment is first carried out and evaluated by the students in small groups and then discussed in plenary.

In addition to subject learning, choice^{2}explore focuses on promoting "joint learning on a common subject" [83]. Contrary to the common recommendation to provide students with different performance levels with differentiated learning materials, in this teaching concept, children in heterogeneous performance small groups work with the same material; this is intended to avoid stigmatization and exclusion and to support joint action.

In order for this learning material to be equally understandable for all students—e.g., also for children with a focus on "reading" or "learning"—the material offers several access methods (Fig. 4.30): Experiment steps or work assignments are formulated in *Easy Language* and also symbolically represented at the same time, so that they can also be grasped non-verbally. Criteria for formulations in easy language can be found under [84]. For symbolic representation, symbols from Annette Kitzinger's *METACOM system* were used [85]. Individual work steps can be "checked off" by the students after completion and are therefore

Fig. 4.28 choice^{2}explore: Verification of the idea "The salt is gone"

Fig. 4.29 choice²explore: Verification of the idea "The salt has become water"

easy to follow even for children with attention deficits. All learning materials are also very clearly structured visually.

The use of *word memory cards* (Fig. 4.31) for key terms or sentence starters and other tools from *Supported Communication* make it easier for students to describe observations and promote collective reasoning. They also support communication during plenary phases [80].

The teaching concept includes five phases (Fig. 4.32), which were named for both the teachers (e.g., *observe phenomenon*) and the students (*Look closely!*) to make the course of the lesson transparent for the children. In addition, each phase was assigned a symbol that refers to the main focus of the phase. This symbol is also found on the corresponding worksheets.

Compared to the teaching concept choice²learn, a teaching phase was added in which the students describe and explain the phenomenon under discussion once again (*Explain it to me!*). For this, they use yellow Lego bricks as a model for water particles and red bricks as a model for the salt, which is "distributed in small particles" (Fig. 4.33).

> **For Reflection**
>
> The analysis of the learning processes in choice²explore by L. Rott provides exciting insights into learning in inclusive science education [81]. It was particularly evident in the actions associated with experimenting (performing an experimental step, observing, explaining, interpreting) that there was a high

4.5 Teaching Processes—Teaching Methods for Achieving Educational … 135

a

Find out!		
Task		✓
[water bottle → water glass]	Pour into a glass of water.	
[water+salt bottle → water glass]	Fill the other glass with "water and salt".	
[potato → water glass] [potato → water glass]	Put a piece of potato in each jar	
[eyes observing]	Observe.	

© Institute for Didactics of Chemistry - Münster Univers - Lisa Rott & Annette Marohn

Fig. 4.30 choice²explore: Design of the learning material (structuring, Easy Language, Symbolic representations—METACOM)

b

Find out!

Draw what you see!

Glass 1: Water	Glass 2: Water and salt (water?)

Think and discuss

If the salt turned into water, there would only be water left in the glass.

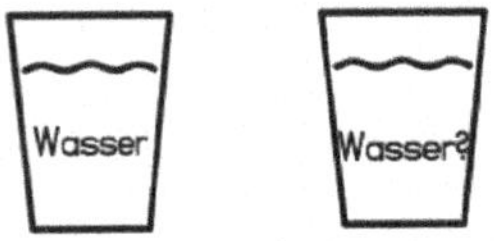

☐ Yes, because _____.

☐ No, because... _____.

© Institute for Didactics of Chemistry - Münster Univers - Lisa Rott & Annette Marohn

Fig. 4.30 (continued)

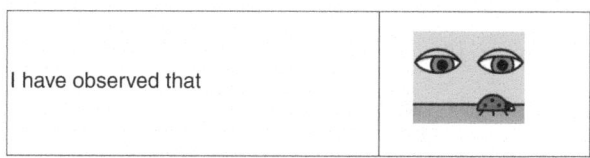

Fig. 4.31 choice²explore: Word memory card

degree of collaboration—a balanced togetherness. Experimenting in science education apparently offers—in addition to all risks—a great opportunity for joint learning in inclusive groups.

The special design of the learning material (structuring, easy language, symbolic support, …) promotes the comprehensibility of the material for all students and supports academic learning. Even children with special educational support needs develop academic ideas and can sometimes even exclude alternative ideas with justification. This is particularly successful when they work in small groups with a high degree of collaboration. Weaker students in particular seem to benefit from learning together.

The use of a simple model (Lego particle model) for visualizing and explaining a process at the particle level has proven to be extremely helpful. Even the children with special educational support needs were able to use the model to describe and explain the facts. The use of models in inclusive science education can thus contribute to developing ideas of the submicroscopic level even in weaker students. Higher-performing children also transferred the particle model to other phenomena or even described limits of the model. Models thus offer a good opportunity for dealing with scientific facts at different levels. ◀

4.5.9 Further Teaching Concepts

The concepts mentioned above represent teaching methods in the "classic" sense: They not only pursue a specific didactic idea, but also structure the lesson into predefined phases. In addition to the teaching methods, however, there are also more general approaches that pursue a didactic goal without specifying a concrete structuring of the lesson.

The **Structure-oriented Chemistry Teaching** by H.-D. Barke is based on approaches by Grosser and Bauer [86, 87]. He advocates that after getting to know the substances and reactions experimentally the *structures* of the involved substances should be illustrated by structure models before the reaction symbols are added. This suggestion is based on the experience that younger students can understand the structure of substances more easily by "grasping" concrete models such as sphere packing and molecule models, rather than through theoretical bonding models. A possible modeling to illustrate a chemical reaction is shown in Fig. 4.34 using the example of the formation of magnesium oxide.

138　　　　　　　　　　　　　　　　　　　　　　　　4　Teaching Objectives

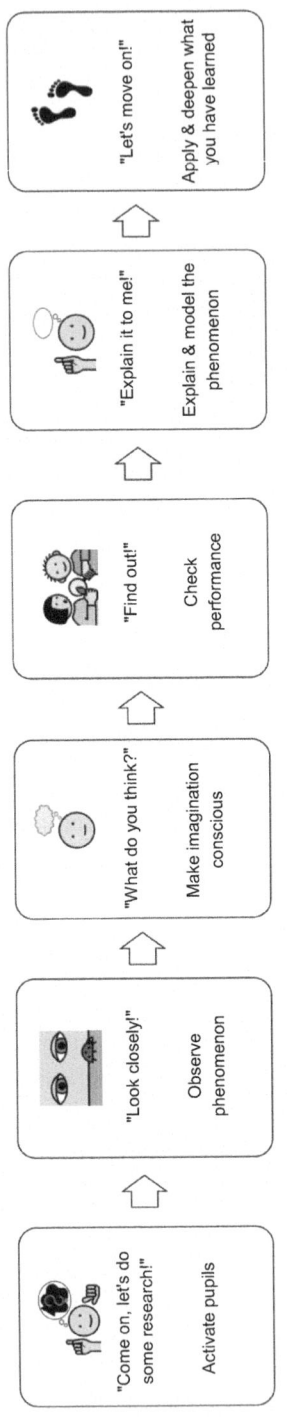

Fig. 4.32 choice²explore: Teaching phases

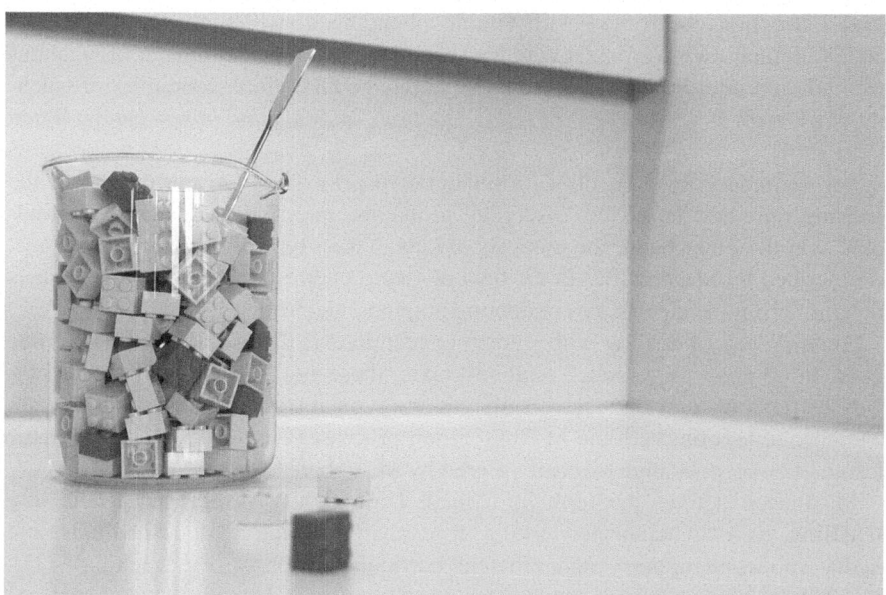

Fig. 4.33 choice²explore: Lego model (Phase: Explain it to me!)

Fig. 4.34 Possible model conception for magnesium combustion according to Barke [88]

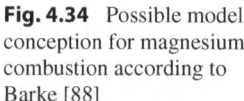 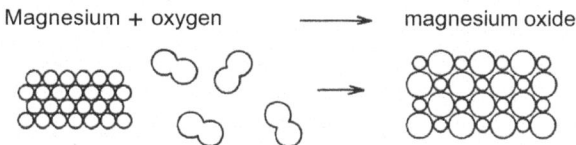

Barke also suggests introducing the concept of ions as early as possible. By mentally combining atoms and ions, it then becomes possible to describe the essential compounds of school chemistry at an early stage (Table 4.7)

A. Flint develops in his approach **Chemistry for Life** a variety of experiment-oriented teaching units that aim for a close link between chemistry and everyday life. According to the principle "from simple to complex", students in secondary

Table 4.7 Linking rules for metal atoms, non-metal atoms, and ions [89]

Linking according to location in the PSE	Particle type	Type of bond	Structure
"left and left"	Metal atoms	Spatially undirected	Metal lattice
"left and right"	Ions	Spatially undirected	Ion lattice
"right and right"	Non-metal atoms	Spatially directed	Molecules, atomic lattice

level I are first confronted with simple phenomena, which they describe and explain in their own words. Only in the second step should the transfer of what has been learned into the chemical technical language take place. Examples of teaching units, such as on the *topic of redox reactions* or *acids and bases*, can be found under [90].

For secondary level II, Flint formulates two principles: On the one hand, the teaching units are based on "everyday problems, questions or astonishing findings". On the other hand, the units should counteract typical student conceptions, as described by Marohn [62] in the field of electrochemistry. An example of this is the unit "From the electricity-conducting potato to electrolysis" [91].

Another way of dealing with student conceptions is the use of **concept change texts**. In contrast to "classic" textbook texts, these not only address the technically correct concepts, but specifically address typical student conceptions. A concept change text on the topic of "dissolution processes" as well as criteria for the design of concept change texts are offered by M. Egbers and A. Marohn [92].

M. Tausch defines the term he coined, **Didactically Integrated Chemistry Teaching**, as a "coherent networking of teaching/learning content, methods, and media into contemporary and efficient curricula for chemistry teaching" [93]. The content includes both chemical subject matter and contexts. He mentions "constructivist learning loops" as an important component in designing teaching sequences, which follow the pattern of *Exploring, Researching, Adapting*, and *Applying*. An overview of the approach and an exemplary representation using the content area "Organic Products—Plastics and Dyes" is provided in reference [93].

The **choice^2 interact** approach by Marohn and Dellbrügge offers interactive learning environments for tablets [94]. Through the "Explain Everything" app, students are provided with learning paths on a subject area that they can work on and supplement in a self-chosen order—in pairs (Fig. 4.35). The learning paths include various elements: *Introduction, Theory, Exercises, Experiments, Research, Contexts, Explainer Videos*, and *Feedback*. The design of the learning environment is based, among other things, on criteria from the Cognitive Theory of Multimedia Learning [95].

In addition to the teaching concepts listed, **hybrid forms** of the various approaches are of course also conceivable. For example, problem-oriented teaching can be integrated into a real-world context. A combination of the history of chemistry and student conceptions is also feasible, as many of the students' thought concepts find a correspondence in the historical conceptions of scientists. An example of this can be found in Marohn [96].

Beyond the methods listed for science teaching, chemistry teaching can also be structured through various social forms or **methods**. These include, for example: teacher lectures, student presentations, classroom discussions, group work, group puzzles, learning stations, individual or partner work, panel discussions, role-playing games, project methods, posters, egg races, and many more. An overview in connection with examples from chemistry teaching can be found, for example, in Kranz and Schorn [97].

4.5 Teaching Processes—Teaching Methods for Achieving Educational … 141

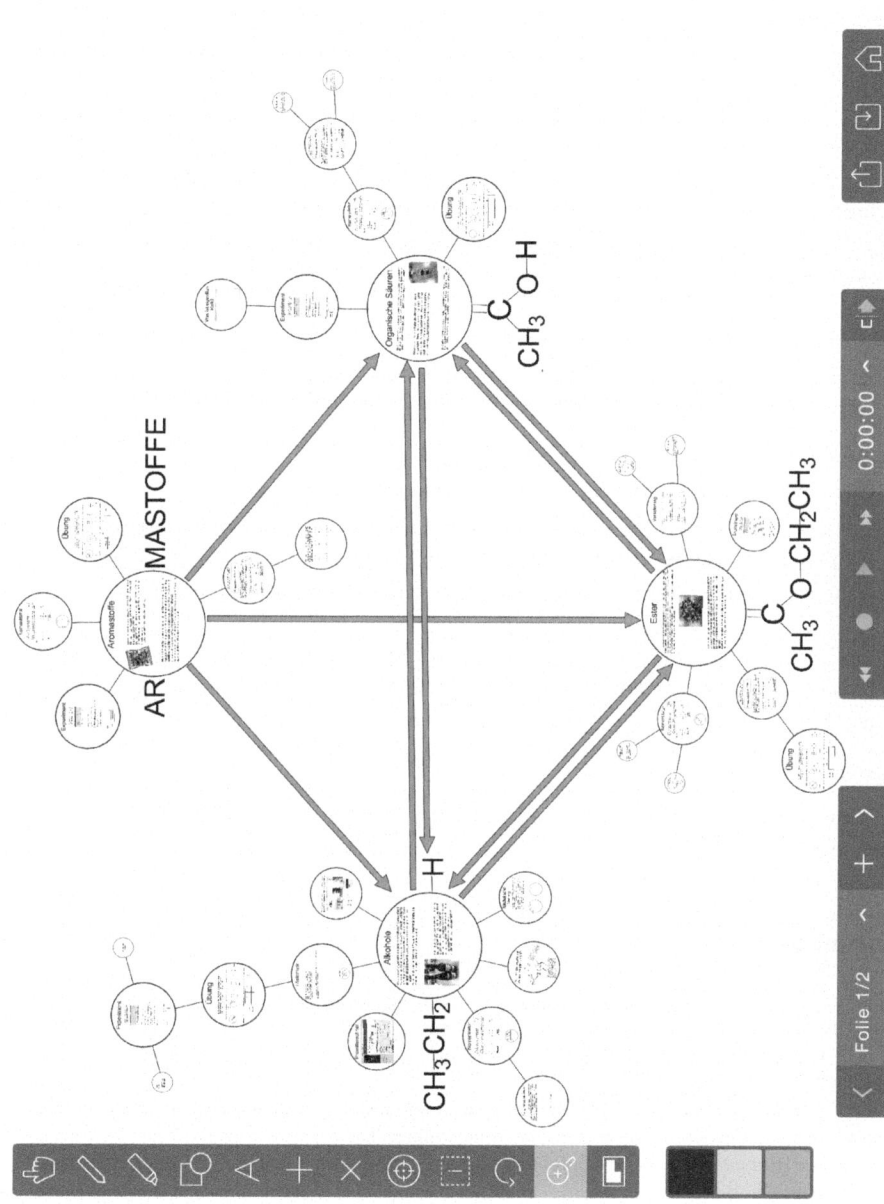

Fig. 4.35 choice²interact: Learning paths on the subject area of flavorings/esters [94]

4.6 Exercise Tasks

A4.1
Reflect on the basis of educational standards and curricula: In your opinion, has the introduction of educational standards achieved a standardization of chemistry teaching?

A4.2
Plan a teaching unit on the topic "Separation Procedures" according to the five thinking stages of the research-developing teaching method. Formulate a specific problem and anticipate the paths students could take to solve the problem. Create a specific materials list for this.

A4.3
Indicate which competencies from the four areas of competence of the educational standards can be promoted through your teaching unit outlined in A4.2.

A4.4
The NRW framework curriculum requires that students in the area of evaluation competence "assess measures and behaviors for maintaining their own health". Choose a suitable context that enables the promotion of this competence, and outline a teaching unit according to the four phases of "Chemistry in Context".

A4.5
Reflect on the advantages and possible disadvantages of a context-oriented chemistry teaching.

A4.6
Select a student conception mentioned in Chap. 2 and develop an idea for a lesson sequence that is oriented towards this student conception.

References

1. Scheible A (1969) Ist unser Chemieunterricht noch zeitgemäß? Math Naturwissenschaftliche Unterr 22:449–457
2. Erhart H (1998) Chemie—einer der unbeliebtesten Unterrichtsgegenstände. Sind die Lehrer schuld daran? Chem Sch 4:29
3. MNU, GDCh, GDCP et al (1998) Mathematische und naturwissenschaftliche Bildung an der Schwelle zu einem neuen Jahrhundert. Chemie konkret, Bd. 5
4. Deutscher Bildungsrat (1971) Empfehlungen der Bildungskommission. Strukturplan für das Bildungswesen. Beltz, Stuttgart
5. Möller Ch (1973) Technik der Lehrplanung. Beltz, Weinheim
6. Bloom BS (1972) Taxonomie von Lernzielen im kognitiven Bereich. Beltz, Weinheim
7. Klafki W (1980) Die bildungstheoretische Didaktik. Westermanns Pädagogische Beiträge 1:32–37

8. Schulz W (1980) Die lerntheoretische Didaktik. Westermanns Pädagogische Beiträge 2:80–85
9. Möller Ch (1980) Die curriculare Didaktik. Westermanns Pädagogische Beiträge 4:164–168
10. Winkel R (1980) Die kritisch-kommunikative Didaktik. Westermanns Pädagogische Beiträge 5:200–204
11. v. Cube F (1980) Die informationstheoretisch-kybernetische Didaktik. Westermanns Pädagogische Beiträge 3:120–124
12. Blankertz H (1973) Theorien und Modelle der Didaktik. Juventa, München
13. Ruprecht H (1976) Modelle grundlegender didaktischer Theorien. Schroedel, Hannover
14. Klafki W (1964) Didaktische Analyse. Schroedel, Hannover
15. Heimann P, Otto G, Schulz W (1965) Unterricht. Analyse und Planung. Schroedel, Hannover
16. Bönsch M (1976) Unterrichtsanalyse. Erziehung und Unterricht, Bd. 10
17. Meyer H-L (1984) Leitfaden zur Unterrichtsvorbereitung. Scriptor, Frankfurt
18. Kultusministerkonferenz (2004) Bildungsstandards im Fach Chemie für den Mittleren Bildungsabschluss. Beschluss vom 16.12.2004. Luchterhand, München, Neuwied
19. Weinert FE (2001) Vergleichende Leistungsmessung in Schulen—eine umstrittene Selbstverständlichkeit. In: Weinert FE (Hrsg) Leistungsmessungen in Schulen. Beltz, Weinheim, Basel, S 17–31
20. Kultusministerkonferenz (2004) Einheitliche Prüfungsanforderungen in der Abiturprüfung Chemie. Beschluss vom 01.12.1989 i.d.F. vom 05.02.2004
21. Ministerium für Schule und Weiterbildung des Landes Nordrhein-Westfalen: Sekundarstufe I. Gymnasium. Chemie. Kernlehrplan. Schule in NRW Nr. 3415. Ritterbach (2008).
22. Pant HA et al (Hrsg) (2013) IQB-Ländervergleich 2012 Mathematische und naturwissenschaftliche Kompetenzen am Ende der Sekundarstufe I. Waxmann, Münster
23. Lück G (2013) Naturphänomene erleben—Experimente für Kinder und Erwachsene. Herder, Freiburg
24. Kahlert J, Demuth R (Hrsg) (2008) Wir experimentieren in der Grundschule. Teil 1/ Teil2. Aulis, Köln
25. Piaget J, Inhelder B (1973) Die Psychologie des Kindes. Walter, Olten
26. Gräber W, Stork H (1984) Die Entwicklungspsychologie Jean Piagets als Mahnerin und Helferin im naturwissenschaftlichen Unterricht. Math Naturwissenschaftliche Unterr 37:193–201
27. Duit R (1996) Lernen als Konzeptwechsel im naturwissenschaftlichen Unterricht. IPN, Kiel
28. Heilbronner E, Wyss E (1983) Bild einer Wissenschaft: Chemie. Chem Unserer Zeit 17:69–76
29. Barke H-D, Hilbing CH (2000) Image von Chemie und Chemieunterricht. Chem Unserer Zeit 34:17–21
30. Müller-Harbich G et al (1990) Die Einstellung von Realschülern zum Chemieunterricht, zu Umweltproblemen und zur Chemie. Chimica Didact 16:150–169
31. Gräber W (1992) Untersuchungen zum Schülerinteresse an Chemie und Chemieunterricht. Chem Sch 39:270–273
32. Wanjek J, Barke H-D (1998) Einfluß eines alltagsorientierten Chemieunterrichts auf die Entwicklung von Interessen und Einstellungen. In: Behrendt H (Hrsg) Zur Didaktik der Physik und Chemie. Leuchtturm, Kiel, S 268–288
33. Harsch G, Heimann R (1998) Didaktik der Organischen Chemie nach dem PIN-Konzept. Vom Ordnen der Phänomene zum vernetzten Denken. Vieweg, Wiesbaden
34. Schmidkunz H, Büttner D (1985) Chemieunterricht im Spiralcurriculum. Naturwissenschaften im Unterricht. Phys Chem 33:19–22
35. Winkel R (1993) Langweilig sein, die ärgste Sünde des Unterrichts. Dtsch Lehrerzeitung 11:9
36. Marohn A (Hrsg) (2016) Unterrichtskonzepte. Praxis der Naturwissenschaften, Chemie in der Schule, 5/65

37. Schmidkunz H, Lindemann H (1992) Das Forschend-Entwickelnde Unterrichtsverfahren—Problemlösen im naturwissenschaftlichen Unterricht. Didaktik, Naturwissenschaften, Bd. 2. Westarp Wissenschaften, Essen
38. Himmerich K, Weiß M (2016) Das forschend-entwickelnde Unterrichtsverfahren am Beispiel "Vorsicht: Rhabarber!—Wie gefährlich ist unsere Nahrung?". Prax Naturwissenschaften Chem Sch 5(65):5–7
39. Marohn A (Hrsg) (2014) Interaktionsboxen. Praxis der Naturwissenschaften, Chemie in der Schule 6/63, S 4
40. Jansen W et al (1986) Geschichte der Chemie im Chemieunterricht—das historisch-problemorientierte Unterrichtsverfahren. Teile 1 und 2. MNU 39(321):391
41. Matuschek C, Fickenfrerichs H, Peper R (1992) Das historisch-problemorientierte Unterrichtsverfahren am Beispiel der Einführung der tetraedrischen Struktur des Methans. Naturwissenschaften Im Unterr Chem 3(13):4–8
42. Berger C, Jansen W, Fickenfrerichs H, Peper R (1987) Die Entdeckung der Alkalimetalle und der Zusammensetzung des Ätznatrons durch Humphry Davy—eine experimentelle Unterrichtskonzeption. Naturwissenschaften Im Unterr Phys 35(30):8–19
43. Busker M, Rosenberg D, Böttger S, Pöhls C, Fittschen U, Jansen W (2016) Der historisch-problemorientierte Unterricht am Beispiel "Metalle und Säuren". Prax Naturwiss Chem Sch 5(65):8–13
44. van't Hoff JH (1877) Über den Zusammenhang zwischen optischer Aktivität und Konstitution. Ber Dtsch Chem Ges 10:1620
45. Kolbe H (1877) Zeichen der Zeit. J Für Prakt Chem 15:472
46. Davy H (1893) Electrochemische Untersuchungen. Vorgelesen in der Königlichen Societät zu London als Bakerian Lecture am 20. November 1806 und 19. November 1807. In: Ostwald W (Hrsg) Ostwalds Klassiker der exakten Naturwissenschaften, Bd. 45. Engelmann, Leipzig
47. Nentwig PM, Demuth R, Parchmann I, Gräsel C, Ralle B (2007) Chemie im Kontext: situating learning in relevant contexts while systematically developing basic chemical concepts. J Chem Educ 84(9):1439–1444
48. Otter C, Pilling GM et al (Hrsg) (2008) Salters Advanced Chemistry: Chemical Storylines. Heinemann, Imprint of Pearson Education Limited, Harlow
49. Parchmann I, Ralle B (2016) Chemie im Kontext—Lernen von und in sinnstiftenden Zusammenhängen. Prax Naturwissenschaften Chem Sch 5(65):14–18
50. Chemie im Kontext: Säuren in der Speisekammer (Sek.I). http://www.chik.de/index2.htm. Zugegriffen: 1. April 2014
51. ChiK-Set Hamburg (2005) Coca Cola—Mehr als ein Erfrischungsgetränk. In: Freie und Hansestadt Hamburg-Behörde für Bildung und Sport (Hrsg) Förderung der Motivation und der Selbstständigkeit im naturwissenschaftlichen Anfangsunterricht. Klasse 6 Chemie. Freie und Hansestadt Hamburg-Behörde für Bildung und Sport, Hamburg, S 33–77
52. Marohn A (2008) Schülervorstellungen zum Lösen und Sieden—Auf der Suche nach "elementaren" Vorstellungen. Math Naturwissenschaftlicher Unterr 61(8):451–457
53. Marohn A (2008) Merksätze, Eselsbrücken und Vereinfachungen im Chemieunterricht—eine kritische Betrachtung. Prax Naturwissenschaften Chem Sch 57(3):46–44
54. Marohn A (2008) "Choice2learn"—eine Konzeption zur Exploration und Veränderung von Lernervorstellungen im naturwissenschaftlichen Unterricht. Zeitschrift Für Didaktik Naturwissenschaften 14:57–83
55. Posner GJ, Strike KA, Hewson PW, Gertzog WA (1982) Accomodation of a scientific conception. Toward a theory of conceptual change. Sci Educ 66(2):211–227
56. Vygotsky L (1978) Mind in Society: The Development of Higher Psychological Processes. Harvard University Press, Cambridge
57. Andriessen J (2006) Arguing to learn. In: Sawyer RK (Hrsg) The Cambridge Handbook of the Learning Sciences. Cambridge University Press, New York, S 443–459. http://dspace.library.uu.nl/bitstream/handle/1874/30943/Andriessen_06_arguing.pdf?sequence=1. Zugegriffen: 10. Jan. 2017

58. Egbers M (2017) Konzeptentwicklungs- und Gesprächsprozesse im Rahmen der Unterrichtskonzeption "choice2learn". In: Marohn A (Hrsg) Lernen in Naturwissenschaften—verstehen und entwickeln, Bd. 1. logos, Berlin
59. Taylor P, Fraser B (1991) Development of an Instrument for Assessing Constructivist Learning Environments. Roundtable at the annual meeting of the American Educational Research Association, Chicago
60. Pintrich PR, Marx RW, Boyle RA (1993) Beyond cold conceptual change: the role of motivational beliefs and classroom contextual factors in the process of conceptual change. Rev Educ Res 63(2):167–199
61. Reinmann G, Mandl H (2001) Unterrichten und Lernumgebungen gestalten. In: Krapp A, Weidenmann B (Hrsg) Pädagogische Psychologie. Beltz, Weinheim, S 603–646
62. Marohn A (1999) Falschvorstellungen von Schülern in der Elektrochemie—Eine empirische Untersuchung. Dissertation, Universität Dortmund
63. Marohn A, Egbers M (2011) Vorstellungen verändern—Lernmaterialien zum Thema "Verdampfen" im Rahmen der Unterrichtskonzeption "choice2learn". Prax Naturwissenschaften Chem Sch 60(3):5–9
64. Schillmüller R, Marohn A (2017) Warum blubbert's in der Brause?—choice2learn in der Sekundarstufe I. Naturwissenschaften Im Unterr Chem 17/28(159):13–18
65. Schillmüller R, Marohn A (2016) choice2learn—Schülervorstellungen verändern am Beispiel "Lösen von Kochsalz". Prax Naturwissenschaften Chem Sch 5(65):18–24
66. Egbers M, Wischerath K, Marohn A (2015) Lernen über Nature of Science im Rahmen der Unterrichtskonzeption choice2learn. Prax Naturwiss Chem Sch 6(64):23–29
67. Marohn A (2012) Wie kommt der Strom durch die Lösung? Choice2learn—Diagnose und Veränderung. In: Wambach H, Wambach-Laicher J (Hrsg) Individualisieren und Aktivieren im Chemieunterricht SII, Bd. 1. Aulis, Hallbergmoos, S 83–93
68. Petermann K, Friedrich J, Oetken M (2009) Orientierung an Schülervorstellungen—Erprobung und Evaluation einer Unterrichtseinheit zum Gesetz der Erhaltung der Masse. Prax Naturwissenschaften Chem Sch 58(8):11–18
69. Friedrich J, Bröll L, Petermann K, Oetken M (2016) Das an Schülervorstellungen orientierte Unterrichtsverfahren—Eine inhaltliche Auseinandersetzung mit Schülervorstellungen im naturwissenschaftlichen Unterricht illustriert am Beispiel des Boyle-Versuchs mit Kohlenstoff. Prax Naturwissenschaften Chem Sch 5(65):25–33
70. Dörfler T (2009) Das an Schülervorstellungen orientierte Unterrichtsverfahren. Beispiel: Neutralisation im Chemieunterricht. Chem Konkret 16:141
71. Jungkamp F, Marohn A (2016) choice2reflect—Kontrovers diskutierte Themen mit Hilfe wissenschaftlicher Prüfkriterien reflektieren—Beispiel Homöopathie. Prax Naturwissenschaften Chem Sch 5(65):38–43
72. Jungkamp F, Marohn A (2016) Homöopathie: Nichts drin—nichts dran? Befähigung zu wissenschaftlicher Reflexion anhand selbst erarbeiteter Prüfkriterien. In: Menthe J, Höttecke D, Zabka T, Hammann M, Rothgangel M (Hrsg) Befähigung zu gesellschaftlicher Teilhabe. Beiträge der fachdidaktischen Forschung. Waxmann, Münster, S 307–311
73. Informationsnetzwerk Homöopathie: www.netzwerk-homöopathie.eu. Zugegriffen: 4. Oct. 2017
74. Eilks I, Marks R, Stuckey M (2016) Das gesellschaftskritisch-problemorientierte Unterrichtsverfahren—erläutert an einem Unterrichtsbeispiel zu Tätowierungen. Prax Naturwissenschaften Chem Sch 5(65):33–37
75. Stolz M, Witteck T, Marks R, Eilks I (2011) "Doping" für den Chemieunterricht und eine Reflexion über geeignete Themen für einen gesellschaftlich relevanten Chemieunterricht. MNU 64(8):472–479
76. Feierabend T, Eilks I (2009) Bioethanol—Bewertungs- und Kommunikationskompetenz schulen in einem gesellschaftskritisch-problemorientierten Chemieunterricht. MNU 62(2):92–97
77. Stuckey M, Witteck T, Eilks I (2013) Chemie die unter die Haut geht: Tätowierungen. PdN-ChiS 62(3):30–34

78. Rott L, Marohn A (2015) choice2explore—Eine an Schülervorstellungen orientierte Unterrichtskonzeption für den inklusiven Sachunterricht. Sache Wort Zahl 154(4):52–58
79. Rott L, Marohn A. Inklusiven Unterricht entwickeln und erproben—Eine Verbindung von Theorie und Praxis im Rahmen von Design-Based Research. Zeitschrift für Inklusion. http://www.inklusion-online.net/index.php/inklusion-online/article/view/325/277. Zugegriffen: 29.05.2018
80. Rott L, Marohn A (2017) Unterstützte Kommunikation. Unterstützte Kommunikation Im Inklusiven Sachunterr 1:49–53
81. Marohn A, Rott L (2018) Naturwissenschaftliches Lernen im inklusiven Unterricht. In A Langner (Hrsg) Inklusion im Dialog: Fachdidaktik – Erziehungswissenschaft – Sonderpädagogik (S 102–108). Heilbronn: Klinkhard
82. Rott L, Marohn A (2015) "Oh mein Gott—man sieht den nicht!" Schülervorstellungen im inklusiven Sachunterricht—Chancen und Umsetzungsmöglichkeiten. Sache Wort Zahl 150(43):87–90
83. Feuser G (2002) Momente entwicklungslogischer Didaktik einer allgemeinen (integrativen) Pädagogik. In H Eberwein, S Knauer (Hrsg) Handbuch Integrationspädagogik (S 280–294). Weinheim: Beltz
84. www.leichte-sprache.de. Zugegriffen: 29. May 2018
85. Kitzinger A. Metacom. Symbolsystem zur Unterstützten Kommunikation. www.metacom-symbole.de. Zugegriffen: 29. May 2018
86. Grosser ChG (1985) Strukturorientierter Chemieunterricht von Anfang an. Naturwissenschaften Im Unterricht-physik/chemie 33:134
87. Bauer H (1985) Die Struktur der Materie. Fachliche Grundlagen. Naturwissenschaften Im Unterricht-physik/chemie 33:139
88. Barke H-D (1980) Die Unverzichtbarkeit der Strukturmodelle für das Verständnis der chemischen Reaktion. Prax Naturwissenschaften-chemie Sch 29:37
89. Sauermann D, Barke H-D (1997) Chemie für Quereinsteiger. Bände 1–6. Schüling, Münster (Siehe auch: www.wikichemie.de)
90. Flint A (2017) Chemie fürs Leben—Sek. I. https://www.didaktik.chemie.uni-rostock.de/forschung/chemie-fuers-leben-sek-i. Zugegriffen: 8. Oct. 2017
91. Flint A (2017) Chemie fürs Leben—Sek.II. https://www.didaktik.chemie.uni-rostock.de/forschung/chemie-fuers-leben-sek-ii/elektrochemie/. Zugegriffen: 8. Oct. 2017
92. Egbers M, Marohn A (2013) Konzeptwechseltexte—eine Textart zur Veränderung von Schülervorstellungen. Chem Konkret 20(3):119–126
93. Tausch M (2016) Didaktisch integrativer Unterricht—Kohärente Inhalte, Methoden und Medien. Prax Naturwissenschaften Chem Sch 5(65):44–48
94. Dellbrügge B, Marohn A (2017) choice2interact—Interaktiv Lernen mit Tablets im Chemieunterricht. In: Meßlinger-Koppelt J, Schanze S, Groß J (Hrsg) Lernprozesse mit digitalen Werkzeugen unterstützen—Perspektiven aus der Didaktik naturwissenschaftlicher Fächer. Joachim Herz Stiftung, Hamburg
95. Mayer R (2005) Cognitive theory of multimedia learning. In: Mayer R (Hrsg) The Cambridge Handbook of Multimedia Learning. Cambridge University Press, New York, S 43–71
96. Marohn A (2008) Ionenbildung durch Strom?—Eine an Schülervorstellungen orientierte und chemiegeschichtlich motivierte Unterrichtskonzeption. Chem Konkret 15(2):75–84
97. Kranz J, Schorn J (Hrsg) (2008) Chemie Methodik. Handbuch für die Sekundarstufe I und II. Cornelsen Scriptor, Berlin

Media

5

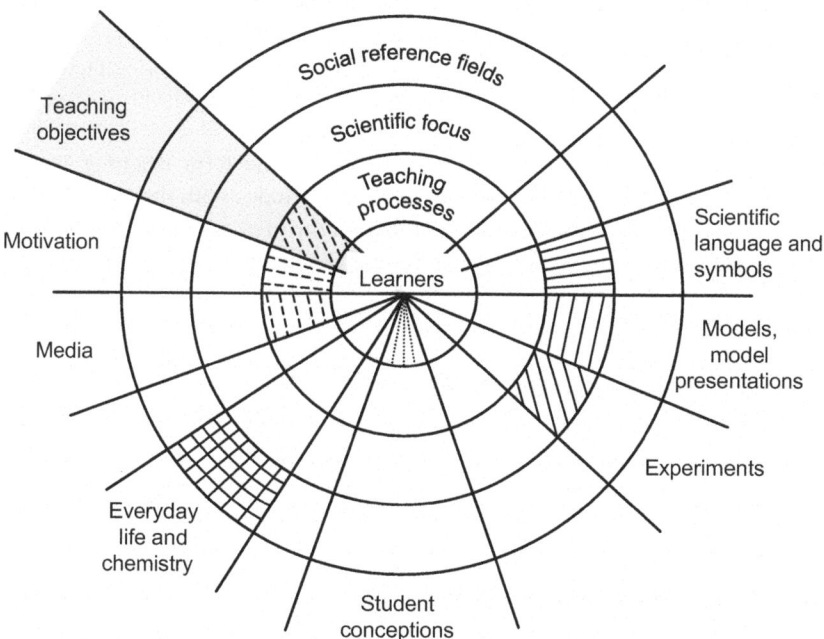

School and teaching are inconceivable without media. In chemistry lessons, the variety of media use is particularly large compared to many other subjects. In addition to "classic" media such as the blackboard and the textbook, experiments, models, and the "new" media are used.

▶ **Media**

Since the term "media" is used in various contexts, a definition of the term is initially required. The scientific discipline relevant to the teaching context is media didactics: "It deals with the functions and effects of media in teaching and learning processes, i.e., with media-mediated learning. Its goal is to promote learning through didactically appropriate design and methodically effective use of media. The selection and use of media should be coordinated with the teaching objectives, content, and methods" (translated from German) [1].

In media didactics, media are understood as "mediators through which signs are transmitted, stored, reproduced, arranged or processed with technical support in communicative contexts [...] and presented in pictorial and/or symbolic form" (translated from German) [2].

To be able to concretize this abstract definition, it is helpful to classify the various characteristics of media. One possibility for classification is the sensory modality of the "symbols". In principle, these "symbols" can be perceived with all the senses available to us. The sense of sight and hearing as well as the sense of touch and smell are particularly important in chemistry lessons (Table 5.1). After that, we distinguish between visual, auditory, audiovisual, haptic, and olfactory sensory experiences. Another distinguishing feature concerns the level of experience (Table 5.1).

In the schedule of teaching media in Table 5.1, the term teaching media is understood comprehensively. A teaching medium includes both the "symbols" that are encoded in the form of letters and numbers, drawings, images, sounds, smells, and "tangible" forms, as well as the technical device with which the "symbols" are presented, e.g., the paper, the CD and the CD player, the presentation slide, the screen and the computer, or even the real object.

▶ **Definition: "New Media" and Multimedia**
In addition to the traditional media, the term "new media" needs to be explained. "New" media are understood to be those media that use digital and computer-technical foundations. Since the term "new media" has been used for more than 40 years, the "new" media are expanded with each technical novelty [3].

Table 5.1 Possible classification of teaching media

Teaching media				
Type of sensory experiences			Type of experience levels	
Visual	Auditory	Audiovisual	Primary experiences	Secondary experiences
Textbook	Tape recorder	Sound film	Animals, plants	Preparations
Blackboard	CD	Video	Substances, crystals	
Replicas				
Transparent	DVD	Television	Experiments	Recordings
Illustration	Radio	Audio-slide series		
Computer	Language lab	Multimedia		

The characterization of new media can be made through the following dimensions [4]:

- **Multimediality:** With computer technology, the processing of large amounts of different data (texts, images, audio data, video data) is realized.
- **Multicodality:** Due to digital technology, a presentation of different types of coding (words, images, numbers) is possible.
- **Multimodality:** The new media combine different sensory experiences.
- **Hypermediality:** This term refers to the possibilities of designing network-like arranged contents that are linked to each other. While in the past only a linear and thus sequential representation of contents was possible, today's technology allows the paths of capturing content to be individualized. Hypermediality is almost self-evident due to the degree of networking of media and the development of the internet.
- **Interactivity:** This term describes the user's ability to control the multimedia application. Especially with regard to the use of multimedia in teaching, adaptivity is a significant, but difficult to realize, property of multimedia applications. Depending on the type of multimedia application, the application adapts to the user. This is particularly important in the context of the demand for individualization and diagnosis of learning processes.

5.1 Teaching Processes: Media and their Functions in Teaching

A clear assignment of functions of individual media for teaching is not possible, as media take on various functions in different teaching situations. Kerres describes **three didactic functions** of media [5]:

1. **Knowledge tool for communication and cooperation:** This function is fulfilled by visual and audiovisual media, as they are "used as tools for the development, collection, preparation and communication of knowledge" [5].
2. **Knowledge presentation:** This function is certainly the most obvious of all functions, which is fulfilled in different ways by all media listed in Table 5.1. Texts, photos, illustrations with different degrees of abstraction, tables, diagrams, experimental observations, models can be used for the presentation of knowledge.
   However, the medium does not in itself act as a mediator that supports the learners' learning processes. Only through active, cognitive engagement initiated by work instructions or learning tasks media become helpful media for teaching. It may be necessary to change the information, for example, in the form of a schematic representation of a blast furnace by omitting details and thus directing attention to what is relevant.
3. **Knowledge transfer and control of learning processes:** This function can be applied especially to audiovisual and multimedia media, as the temporal control

is decisive for the design of learning processes. The learner and the teaching content must be connected via the medium. To realize such a coupling, the teaching progress would have to take into account each personal learning process.

Among the media, especially multimedia applications with a high degree of interactivity and adaptivity fulfill this didactic function excellently. The often associated hope of increased motivation of learners with new media does not prove true in empirical studies, as the curiosity effect wears off very quickly [5].

This rather abstract categorization of the didactic functions of media according to Kerres is to be supplemented by a further categorization according to Martial, who formulated the following functions of media [6]:

- Information mediation
- Attention guidance
- Activation of learners
- Stimulation and control of thought processes
- Structuring of thought processes
- Support of information processing
- Accentuation of information
- Elementarization and illustration of information
- Structuring of learning content
- Generalization and/or abstraction of learning content
- Learning of thinking and working techniques
- Promotion of memorizing performance
- Enabling of experiences
- Mediation of feedback
- Control of results
- Diagnosis of competence increase
- Practice
- Differentiation and individualization of learning processes
- Media education

In the following, the teaching media relevant for chemistry lessons (Table 5.2) are described and aspects of chemistry didactics and media didactics are addressed.

Table 5.2 Classification of subject-appropriate media for chemistry lessons

Media for chemistry lessons			
Visual	Audiovisual	For experiments	For models
Textbook	Sound film	Experimental equipment	Structure models
Blackboard	Video, DVD	Measuring devices	Model drawings
Presentation slide	Computer, Television	Apparatus	Model experiments
Illustration	Multimedia	Projections	Functional models
		Computer support	Computer models

5.1.1 Textbook

The textbook is certainly considered a classic among teaching media and should represent a supportive medium for both the learner and the teacher. Textbooks must be approved by the state government of the respective federal state in Germany if they are to be officially used in teaching. A commission of the ministry examines whether the requirements of the core curricula of the federal state are met. This applies to both the individual competence areas and competences as well as the technical content. The textbook may indeed contain content that goes beyond the core curricula. If the commission's result is positive, the book is added to the list of textbooks approved for the state [7]. At schools in Germany, the school conference then decides on the introduction of a specific textbook based on the proposal of the chemistry conference. If a specific chemistry book is to be considered a useful teaching medium by the colleagues of the subject conference, important criteria for the book need to be developed, which should be aligned with the school's internal curriculum. A working group of the Chemistry Teaching Division of the German Chemical Society (GDCh) has developed a criteria catalogue to assist in the selection of a textbook [8]. The analysis grid is based on the competence areas of the educational standards of the Conference of Ministers of Education for the intermediate school leaving certificate in chemistry [9] (see overview).

Excerpt from the analysis grid for chemistry books of the textbook working group of the Chemistry Teaching Division of the GDCh [8]
Competence area "Technical knowledge":

- Basic prerequisite (curriculum-compliant, cumulative knowledge acquisition)
- Methodical didactic procedures (problem-oriented, research-developing)
- Handling of technical terms
- Content focus and summary
- Clarity and comprehensibility of texts
- Inclusion of illustrations
- Representation levels
- Structure and design of the textbook
- Task formats
- Offers for reflection

Competence area "Knowledge acquisition":

- Use of experimental investigation methods
- Promoting scientific curiosity
- Working with and thinking in models
- Consequences of knowledge acquisition

Competence area "Communication":

- Methods and social forms
- Materials as a basis for communication
- Use of technical language

Competence area "Evaluation":

- Topic selection
- Materials as a basis for controversies
- Task formats
- Methodical arrangements

5.1.2 School Board

A board is almost in every classroom and usually in a prominent place, so that it can be easily seen from all student seats. It is therefore often used, and one should be aware of the importance of its function as a medium. Often, the board picture is also a model for the students' notebook entry—for this reason, it needs to be carefully planned and well structured, even if the board writing often has to be done spontaneously.

Especially in experimental instruction, the blackboard is indispensable when the individual steps from problem definition to problem solution need to be traced. In this case, the **blackboard diagram** can be divided into the following sub-points:

1. Lesson topic
2. Problem definition
3. Planning and execution of the experiment
4. Observations and measurements
5. Evaluation and error discussion

Graf [10] provides further arguments for the "key functions" of the school blackboard:

> Since the graphical representation is usually developed step by step in front of the students' eyes or (which is extremely useful from a learning psychology perspective) even together with the students, and thus the blackboard design—and at the same time the learning object—is sequenced into manageable learning steps, the facts can be well remembered by the students. If you have a folding blackboard available in the chemistry room, it is advisable to use one or both foldable side boards as a "surface quarry area" to fix, for example, the students' questions and assumptions or to record central technical terms (known or newly introduced). (translated from German)

5.1 Teaching Processes: Media and their Functions in Teaching

The **advantages and disadvantages of the blackboard** are listed by Graf according to the overview.

Advantages and Disadvantages of the Blackboard according to Graf [10]
Some advantages of the blackboard include:

- Available in every classroom
- Quickly usable (no darkening, no electrical connection needed)
- Methodically versatile
- Targeted adaptation of the blackboard image to the students' level of development
- Differentiated coordination of the blackboard image with the lesson (progress of the lesson)
- Gradual development of the blackboard image with the participation and in front of the students
- Emphasis of important elements or focal points of the lesson without much effort
- The teacher's role model in designing the blackboard image as a learning aid
- Successive addition of the blackboard image (also beyond the individual lesson) possible
- Various design possibilities of the blackboard image (structure of the blackboard image ≡ structure of the lesson, mind maps next to experiment sketches, pre-made magnet models, experiment observations and interpretations etc.)
- Participation opportunities for students
- Clearly visible from every student's seat
- Corrections to the blackboard image are relatively easy to make
- High contrast, i.e., almost every color is easy to see (exception: thin pastel chalk and dark colors)
- Concentration of the students' attention "on one point"
- Stimulation of the students for active participation and follow-up work
- From an ecological balance point of view, the use of the blackboard is undoubtedly quite favorable
- Many different blackboard chalk colors available (aesthetic design of the blackboard image)

Some disadvantages of the blackboard include:

- Limited surface
- Wet boards are hardly usable
- Graphical representation on the blackboard is only available for a limited time

- Graphical representation must be carefully planned (proportions, relationship of sketch and text etc.)
- Sketching complex experimental setups and technical facilities takes a lot of time
- Drawing on the board is a skill that needs to be learned
- While the teacher is making a blackboard sketch, he turns his back to the class and covers part of the board, even if he stands a little to the side of the blackboard sketch.
- The board often needs to be cleaned at the beginning of the lesson
- Even the correct cleaning of a board needs to be learned, takes time and requires care.
- Large blackboard surface tempts to write too much text
- Sliding blackboard surfaces rather correspond to a DIN-A4 notebook double page in landscape format
- Transferring the blackboard image to the student's notebook takes a lot of time for some students
- If students write on the board, they do not yet have the text or the illustration in their folder or notebook.
- The blackboard is only two-dimensional, i.e., for example, the "methane tetrahedron" is difficult to represent
- Soiling of hands and clothing (when drawing or wiping)
- Blackboard writings that are developed over the entire approx. 2 m wide sliding board during the lesson are usually only transferable to the notebook with considerable difficulties.
- Copying the pre-made blackboard image makes students dependent

5.1.3 Presentation Slides (via Overhead Projector or Computer/Tablet and Projector)

Similar to the blackboard, one often encounters a projector for overhead projection in most classrooms. Initially, one can use the overhead projector like a blackboard, therefore the same applies to the projected slide as to the blackboard. The following advantages arise compared to the blackboard:

- You write facing the class.
- The available writing area is arbitrarily large.
- The slide image developed in a lesson can be projected again for repetition in the next lesson.
- Own slides can be supplemented by suitable slides from the teaching material trade.

Many science classrooms today are equipped with a ceiling-mounted projector in addition to a blackboard and overhead projector, to which a fixed computer is

5.1 Teaching Processes: Media and their Functions in Teaching

connected or to which a laptop can be quickly connected. As a result, overhead projectors are used less and less. A still common use of the overhead projector is the presentation of experiments in projection (Sect. 5.6: V5.4 and V5.5).

When using the overhead projector, a few disadvantages must be considered:

- The slide pens dry out easily and therefore must always be closed again after writing.
- The projector's lamp dazzles the teacher while writing.
- The projector's lamp can burn out (if necessary, a spare lamp can be switched on).

A possible failure of the technical systems must always be considered when using a projector and computer.

The diverse design possibilities of overhead slides and presentation slides can lead to students being overwhelmed with a flood of information and visual stimuli. A recipient-appropriate adaptation of the information flood, e.g., by supplementing detailed information, can be achieved by performing a custom animation of the presentation slides in the presentation programs. In this way, details can be added or faded out again. The service of textbook publishers to distribute CDs or DVDs with the image material from the books is a helpful support in the preparation of presentation slides.

When designing presentation slides on the computer, in contrast to the overhead slide, spontaneous reactions to student comments cannot be made. However, the handwritten addition and then even the digitization of the additions can be realized today with so-called interactive whiteboards (Sect. 5.1.8).

Chemical, oil, or energy corporations often offer ready-made presentations on their products on their websites. Specifically for chemistry lessons, the Fund of the Chemical Industry (FCI) provides extensive media packages with slide sets and accompanying texts [11]. Topics include the wonder world of nanomaterials, renewable raw materials, biotechnology: smallest helpers—great opportunities, textile chemistry.

What is shown on slides can encompass the entire repertoire of visual, but also audiovisual, information carriers. In addition to texts, tables, diagrams, digitized original texts such as newspaper reports and patent specifications, photos, schematic representations, symbolic representations such as reaction equations and reaction mechanisms, molecule representations, etc. can be presented. If the slides are linked with hyperlinks to other media (videos, animations, homepages, audio files), then the presentation slide becomes a medium that is defined as multimedia.

At this point, it should be noted that copyright must be observed when using foreign materials. Information on this can be found on websites such as Teacher-Online [12].

5.1.4 Newspaper Report

Teaching on current topics of everyday life and the environment can be made interesting if current newspaper reports are used as a basis and interpreted. It is often also motivating to uncover factual errors by journalists and to clarify correspondingly incorrectly described facts in class discussion. For this reason, it would be advantageous to compile a collection of such newspaper reports and keep them ready for use in class. Haupt [13] has been collecting articles from the newspapers of the Oldenburg region since the 1970s and sorting them by topic: cartoons, specific substances, pollution of water, air and soil, chemical technology and chemical accidents, food and genetic engineering, energy, radioactivity, etc.

5.1.5 Videos, Films, Online Appearances

Platforms such as YouTube, online presences of knowledge shows such as Quarks and Co. [14] and research institutes such as the Helmholtz Center Berlin for Materials and Energy [15] and many others provide videos on a wide range of topics. Since computers with internet access are now standard equipment in many science rooms, the technical hurdles for using film material have become lower. In the context of lesson preparation, the material must be reviewed in terms of the selection of clips from videos and films, the technical accuracy of the content, and appropriate language and presentation of the content.

In addition to many freely accessible media, the Media Institute of the States FWU in Germany, among others, offers a large number of videos with accompanying information in its media library, which can be accessed through a purchasable school license. The variety of topics covers all mandatory content of chemistry teaching in SI and SII [16].

Free films from federal ministries, knowledge shows, and large chemical companies are often suitable for use in chemistry lessons, e.g., films about the future of electromobility [17]. These media usually cover topics from various perspectives and thus in an interdisciplinary manner, reflecting current discussions in politics, science, and society. Feature films can also be used in class as a motivating medium (Sect. 5.4).

5.1.6 Computer, Tablet

The computer has evolved into a universal medium (Table 5.3): Texts are scanned and processed, tables and graphics are created, drawings or photos are inserted and altered. If appropriate programs are available, simulations of real processes can be followed, measurement values can be recorded and processed in the shortest possible time and stored, programs can be developed and questions answered in collaboration with computer science classes. With access to the Internet, teachers and students are able to retrieve information and data on almost all substances, manufacturing processes and environmental issues from all over the world, make contact with other institutions via e-mail or present their own projects to the public via the homepage.

5.1 Teaching Processes: Media and their Functions in Teaching

Table 5.3 Different possibilities for the use of computers in chemistry lessons

Use of the computer		
Common programs on the hard drive	Drives for CD-ROM	Internet access
Word processing	Simulations	Databases
Spreadsheet	Data acquisition	E-mail
Presentation programs	Programming	Homepages
Image editing	Video editing program	Social Media
Formula symbol programs		

In dealing with the use of computers in teaching, a distinction must be made between the medium itself (hardware and software) and the didactic function of media use and the concrete action with the presented information (images, video, simulations, animations) [18]. However, this can only be realized in a specific case. Here, therefore, mainly current technical innovations can be mentioned and exemplary possibilities for use can be shown. These developments certainly include the ubiquity of digital media in students' everyday life. The size and weight of computers and laptops have now been reduced to such an extent that these media can also be used mobile. While previously one usually had to decide whether the chemistry lesson should take place in a science classroom or in the computer room at school, teaching is becoming more flexible, especially through the use of tablets. When using tablets in teaching, many questions arise that need to be discussed with all participants in the context of the discussion about media education in schools.

The cost reduction for tablets in recent years has also affected the media equipment in schools. At some schools, entire classes are equipped with tablets, so that the tablet can be used as a digital tool in all subjects. In chemistry lessons, the tablet can be used in conjunction with apps for various functions, some of which are mentioned here [19, 20]:

- Research on the Internet
- Use general, cross-curricular apps like mind map programs, spreadsheets, etc.
- Videograph experiments
- View animations and dynamic representations at the submicroscopic level
- Draw molecule models
- Use chemistry-specific apps like a virtual periodic table of elements

An alternative concept to school-owned tablets in teaching is behind the abbreviation BYOD: Bring your own device. Learners use their own smartphone or tablet as a digital tool in school lessons. In studies on media use in teaching and on the media competence of teachers, teachers mention various reservations about BYOD [21]. There are reservations especially with regard to legal uncertainties such as data protection and usage rules as well as the insufficiently developed infrastructure in schools [22, 23].

5.1.7 Multimedia

Following the definition in the first part of the chapter, multimedia also includes presentations that combine purely visual information with audiovisual or auditory media. In this context, the term **hypermedia** is mentioned, which includes the terms hypertext and multimedia. Hypermedia connect hyperlinks and multimedia applications [24]. The links enable different processing paths. It is of great importance that the user receives an overview and orientation, for example, through clear navigation using a sitemap. Multimedia applications have the great advantage that they combine the use of texts in image and sound, images with or without commentary, film sequences or model animations, and can switch quickly from one application to another, repeat them or skip another. It is possible for each student to interactively follow their individual learning steps on their PC according to their learning progress, or for a whole group of students to be taught by the teacher in lockstep using a projector.

The working group around Tausch offers animations on a variety of mandatory content for chemistry lessons in SI and SII as well as innovative topics for chemistry lessons under [25]. In the animations, video clips of experiments, schematic representations of the experiments, and models for the submicroscopic level complement each other to form a multimedia application, into which task formats for knowledge testing are partly integrated. In an animation on Otto fuels, students can interactively simulate the large-scale synthesis of methyl tert-butyl ether, an anti-knock agent for fuels, and determine the optimal conditions for the synthesis via parameters such as molar ratio, temperature, and pressure [26].

In the $CHEM_2DO^®$ project, a cooperation project of several chemistry education departments and Wacker Chemie AG, animations are being created for experiments with silicones and cyclodextrins [27]. The detailed and professionally designed animations are characterized by a close interlocking of experiments and thus the macroscopic level and the submicroscopic level. On the submicroscopic level, models of different degrees of abstraction are used, so that the experiments can be located in both secondary level I and secondary level II with reference to the basic concepts from German school curricula. In Fig. 5.1a, you can see a simple puzzle model with which the synthesis of a silicone rubber with elastic properties can be illustrated on a simple particle level [28]. For secondary level II, ball-and-stick models are used for the same experiment, with which the platinum-catalyzed reaction between a silane molecule and a vinyl group in detail and the polyaddition to a three-dimensionally linked polymer can be explained on a submicroscopic level (Fig. 5.1b).

5.1 Teaching Processes: Media and their Functions in Teaching 159

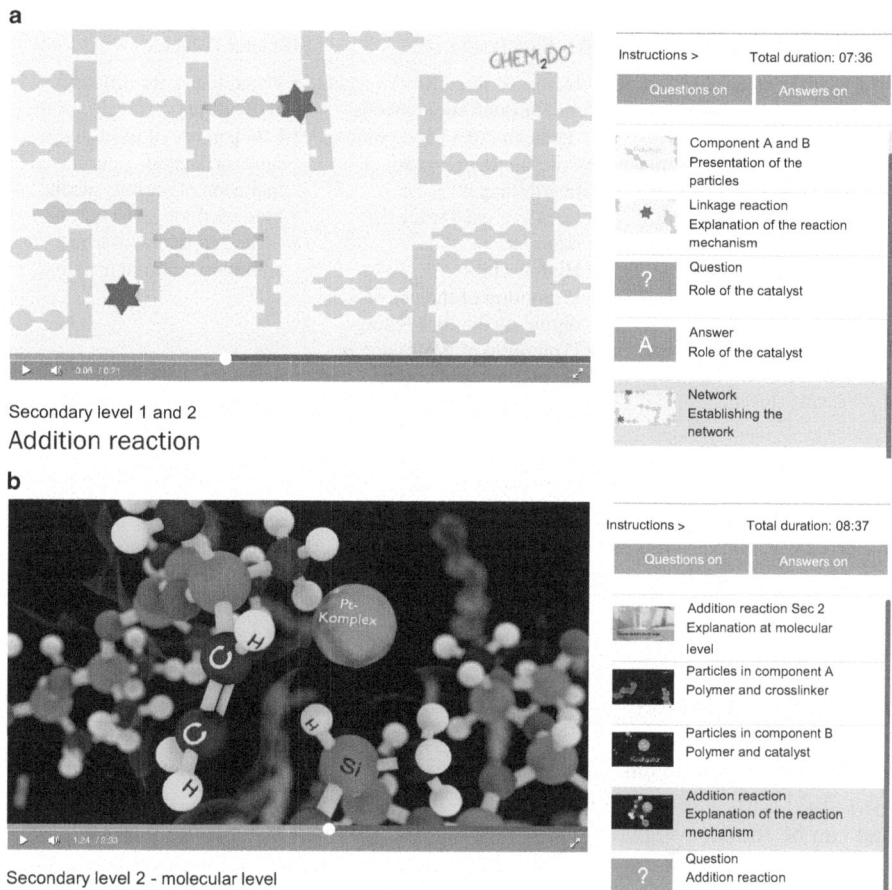

Fig. 5.1 Screenshot from the animation for the synthesis of a silicone rubber (**a**) for secondary level I and (**b**) for secondary level II [28]

5.1.8 Interactive Whiteboard

In addition to a conventional blackboard, interactive whiteboards are increasingly being installed in science classrooms. These consist of a white presentation surface onto which the screen of a connected laptop is projected. The interactivity lies in the ability to operate the computer via the presentation surface using fingers or a special pen. For example, prepared presentations can be supplemented with handwritten notes or animations can be shown by the teacher or students. Some didactic potentials of using an interactive whiteboard are summarized in Table 5.4.

Some of the potentials mentioned in Table 5.4 may also turn out to be overestimated potentials. Essential for meaningful use in teaching is the further training of teachers in handling the interactive whiteboard and the software. Sieve describes

Table 5.4 Potentials of using interactive whiteboards in teaching [29]

Learning Potentials	Teaching Potentials	General Potentials
• Increase in motivation • Increased participation opportunities • Promotion of creative student presentations • Easier access to teaching resources to offer options for different learning styles	• Increased flexibility of teachers through spontaneous inclusion of various sources • Printing and securing of board images • Recall of used teaching materials • High usability • Promotion of the use of digital media in teaching • Promotion of professional development of teachers	• Versatile use for different grade levels • Efficient use of teaching time through uncomplicated inclusion of various media • Increased job satisfaction • Increased interaction and communication in teaching

some examples of a learning-promoting use of the interactive whiteboard in chemistry lessons: planning experiments with drawing elements for experimental apparatus, evaluating experiments in connection with the use of data acquisition systems, using images, animations and molecule models, and learning games for chemistry lessons [30].

5.1.9 Experiments

Experimental equipment, measuring devices and apparatus (see overview in Sect. 5.1.1) are depicted in experimental instructions of teaching or school books and can be found, compared and ordered in the catalogs of teaching material companies. When assembling apparatus, certain psychological perception rules must be observed (Sect. 6.2). Furthermore, reference is made to chapter 6, which makes the experiment a central concern. A few selected examples are intended to present aspects of the use of media in the demonstration and execution of experiments. The additional media should fulfill many functions: from better visibility of observations in experiments to modeling of processes in the experiment. It should not be forgotten that the experimental materials are also media by definition.

5.1.10 Backgrounds and Light Wall

With a very simple method, the visibility of observations in demonstration experiments but also student experiments can be improved: The conscious choice of a white, black or monochrome background helps to make turbidity or color changes more visible (Sect. 5.5: V5.1). Many secondary information such as the course of clamps, stands and muffles can be faded out by a monochrome background

(Sect. 6.2). For colored solutions, illuminating the solutions from the back can contribute to a clear observation of color changes (Sect. 5.5: V5.2).

5.1.11 Camera Use

To enlarge experimental setups and to present detailed observations such as the size and ascent speed of gas bubbles, gooseneck cameras or document cameras can be used in combination with a projector (Sect. 5.6: V5.3). Compared to document cameras, webcams (with autofocus) or even the cameras of tablets and smartphones are cheaper alternatives. The functions of the different cameras should be carefully considered when discussing the purchasing of such media. Features such as autofocus, high zoom capability, and an independent light source on the camera are helpful in presenting experiments. When using camera projections of experiments, it should be taken into account that the students must see the experimental setup in duplicate and must distinguish between the representations.

5.1.12 Projections

Another type of projection is possible with the overhead projector. Experiments in Petri dishes are thus displayed in transmitted light (Sect. 5.6: V5.4). Several advantages are utilized: On the one hand, the amount of chemicals is very low, on the other hand, the analysis of observations can be started on the projection of the real object by placing a foil underneath.

The disadvantage is that opaque solids such as electrodes or precipitates always appear black and cannot be displayed in their true color.

The advantage of minimizing the amount of chemicals is also mentioned when using special cuvettes for use on the overhead projector (Sect. 5.6: V5.5).

5.1.13 Magnetic Whiteboards

Often, the observations in demonstration experiments are not clearly visible to all learners. The use of white, black, or colored backgrounds can sometimes quickly remedy this (Sect. 5.6: V5.1). The use of magnetic whiteboards and magnets, with which the devices can be attached to the whiteboard via clamps, simplifies the execution of many experiments. This proven in use system helps in considering the perception rules in psychology (Sect. 6.2) and not only enables the presentation of experiments but also supports the analysis of the experiments, as the experimental setup can be labeled. Grofe and Brand have developed a whole system of techniques for use in chemistry lessons [31, 32].

5.1.14 Computer Use

When using computers for experiments, it is not intended to replace experiments with images on the computer screen: If it is viable, the real experiment should take place either in form of a demonstration or a student experiment. As soon as it comes to the recording and evaluation of many measurement data, the creation of tables and graphs, the comparison of different measurement curves, this can be done by the computer. This **computer-assisted** data acquisition can have multiple functions:

1. The large display can show measurements in large numbers that are easy for all students to see and can be followed during the measurement process, such as the decrease in mass measured on a balance during the evaporation of a volatile solvent or the measurement of the pH value or the electrical conductivity during an acid-base titration.
2. Very slow and very fast reactions can become more vivid with the help of the connected PC than in the traditional way. For example, the very rapid increase in pH values during neutralization becomes clear directly on the screen.
3. Series of measurement data of the same fact, such as analyses of pollutants in water or air samples at different times of the day, can be realized with little effort and the results of the measurements can be compared using data processing programs.
4. Measurement data obtained through a real experiment can be processed into tables or graphs, also averages, deviations, compensation curves determined or error calculations carried out and marked on the screen.

5.1.15 Data Acquisition Systems, Handheld Devices

For these purposes, educational supply companies offer various systems. At this point, the multifunctional system All-Chem-Misst by AK Kappenberg [33] should be mentioned, which is particularly suitable for demonstration experiments and projects due to its purchase price. For student experiments, systems are suitable where various sensors can be connected to a handheld device. An overview of handheld devices and sensors and a comparison regarding usability, application possibilities, sensor variance, software and support have been compiled by Schrader and Schanze [34].

A recent development serves both the requirement for low material costs and the promotion of technical understanding. With microcontrollers like the Arduino board or the Raspberry Pi, a hardware product and the programming are made open source [35–37]. Various sensors can be connected to the microcontrollers, so these systems can be used in student experiments in science lessons [38].

5.1.16 Models

Real models, model drawings, model experiments or functional models are important media for any chemistry lesson, especially when initial model ideas for the association of smallest particles (particle model) or for atoms and ions were introduced. Such models of this kind and their reflection are the content of Chap. 7. Model calculations or **model representations** that can be carried out with the computer are special media of chemistry lessons, which should be explained at this point. The following should be mentioned as examples from the multitude of current and future programs:

- Standard drawing programs such as Chemdraw, ChemSketch, etc. are suitable for drawing and printing the structure of molecules in two or three dimensions. In these programs, molecules can be drawn in various representations and also displayed in three dimensions with the 3D viewer.
- For a representation independent of formula drawing programs, plug-ins for internet browsers can be used. Schmitz offers helpful instructions for creating such molecular models on the screen as well as a large selection of ready-to-use molecular models on his homepage [39].

5.1.17 Experiment Kits

There are experiment kits or experiment sets on the teaching material market, which, in addition to chemicals and equipment, also provide school-appropriate experiment instructions and working materials as copy templates [40]. WACKER Chemie AG offers a kit for experiments with silicones and cyclodextrins named $CHEM_2DO$, which is given free of charge to the teacher after a further training of teachers [41].

5.2 Scientific Focus: Objective Adequacy Appropriateness of Media

Whether a medium is suitable for use in chemistry lessons depends on many factors. Some evaluation criteria can be used as a guide for the decision of a medium.

Valuation Criteria
Media should be used appropriately in relation to the respective learning situation:

- the cognitive development stage of the learners,
- prior knowledge and interests of the learners,
- the intended goals of the lesson,
- the planned social forms of the lesson,
- aspects of gestalt and perception psychology,
- the technical feasibility and mastering of the technical devices.

Both the learning materials industry and textbook publishers deliver a multitude of graphical representations, slide sets, videos, drawings and images—often they are different media for the same fact. Before using such media, their factual appropriateness must be reflected. The following criteria for reflection can be found:

Accuracy

Before using many media, it must be checked to what extent they are factually correct or how they can be brought up to date with the latest scientific ideas. For example, if the use of images for the construction of the atomic shell is planned, which show isolated spheres as models for electrons and orbits for their movement in the sense of Bohr's idea ("sighting discs"), additional information must be prepared that start from the wave-particle dualism and bring terms such as occupancy probabilities or energy levels into the discussion. Or corresponding films must be shown and explained in addition, which avoid a fixation on the "sighting disc model". Since the area of protons, neutrons and electrons is fundamentally not to be visualized, it can also be dispensed with to use of such flawed media.

Didactic Reduction

Many facts must be didactically reduced according to the learning group. The following aspects should be included in the considerations for didactic reductions: The factual representability of elementary processes as well as the connectivity of contents to scientifically valid statements that have been didactically reduced must be guaranteed, while representations of contents must be adapted to the cognitive structure of the learners.

In iconic representations of the processes in a galvanic element, various didactic reductions are made. The water molecules of the electrolyte solutions and also the hydrate shell of the ions are often not represented. This reduction allows the focus on the elementary particles in the processes at the electrodes and is generally accepted. Another reduction concerns the representation of the electrodes, which are usually represented as compact blocks, and the electrical conductors of the external circuit. The teacher must decide for himself whether this reduction is still permissible from a technical point of view and whether it facilitates or rather complicates the learning process.

Real Objects, Real Processes

If the opportunity exists to use real objects instead of illustrations or transparencies for a factual situation, then real objects should be given preference initially. However, the real objects should be combined with photos or model drawings to achieve an optimal learning effect with these additional media. For example, when dealing with the topic of "car battery", not only is the demonstration of the real object possible and necessary, but an experiment on the real process can even be

5.2 Scientific Focus: Objective Adequacy Appropriateness of Media

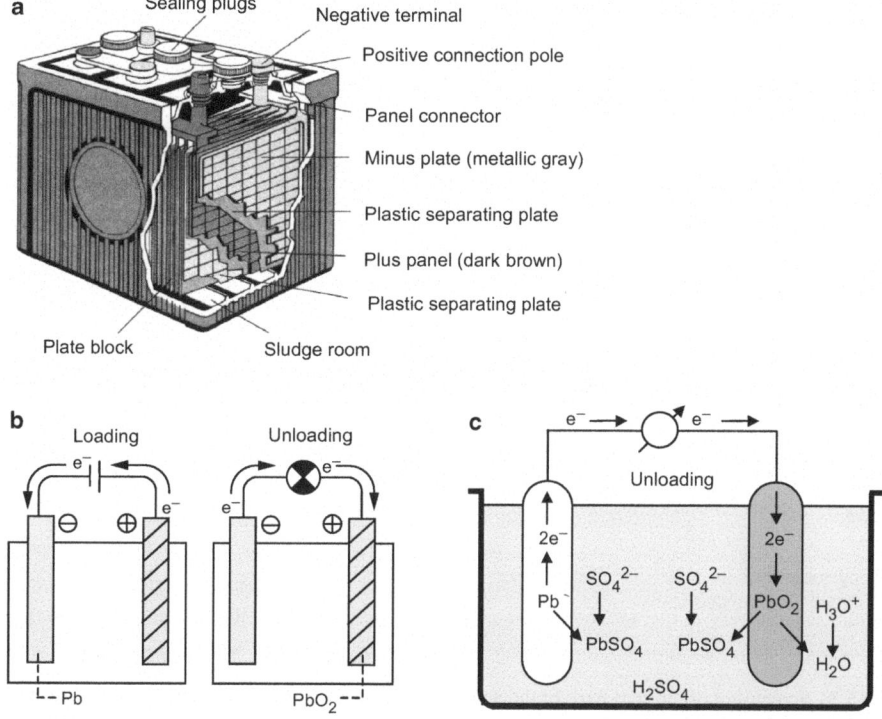

Fig. 5.2 a Schematic structure of the real object, b Sketch of the experimental setup, c Representation of the processes at the electrodes during discharge on the topic of "car battery"

shown (Sect. 5.6: V5.7). Only after the real object and experiment should slides and images be offered as additional media for explanations (Fig. 5.2).

Separation of Different Levels of Abstraction
In all the mentioned criteria, it must be considered whether the levels of representation and thus also the material and particle level are clearly distinguished from each other (Sect. 7.2.1). The separation of the levels is already evident in the overview using the example of an animation of host-guest complexes with cyclodextrins, which gives an overview of the learning tools in the animation [42] (Fig. 5.3). In addition to a video recording of the experiment, which can be used to repeat observations, a simple animation of the experiment is offered, which is supplemented by an animation at the molecular level using different model representations of the cyclodextrin molecule. In the learning tools on the left side, the molecular structures of the reactants—in the experiment β-cyclodextrin and phenolphthalein—can be explored in different molecular representations.

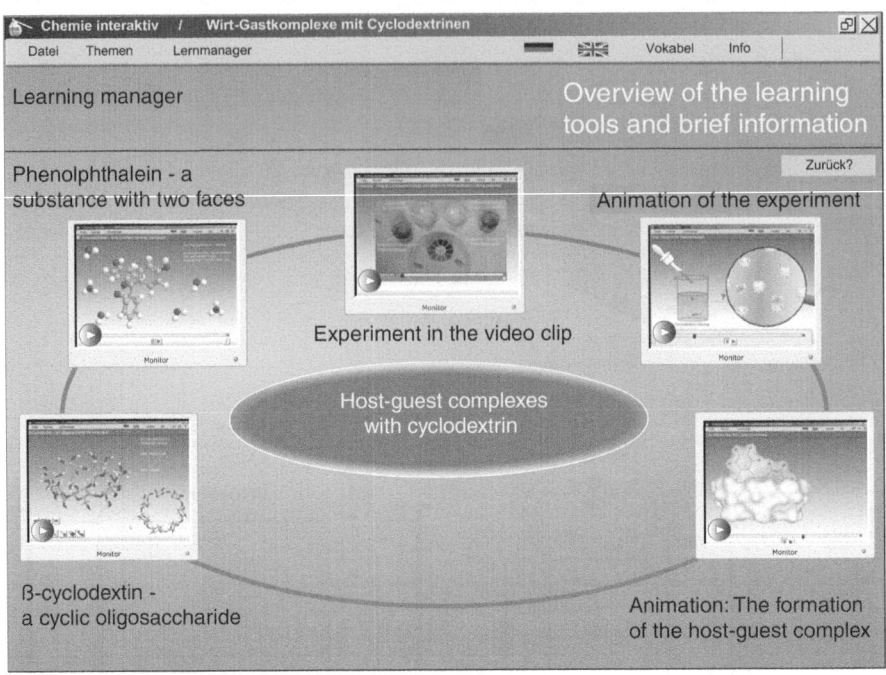

Fig. 5.3 Screenshot of the start page of the animation "Host-Guest Complexes with Cyclodextrins" [42]

5.3 Learners: Media Competence and Media Production

For students, media such as smartphones or tablets are part of everyday life. They therefore already bring a lot of skills and abilities in dealing with media. In school, other aspects should be addressed based on this knowledge to promote the media competence of the students.

In the decision of the Conference of Ministers of Education and Cultural Affairs "Media Education in School", schools should be given guidance for the implementation of media education and the development of media competence. Media competence is understood here as knowledge, skills and abilities

> that enable appropriate, self-determined, creative and socially responsible action in the media-imprinted world. It also includes the ability to move responsibly in the virtual world, to understand the interaction between the virtual and material world and to recognize not only the opportunities but also the risks and dangers of digital processes (translated form German) [43].

Thus, media competence encompasses the four dimensions of media criticism, media knowledge, media use and media design [44] (Fig. 5.4).

5.3 Learners: Media Competence and Media Production

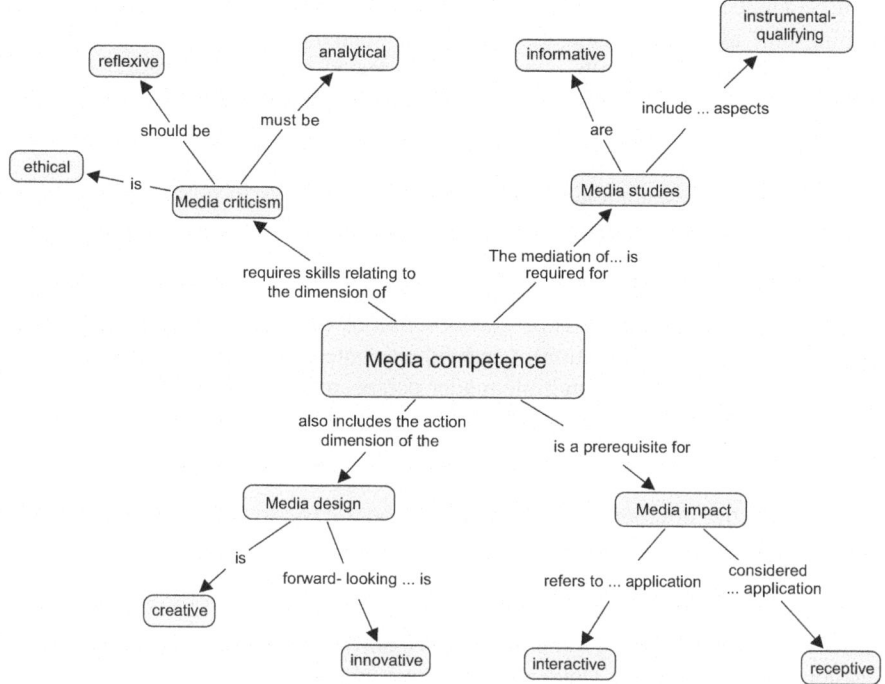

Fig. 5.4 Concept map of media competence, created with a Cmap Tools program [45]

The following are some possibilities for media design in chemistry lessons. Most students will keep a notebook or similar in chemistry lessons, in which, in addition to blackboard notes, mainly experimental protocols and completed worksheets will be found. This cannot be considered as media production. The following products could be created in chemistry lessons:

Posters
When designing a poster, many aspects need to be considered. In doing so, students acquire competencies in various areas or apply their competencies. In class, criteria for a good poster should be developed together. Important criteria can be: title, readability, structuring, use of images with captions, selection of content, preferably short texts, illustration of numerical material with diagrams, illustration of the submicroscopic level with models. A topic suitable for chemistry lessons in secondary level I addresses the problem "Salts and Health" from very different facets [46].

Concept Maps
A concept map consists of a network of terms that are linked to each other via relations (Fig. 5.4). This distinguishes a concept map significantly from a mind map, in which terms are linked to each other, but the relations in between remain undefined. Thus, the design of a concept map places significantly higher demands

on the students. Concept maps can be used in class both at the beginning of a teaching unit as an advanced organizer, accompanying but also at the end of a teaching unit [47]. Especially concept maps created individually by the students can help teachers in diagnosing the competencies of their students.

Presentations
When designing presentations, in addition to learning how to use the program on the computer, many aspects need to be considered that are also important when creating a poster. So that the students can mainly deal with the content and the presentation of the content and not use the available time for the design of the medium, a step-by-step processing makes sense, as suggested by Stahl: basic conception of the medium, creation of individual content modules, determination of the overall structure, taking multiple reader perspectives, integration of the content modules into a product [48].

Films
If you go on the internet looking for films about experiments, you will quickly find what you are looking for. In addition to offers from teachers, universities and teaching material companies, you will also find videos that were recorded in class by students, for example with a smartphone. Often the videos were created without any thought to the camera perspective, the setting, the use of the zoom function, etc. This type of video can also have its justification. Since students can already handle the technology today, it seems sensible to integrate the aspect of film education into subject teaching. In a production-oriented film-integrative teaching unit, in addition to a limitation to subject content and thus a focus, both aspects of media education and many practical and information technology prerequisites need to be considered. A certain competence of the teacher in the field of film education and in dealing with camera, editing program, etc. is necessary [49].

5.4 Social Reference Fields: Mass Media

The influence of media on learners—in the sense of mass media—is often stronger than educational media can be, due to their technical perfection and constant repetition. Mass media such as newspapers, magazines, radio and television broadcasts, and the internet act as **extracurricular information carriers** in both a positive and negative sense.

In relation to chemistry-related content, extracurricular media can have a **positive** effect by:

- **Stimulating interest and curiosity.** If students find reports in newspapers about the importance of memory metals for the construction of engines and many other technical objects, the phenomenon of the "shape memory" of certain alloys may be so interesting that it is addressed in chemistry lessons and discussed using relevant articles from professional journals.

- **Encouraging critical engagement**. Newspaper reports and television broadcasts about climate issues or innovative applications such as OLED display technology can be an opportunity to objectively prepare this topic in class and then critically engage with these reports, including finding and correcting factual errors in the reports.

However, mass media also have a **negative** impact by:

- **Transmitting negative attitudes.** One-sided reports about toxic substances in food or about unwanted side effects of drugs quickly lead to blaming "chemistry" for this—with the effect that students adopt the negative attitude towards chemistry from the journalists and thus limit their willingness to learn in chemistry lessons.
- **Conveying false ideas.** Newspaper and television reports about energy*consumption* such as fuel*consumption* or electricity*consumption* reinforce the everyday notion that energy can "disappear". When the principle of energy conservation is taught in class and the *conversion* of certain forms of energy into other forms of energy is discussed, this teaching often contradicts the "destruction concept" conveyed by the media (Chap. 2),
- **Misleading through advertising and marketing.** Many commercials—for example for detergents—often argue in a magical-animated way when substances are attributed the ability to perform "miracles". The laundry becomes "whiter than white" or "deeply clean"; detergents have the "power of the white giant" or can make things "citrus fresh". When a newspaper [13] reports, "Organic detergent must not contain any chemicals", the reader's deception is perfect: On the one hand, the supposed, completely nonsensical contrast "biology good—chemistry bad" is highlighted, on the other hand, it is deliberately obscured that every detergent must contain washing-active substances and thus chemicals.

Because of this deception by the mass media, chemistry lessons have the great task of educating young people well and enabling them to critically examine the information from the mass media based on their own knowledge and skills, to interpret it independently and to draw their own conclusions accordingly.

To achieve this goal, young people need competencies from all areas of competence. Two examples will be used to demonstrate how mass media can be effectively used in chemistry lessons.

5.4.1 Webquest

The use of WebQuests represents an action-oriented method for using the internet in teaching. Derived from the word "Quest", which was used in the Middle Ages to describe knightly adventure journeys, the method of WebQuests represents a journey through web-based information sources.

Starting points for a WebQuest can be various tasks. Depending on the task, different intentions underlie, which the inventor of the WebQuest Dodge has summarized in a taxonomy [50, 51]. To cope with the task, the teacher selects certain sources, mainly from the internet, which contain important information for solving the task. The selection of sources makes the difference to a free information search on the internet. When working on WebQuests, the focus is not on promoting the competence to select information sources themselves and to assess their content qualitatively. The students must extract and process the essential information from a resource pool that the teacher has selected with regard to the quality, usability and meaningfulness of the information as well as the suitability for the recipients.

The result of the WebQuests can be designed and presented in various media forms (journalistic texts, recommendation letters, expert opinions, posters, flyers, wall newspaper, internet presentation, brochure, role play, etc.).

The process of a WebQuest in class is divided into the following sequences [52]:

1. **Introduction:** The students are made aware of a question by pictures, texts, stories.
2. **Task:** After the procedure of a WebQuest has been explained, the task is given by concrete work instructions, which are integrated together with the links to the internet sources into a presentation or into a special format for WebQuests [53]. Usually, the class/course is divided into groups. The processing of the question can be done in a division of labour, so that the task for the individual groups also allows for internal differentiation.
3. **Elaboration:** Clear instructions for the procedure (reading information, documenting sources, selecting, preparing and presenting information, presenting results, discussing and evaluating results) help in organizing group work. The teacher now takes on the role of process consultant and uses the time to observe, for example, the distribution of work in the group (also as a basis for evaluating group work and individual supporting opportunities).
4. **Presentation:** The presentation of the results from the group work takes place in the plenary (social form and mode depends on the agreed type of product).
5. **Discussion:** Especially with division of labour processing of the WebQuest, another phase is necessary in which the results are related to each other or differentiated from each other, e.g. in the design of an expert opinion or a recommendation for action or in the search for a proposal for agreement.
6. **Evaluation:** The teacher evaluates the group work according to previously defined assessment criteria.

5.4.2 Movie Scenes

Movie scenes from feature films play a central role in the chemistry didactic research project "ChemCi" (Chemistry and Cinema), serving as anchor media for a problem-oriented chemistry class with experiments [54]. As part of the project, teaching concepts and experiments for scenes from the films "Apollo 13", "The Boat", "Perfume", "Dante's Peak" and others have been developed so far [55]. The integration of exciting film sequences enriches the teaching and increases the learners' interest in scientific questions. The question of how realistically the scenes in the feature films are depicted in terms of the subject matter leads to the mediation of both concept-related and process-related competencies, especially from the area of evaluation.

5.5 Exercise Tasks

A5.1
Media can have various functions in chemistry lessons. Choose three functions and explain the use of specific media based on situations of your choice according to the chosen functions.

A5.2
A medium equally important for teachers and students is the textbook. Choose three current textbooks and evaluate them based on important criteria for the use of textbooks in chemistry lessons.

A5.3
The blackboard and overhead projector, or nowadays often a fixed installation of computer and projector, are the most important media in the classroom. Name advantages and disadvantages for their use.

A5.4
The computer is gaining more and more importance as a medium in teaching due to its universal usability. Describe three teaching situations in which you would use the computer sensibly.

A5.5
Watch the animation "The Synthesis of Sodium Chloride" [25]. Assess which characteristics of multimedia the animation exhibits.

5.6 Experiments

V5.1 Use of white or black backgrounds for experimental setups

Problem
Demonstration experiments often cannot be observed sufficiently well by all students. To improve the observation possibilities, the use of white, black or colored backgrounds can represent a simple optimization of demonstration experiments.

Material
Funnels, silicone hose pieces, glass tubes, U-tube, crystallizing dish, gas washing bottle, water jet pump or aquarium pump; gas cartridge burner, ice water, white copper sulfate, lime water; white cardboard, black cardboard

Procedure
The funnel is attached with the wide opening facing downwards. The U-tube, whose curvature is immersed in ice water in the crystallizing dish, the gas washing bottle and the water jet pump are connected to the funnel via glass pieces. A spatula tip of white copper sulfate is added to the U-tube. The gas is ignited under the opening of the funnel and the combustion products are sucked through the apparatus with the pump connected to the gas washing bottle. A white background is placed behind the U-tube and a black background behind the gas washing bottle.

Observation
A colorless liquid condenses in the U-tube, which can be identified as water by the blue coloration of the white copper sulfate. The whitish cloudiness of the lime water in the gas washing bottle can be observed very well against a black background.

V5.2 Presentation of colored solutions in front of a light wall (based on [56])

Problem
The color differences of solutions cannot be clearly observed even when using a white background. If the test tubes are presented in front of a light wall, the distinguishability of colors or color intensities improves.

Material
Test tubes, test tube rack, pipettes, stand material; aqueous solutions of different pH values (pH 1-14), raspberry or red cabbage juice or anthocyanin-containing aqueous plant extracts; light wall e.g. from Hama [56]

5.6 Experiments

Procedure
The differently acidic or alkaline aqueous solutions are presented in the test tubes. In front of the glowing wall, a few drops of the anthocyanin-containing solution are pipetted to the prepared solutions and the color differences are observed. With known coloration of the pH indicator based on anthocyanins, the solutions can be sorted according to increasing pH values. Conversely, a color scale of the pH indicator based on anthocyanin can be created with aqueous solutions of known pH values.

Observation
The color differences can be seen very well in front of the light wall.

V5.3 Extension of experimental setups and details with camera and projector

Problem
Some experimental setups do not allow for arbitrarily large dimensions, so observations cannot be made visible for a large number of learners. The transmission of the experimental setup via a camera onto a large screen or using a projector onto a large presentation area allows detailed observations for all.

Material
Beaker, razor foils, crocodile clips, double-drilled cork stopper or electrode holder, electrical cable, flat battery (4.5 V), electric motor, multimeter; potassium hydroxide solution, c = 0.1 mol/L, digital camera including a tripod [56] or a document camera (e.g. Elmo document camera [57])

Procedure (based on [58, 59])
Roll up a razor foil and fix it with a crocodile clip. These two low-cost platinum electrodes are immersed in the potassium hydroxide solution in the beaker and connected conductively to the flat battery via the cables. After a few seconds, disconnect the battery and replace it with the electric motor or the multimeter.

Observation
Colorless gas bubbles form very quickly on the surfaces of the razor foils, which differ in their size and ascent speed from the solution. If the circuit is interrupted, the gas bubbles remain stuck in the perforated razor foil. If the circuit is now closed via the electric motor, it rotates for a few seconds. If the voltage of the charged model fuel cell is measured, a voltage of 1.2 V can be observed. The formation of the gas bubbles and the differences in size and ascent speed at the two

electrodes can be well observed via the camera. A black background of the apparatus (see V5.1) supports the visibility of the gas bubbles.

V5.4 Projecting experiments in petri dishes on the overhead projector (based on [60])

Problem

As soon as chemical reactions take place in transparent solutions and the substances formed—such as in electrolysis—only occur in the smallest quantities, the phenomena can be projected with the overhead projector. The colors of the solutions are transferred in the projection, but not the color of solid substances such as zinc. To help evaluate the experiment, a transparent foil under the petri dish can be labeled with the technical terms for the electrolysis cell and, for example, the electron flow can be clarified with arrows. This experiment can be transmitted even better under the document camera, as then the colors of the electrodes and solid deposits are also displayed (see V5.3).

Material

Petri dish, two platinum wires, crocodile clips, cables, power source; zinc bromide solution; overhead projector or document camera, transparent foil, foil pens

Procedure

The two platinum wires are attached as electrodes to the edge of the petri dish with clamps and connected to the DC poles of the power source via the cables. The bottom of the dish is covered with zinc bromide solution and a DC voltage of about 3-4 V is applied.

Observation

The brown deposit at the positive pole is immediately noticeable, and the formation of crystals at the negative pole is noticed a little later. With a suitable voltage, a constantly growing "zinc tree" is observed: This observation can only be made accessible to a larger audience through projection.

V5.5 Special cuvettes for projection on the daylight projector (based on [61])

Problem

Small portions of gas—especially those to be measured quantitatively—are better observed when the corresponding apparatus is projected in large format. An example is the quantitative execution of the analysis of water.

Material

Special Hofmann cuvette (electrolysis), power source; sulfuric acid, c = 0.5 mol/L

Implementation

The cuvette is filled with sulfuric acid solution on the switched-on work surface of the overhead projector, the power source is connected and a direct voltage of 5 V is applied, so that a constant gas development occurs. If the gas volumes are large enough, hydrogen and oxygen can be detected through the known reactions.

Observation

Gas bubbles form from the electrodes; at the negative pole, a gas volume twice as large as at the positive pole is achieved. Hydrogen appears at the negative pole, oxygen at the positive pole.

V5.6 Recording and processing of measured values with data acquisition systems (in accordance with [62])

Problem

There are reactions so fast that they cannot be recorded with traditional methods. If, for example, the pH jump around the equivalence point is to be recorded in the neutralization reaction of an acidic with an alkaline solution, pH-metric titration with a data acquisition system is a very good option. However, the students should first carry out a titration and the evaluation up to the titration curve in the traditional way in order to learn the basics and the procedure and then be able to compare them with the data acquisition and with the evaluation in the associated program.

Materials

Beaker, magnetic stirrer, stirring fish, burette, funnel, full pipette, pH electrode; hydrochloric acid, c = 0.1 mol/L; sodium hydroxide solution, c = 0.1 mol/L; data logger, e.g. All-Chem-Misst II [33]; evaluation software, e.g. AK Analytik 11 [63]; computer

Implementation

The pH meter is calibrated and connected to the data acquisition system. The data acquisition system is connected to the computer and the evaluation software is started there. Then the measurement parameters are set on the computer. 10 mL of hydrochloric acid is pipetted into the beaker with a full pipette and the burette is filled with sodium hydroxide solution. The pH electrode is immersed in the hydrochloric acid solution. After adding 0.5 mL of sodium hydroxide solution each time, a measured value is stored by pressing a button, which is simultaneously displayed in the diagram on the computer screen. After the titration is completed, the equivalence point is determined using the three-straight line method with the help of the evaluation software.

Observation

At the beginning of the titration, the pH value 1.0 is displayed. The computer screen builds up the titration curve during the titration and shows the pH jump from about pH 3 to pH 10 after about 10 mL of sodium hydroxide solution has been added. After adding a few more milliliters, the curve approaches the pH value 13.

V5.7 Model experiment as a medium for the lead accumulator

Problem

The processes of charging and discharging a lead accumulator cannot be directly observed on the car battery. The model experiment reproduces how charging and discharging take place and how both lead electrodes behave in each case. Finally, the explanations can be made through other media such as illustrations from textbooks or on worksheets.

Materials

Beaker, power source, voltmeter, electric motor (2 V), cables, crocodile clips; two lead plates, sulfuric acid solution (w = 20%)

Implementation

The beaker is half filled with sulfuric acid solution. Both lead plates are placed and fixed in it in such a way that they do not touch each other; they are connected to the power source with cables (Fig. 5.2). The direct voltage is to be regulated so that gas development can be observed for a few minutes. The power source is replaced by the voltmeter and the voltage between both plates is measured. Then the electric motor is connected.

Observation
A deep brown substance forms on one lead plate during the power supply, as can be seen on one plate of a battery cell in the car battery. After the power supply has been stopped, a voltage of 2 V can be detected between both plates. The electric motor connected afterwards initially turns very quickly, becomes slower and slower and finally stops.

▶ **Note** The sulfuric acid solution is contaminated with lead sulfate and is toxic. It should be filled into a storage bottle as much as possible, labeled, and kept for use in the same experiment. Otherwise, the solution should be disposed of in the heavy metal waste container.

References

1. Issing LJ (1987) Medienpädagogik im Informationszeitalter. Studienverlag, Weinheim
2. Tulodziecki G, Herzig B (2004) Mediendidaktik. Medien in Lehr- und Lernprozessen. Handbuch Medienpädagogik, Bd. 2. Klett-Cotta, Stuttgart
3. Sofos A, Kron FW (2010) Erfolgreicher Unterricht mit Medien. Logophon, Mainz
4. Stadtfeld P (2011) Tradierte Lehrmittel, neue Medien, „Moderner" Unterricht – systematische Betrachtung und praktisches Modell. Bild Erziehung 64(1):69–84
5. Kerres M (2000) Medienentscheidungen in der Unterrichtsplanung. Zu Wirkungsargumenten und Begründungen des didaktischen Einsatzes digitaler Medien. Bild Erziehung 53(1):19–38
6. von Martial I, Ladenthin V (2005) Medien im Unterricht. Grundlagen und Praxis der Mediendidaktik. Schneider Verlag Hohengehren, Baltmannsweiler
7. Ministerium für Schule und Weiterbildung des Landes Nordrhein-Westfalen (Hrsg) https://www.schulministerium.nrw/zulassung-von-lernmitteln-nrw (accessed on 24. Feb. 2025)
8. Jansen C, Pfangert-Becker U, Raguse K, Schultheiß-Reimann P (2012) Gütekriterien für Schulbücher im Chemieunterricht. Chemkon 20(1):41–44
9. Sekretariat der Ständigen Konferenz der Kultusminister der Länder in der Bundesrepublik Deutschland (Hrsg) (2005) Bildungsstandards im Fach Chemie für den Mittleren Schulabschluss. Luchterhand, München, Neuwied
10. Graf E (1997) Die Wandtafel im Chemieunterricht – ein noch zeitgemäßes Unterrichtsmedium? NiU-Chemie 38(8):24–29
11. Fonds der Chemischen Industrie (Hrsg). https://www.vci.de/fonds/schulpartnerschaft/seiten.jsp (accessed on 24. Feb. 2025)
12. Lehrer-Online GmbH (Hrsg) https://www.lehrer-online.de/fokusthemen/dossier/do/urheberrecht-in-schule-und-unterricht/ (accessed on 24. Feb. 2025)
13. Haupt P (2014) Die Chemie im Spiegel einer Tageszeitung. Band 1–10. BIS Verlag, Oldenburg
14. WDR-Sendung Quarks und Co. www1.wdr.de/fernsehen/wissen/quarks/index.html (accessed on 24. Feb. 2025)
15. Helmholtz-Zentrum Berlin. https://www.helmholtz-berlin.de/zentrum/presse-newsroom/mediathek/video/portraits_de.html (accessed on 24. Feb. 2025)
16. Medieninstitut der Länder FWU. www.fwu-mediathek.de/. (accessed on 24. Feb. 2025)
17. Bundesministerium für Umwelt, Naturschutz, Bau und Reaktorsicherheit: Informationen und Medien zur Elektromobilität. https://www.bmuv.de/themen/verkehr/elektromobilitaet/ueberblick-elektromobilitaet (accessed on 24. Feb. 2025)
18. Eilks I, Krilla B, Flintjer B, Möllencamp H, Wagner W et al (2004) Computer und Multimedia im Chemieunterricht heute – Eine Einordnung aus didaktischer und lerntheoretischer Sicht, GDCh, FGCU, AG Computer im Chemieunterricht

19. Sieve B, Schanze S (2015) Lernen im digital organisierten Chemieraum. NiU-Ch 26(145):2–7
20. Krause M, Eilks I (2014) Tablet-Computer im Chemieunterricht. PdN-ChiS 63(4):16–21
21. BITKOM (Bundesverband Informationswirtschaft,Telekommunikation und neue Medien e. V.) (2015) Ergebnisse repräsentativer Schüler- und Lehrerbefragungen zum Einsatz digitaler Medien im Schulunterricht. https://www.bitkom.org/Bitkom/Publikationen/Digitale-Schule-und-vernetztes-Lernen.html (accessed on 24. Feb. 2025)
22. Murauer R (2017) BYO[m]D – Bring your own [mobile] device – Eine empirische Analyse der, aus Sicht der Lehrkräfte, erforderlichen Rahmenbedingungen für die Implementierung von schülereigenen Smartphones und Tablets im Unterricht. Dissertation, Hamburg
23. Führer M (2016) [CHEM2DO®]digital – Mobiles Zusatzangebot für das Experimentieren mit Siliconen und Cyclodextrinen im Chemieunterricht mit neuentwickelten Inhalten der digitalen DiSiDo-Cy. Masterarbeit. Erlangen-Nürnberg
24. Nick S, Andresen J (2001) CHEMnet – Ein Hypermedia-Framework. PdN-Chis 50(7):5–8
25. Bergische Universität Wuppertal, Arbeitsgruppe Prof. Dr. M. Tausch: Flash-Animationen unter https://chemiedidaktik.uni-wuppertal.de/de/forschung/forschung-tausch/ (accessed on 24. Feb. 2025)
26. Woock M. Animation „MTBE im Ottokraftstoff". https://chemiedidaktik.uni-wuppertal.de/de/digitale-medien/animationen/weiteres/ (accessed on 24. Feb. 2025)
27. Wacker Chemie AG: Schulversuchskoffer CHEM2DO® - interaktive Lerntools. Videos auf YouTube: https://www.youtube.com/results?search_query=chem2do (accessed on 24. Feb. 2025)
28. Kröger S, Hock K, Tausch M, Anton M, Bader A, Zdzieblo J (2017) CHEM2DO®-Schulversuchskoffer – ein Kooperationsprojekt von Wirtschaft, Fachdidaktik und Lehrerfortbildungszentren. CHEMKON 24:241–250
29. Irion T (2010) Interaktive Whiteboards: Was sollten Lehrkräfte wissen und können? Computer+Unterricht 20(78):16–18
30. Sieve B (2014) Interaktive Whiteboards – Beispiele für den lernförderlichen Einsatz im Chemieunterricht. PdN-ChiS 63(4):5–9
31. Grofe T, Beyer-Evensen B (2010) Chemische Experimente an Magnettafeln – Eine faszinierende Technik mit vielen Vorteilen. PdN-ChiS 59(4):16–18
32. Brand B-H. Magnetisches Stativsystem. www.bhbrand.de/magnethalter-und-whiteboard/index.html (accessed on 24. Feb. 2025)
33. Kappenberg F. Der ALL-CHEM-MISST II. www.kappenberg.com/pages/messgeraete/geraete_vom_ak/all_chem_misst/acm2.htm (accessed on 24. Feb. 2025)
34. Schrader F, Schanze S (2012) Digitale Messwerterfassungsgeräte. Ein kriterienorientierter Überblick. PdN-ChiS 61(4):42–48
35. Assaf D (2014) Maker Spaces in Schulen: Ein Raum für Innovation (Hands-on Session). In: Rummler K (Hrsg) Lernräume gestalten – Bildungskontexte vielfältig denken. Waxmann, Münster, S 141–149
36. Arduino. www.arduino.cc/ (accessed on 24. Feb. 2025)
37. Raspberry Pi. www.raspberrypi.org/ (accessed on 24. Feb. 2025)
38. Wejner M, Wilke T (2017) Low Cost – High Value: Minicomputer als leistungsstarke Messsysteme für MINT-Fächer. Poster, GDCh Wissenschaftsforum Chemie, Berlin
39. Schmitz R-P. Design von Molekülen, Kugelstabmodellen und Reaktionen. https://chemiedidaktik.uni-wuppertal.de/de/forschung/forschung-tausch/chemie-interaktiv/ (accessed on 24. Feb. 2025)
40. Hedinger: Experimentierkoffer und Experimentiersets. www.der-hedinger.de (accessed on 24. Feb. 2025)
41. Wacker Chemie AG. Wacker-Schulversuchskoffer CHEM2DO. www.chem2do.de/c2d/de/service/service.jsp. Zugegriffen: 9. Okt. 2017
42. Krees S. Animation zu Wirt-Gast-Komplexen mit Cyclodextrinen. https://chemiedidaktik.uni-wuppertal.de/de/digitale-medien/animationen/weiteres/ (accessed on 24. Feb. 2025)

43. KMK (2012) Beschluss der Kultusministerkonferenz: Medienbildung in Schulen. www.kmk.org/fileadmin/veroeffentlichungen_beschluesse/2012/2012_03_08_Medienbildung.pdf (accessed on 24. Feb. 2025)
44. Baacke D (2008) Was ist Medienkompetenz? In: Lauffer J, Röllecke R (Hrsg) Mit Medien bilden – der Seh-Sinn in der Medienpädagogik. Dieter Baacke Preis Handbuch, kopaed, München
45. CmapTools knowledge modelling kit. http://cmap.ihmc.us/ (accessed on 24. Feb. 2025)
46. Tausch M, von Wachtendonk M (2010) Chemie 2000+ Sekundarstufe I. C.C. Buchner, Bamberg
47. Schanze S, Brüchner K (2005) Computergestütztes Concept Mapping. Eine Methode zur Unterstützung selbstgesteuerten Lernens. NiU-Chemie 16(90):16–17
48. Stahl E (2010) Die Rolle motivierender Medien im naturwissenschaftlichen Lernprozess am Beispiel der Medienproduktion. PdN-ChiS 59(4):19–22
49. Dirkmann R (2014) Entwicklung einer produktionsorientierten filmintegrativen Unterrichtsreihe zum Thema Ionenbindung und Ionenkristalle. Masterthesis, Münster
50. Dodge B. WebQuest Taskonomy: A Taxonomy of Tasks. webquest.sdsu.edu/taskonomy.html (accessed on 9. Okt. 2017)
51. Wagner W-R (2007) WebQuest. Ein didaktisches Konzept für konstruktives Lernen. Computer+Unterricht 17(67):6–9
52. Nolte M (2005) WebQuests. Eine handlungsorientierte Methode zum Interneteinsatz im Chemieunterricht. Naturwissenschaften Im Unterr Chem 90(16):1215
53. Lehrer-Online: WebQuest-Bau leicht gemacht. www.lehrer-online.de/artikel/fa/so-kommen-webquests-in-den-unterricht/ (accessed on 24. Feb. 2025)
54. Friedrich J, Kunze N, Rubner I, Oetken M (2010) Chemistry and Cinema. Das Projekt ChemCi: Eine Unterrichtseinheit zum Themenfeld Atmung – inszeniert und illustriert mit Szenen aus den Spielfilmen „Das Boot" und „Apollo 13". PdN-ChiS 59(4):6–12
55. Pädagogische Hochschule Freiburg. Chemiedidaktik. www.ph-freiburg.de/chemie/forschung/chemistry-and-cinema-chemci.html (accessed on 9. Okt. 2017)
56. Brand H-B, Grofe T. Ergänzungen zum Magnettafelwagen. www.bhbrand.de/downloads/magnettafelwagen-ergaenzungen-systematisch.pdf (accessed on 24. Feb. 2025)
57. ELMO. Dokumentenkamera elmo-germany.de/ (accessed on 24. Feb. 2025)
58. Tausch M, von Wachtendonk M (2007) Chemie 2000+ Gesamtband SII. Buchner, Bamberg
59. Seesing M, Bohrmann C, Tausch M (2002) No-cost Brennstoffzelle. PdN-ChiS 51(6):43–44
60. Full R (1996) Lichtblicke – Petrischalenexperimente in der Overhead-Projektion. ChiuZ 30(6):286–294
61. Hedinger. www.der-hedinger.de/produkte/projektionsmodelle/artikel/108324.html (accessed on 9. Okt. 2017)
62. Kappenberg F. Titration von Salzsäure und Natronlauge pH-Wert-Messung. www.kappenberg.com/experiments/ph/pdf-aka11/f03.pdf (accessed on 24. Feb. 2025)
63. Kappenberg F. Software AK Labor und Analytik 11. www.kappenberg.com/pages/aklabor/einfuehrung.htm (accessed on 24. Feb. 2025)

Experiments

6

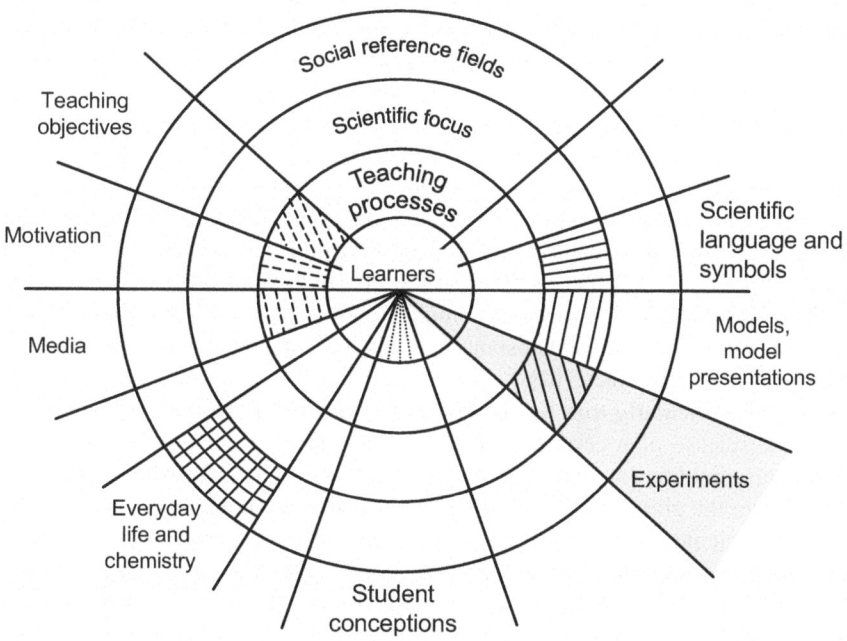

Entirely educated people regard chemistry as an experimental art brought into rules, useful for making soda and soap, fabricating better steel, and delivering good solid colors on silk and cotton [1].

As Liebig already complained in 1840, laypeople often believe that chemistry is an art of trial and error, which occasionally produces a new product by chance. It is not readily recognized that a classical experiment in the natural sciences is preceded by an idea or even a philosophy. To explain this context, the sharpening of the concept of an element will serve as an example.

Example "Idea of an Element"
A preliminary hypothesis about chemical elements already appeared in the philosophy schools of ancient Greece: Empedocles "speaks of the four roots of all things, which he conceives in a mythical way as deities. These four deities represent fire, earth, air, and water, from which all beings are thought to be composed. The concept of the element goes back to him" [2]. Upon closer study of this idea, it becomes apparent that the Greek philosophers had erected such and other thought structures without any experiment: Althoughachievements of the time were certainly based on a series of careful and orderly observations of nature, there can be no talk of experimenting in the scientific sense here [2].

The alchemy of the Middle Ages was dominated by the idea of making gold from base metals: In the processing of ores, small amounts of gold or silver were accidentally separated, and attempts were made in many ways to obtain gold and silver through "metal refinement" or "transmutation". In these "trial arts", it was observed that metallic copper is deposited when a vitriol solution—today we call it copper sulfate solution—is mixed with iron. However, the alchemist van Helmont found out that he could also produce copper from vitriol solution in other ways, that copper was "somehow" contained in the vitriol solution and only had to be separated. Boyle took up this idea of decomposition in his work "The Sceptical Chymist" to arrive at his theory: Elements are basic substances that cannot be further decomposed. "Boyle repeatedly emphasizes that it is not appropriate to consider theories about chemical substances as valid as long as they are not supported by experimental experience" [2].

This was particularly true for the phlogiston theory of combustion, which postulated the release of a certain "substance phlogiston" from the fuel—however, experiments could never verify this theory. It was only Lavoisier who succeeded in overthrowing the phlogiston theory and showing that the pure metals are the indecomposable elements, while the "metallic limes" in the experiment provide metal and oxygen and therefore do not represent elements. Lavoisier clearly recognized that while decomposable substances are not elements, the reverse conclusion does not apply: Substances that were initially considered indecomposable might prove to be decomposable through improved techniques. Therefore, Lavoisier cautiously stated (Chap. 13): "Certain substances are chemically determined as elements, as long as and as long as we have no means to decompose them further. They are elements for us and our point of view, pour nous, à notre égard" [2].

This preliminary concept of an element was the basis of thinking and acting in the time of Lavoisier; experiments were derived from the hypothesis of a possible decomposability and confirmed this in the case of success. The experiment gained the scientific significance of an instrument that decides on the validity of laws and theories!

Another major step occurred in 1808 when John Dalton published "A New System of Chemical Philosophy". At the end of the 18th century, Jeremias Benjamin Richter had published the laws of constant mass ratios in his "Contributions to Stoichiometry". Building on this, Dalton had the grand idea of assigning a specific type of atom to each element, which differs in atomic mass from other types of atoms. He proposed the first atomic mass table of natural sciences in 1808 (Fig. 6.1)—however, six of the twenty listed "elements" later turned out to be chemical compounds: for example, "magnesia", today's magnesium oxide, was considered an element.

With this hypothesis of specific atomic masses, a whole field of research was opened up for chemistry: The quantitative composition of many compounds from their elements was explored, corresponding models and chemical symbols were subsequently developed. Even though Dalton still considered "magnesia, lime, soda, potash, strontian and barytes" as elements in his time, even though the atomic masses turned out to be completely insufficient, the idea of atomic masses was so far-reaching that chemistry could only become an exact science in this way.

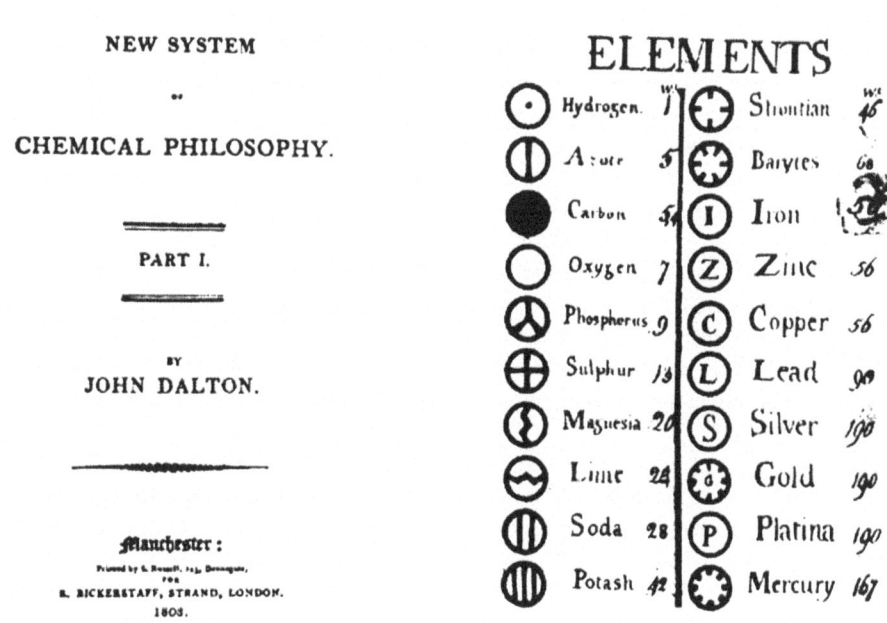

Fig. 6.1 First preliminary atomic mass table by Dalton from the year 1808 [2]

6.1 Scientific focus: Experiment, Experimental Skills, Safety

In line with the introduction, chemistry has developed as an experimental natural science. It is therefore necessary to reflect on functions of experiments in the scientific discipline, in what way these also apply to chemistry lessons and what additional functions are added for lessons.

6.1.1 Experiment and Process of Knowledge Acquisition

One of the most important functions of the experiment in chemistry is the empirical acquisition of knowledge through the *formulation and experimental testing of hypotheses*. In the introduction, the hypothesis of the concept of elements is outlined as a historical example, which has been tested over centuries before the idea of John Dalton was confirmed by the universally reproducible measured values of atomic masses.

First, this empirical method of knowledge should be formalized through essential steps of knowledge (Fig. 6.2). Often one hears inaccurately that only a single phenomenon could lead to the postulation of a hypothesis, and that a single experiment should suffice to confirm a hypothesis. Multiple observations or measurements are always necessary, which are first ordered, collected or compared and put into a suspected context. This is followed by the development of a general hypothesis. If enough individual cases are then derived from the hypothesis and experimentally tested, a decision is made about verification or falsification. Repeated confirmation of the hypothesis leads in some cases to a new theory or a new law, from which predictions or forecasts become possible. The following three examples illustrate this.

Torricelli, together with Galilei, developed the *hypothesis on air pressure* based on considerations of "horror vacui" (Chap. 2). In 1644, he carried out the well-known mercury experiment in the laboratory for the first test of the hypothesis (Fig. 2.1b) and was able to verify the hypothesis through further experiments—such as air pressure measurements at different heights of a mountain—and arrive at the theory of air pressure: air is a substance with specific density, the surrounding atmosphere has a certain air pressure, which decreases with increasing height above the ground.

Kekulé, starting from empirical analyses of mass ratios of elements in many organic substances, dealt with the description of these analysis results by formulas. In doing so, he always thought about the spatial structure of corresponding molecules made of atoms. The attempt to describe the benzene molecule with his means initially failed, then in 1865 he put forward the *hypothesis of the ring-shaped molecular structure* (Fig. 6.3a). In order to save the four-valency model of the C-atom postulated by him—which was by no means generally accepted at that time—Kekulé initially described the benzene molecule with alternating single and double bonds, but suspected that all the bonds are equivalent. Starting from

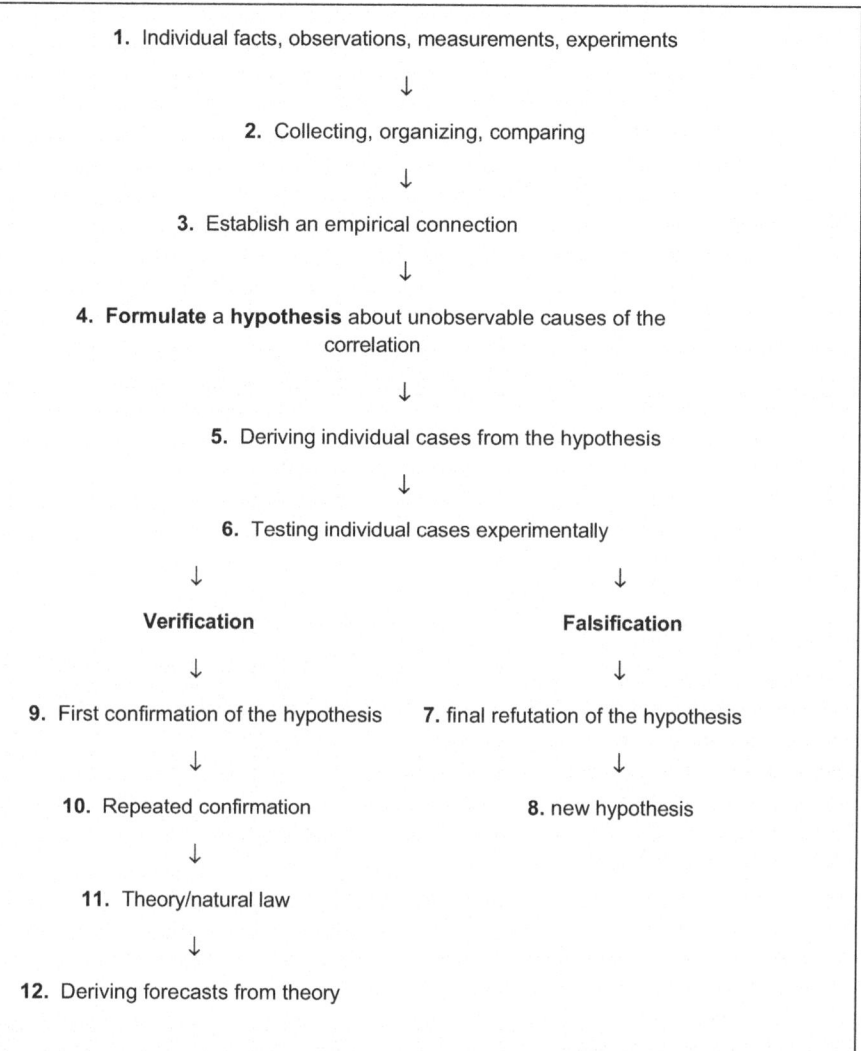

Fig. 6.2 Sequence of steps in the empirical method of knowledge

this hypothesis, he derived many individual cases and corresponding experiments and showed that there can only be a single monosubstituted molecule C_6H_5Cl, that only exactly three disubstituted molecules $C_6H_4Cl_2$ can be realized and that only three trisubstituted molecules $C_6H_3Cl_3$ exist (Fig. 6.3b): All six bonds in the benzene molecule are therefore equivalent. With today's instrumental methods, it is shown that the distance between adjacent C-atoms is always the same, that the measured value lies between the bond lengths of single and double bond.

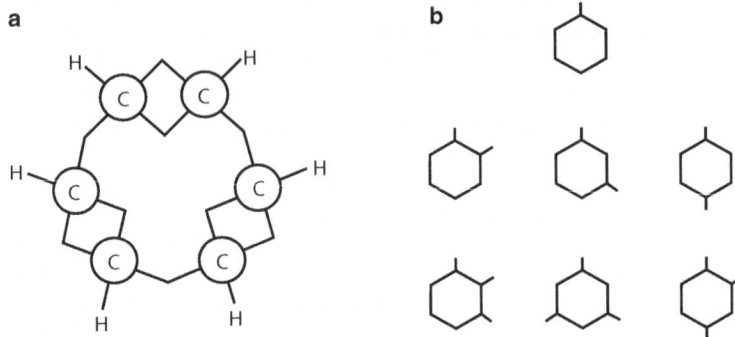

Fig. 6.3 Kekulé's hypothesis of the structure of the benzene molecule and some substituted derivatives [3]

In addition, Kekulé and van't Hoff postulated the spatial tetrahedron model for the structure of the methane molecule. All individual cases derived from the hypotheses could be experimentally confirmed: Both hypotheses—of the four-valency model of the C-atom and of the structure of the aromatics—were gradually sharpened into theories that are still valid today.

Since the discovery of X-rays in 1895 by Konrad Wilhelm Röntgen, many scientists have been feverishly working on the *hypothesis of X-ray diffraction*. Max von Laue finally proved it in 1912, and he himself commented on his decisive thought as follows (Fig. 6.4):

> The history of the discovery of X-ray interference truly illustrates the value of the scientific hypothesis. Many have long before Friedrich and Knipping sent X-rays through crystals. But their observations were limited to the directly transmitted beam, on which, apart from the weakening by the crystal, there was nothing remarkable to see; the much less intense diffracted rays escaped them. Only the hypothesis of the space lattice in crystals and their diffraction properties brought the idea to explore its surroundings [4].

6.1.2 Data Acquisition

In addition to the arguably most important function of hypothesis testing, another function of the experiment is to perform measurements to describe *substances and their properties*: They enable comparison with properties of other substances. The measurements traditionally refer to densities, melting and boiling temperatures, in order to obtain a first characterization of substances with three parameters each. Furthermore, solubility in water and other solvents, refractive index, viscosity, optical activity, thermal conductivity, electrical conductivity, pH dependencies, redox potentials, MS-, IR- and NMR-spectra etc. can be added.

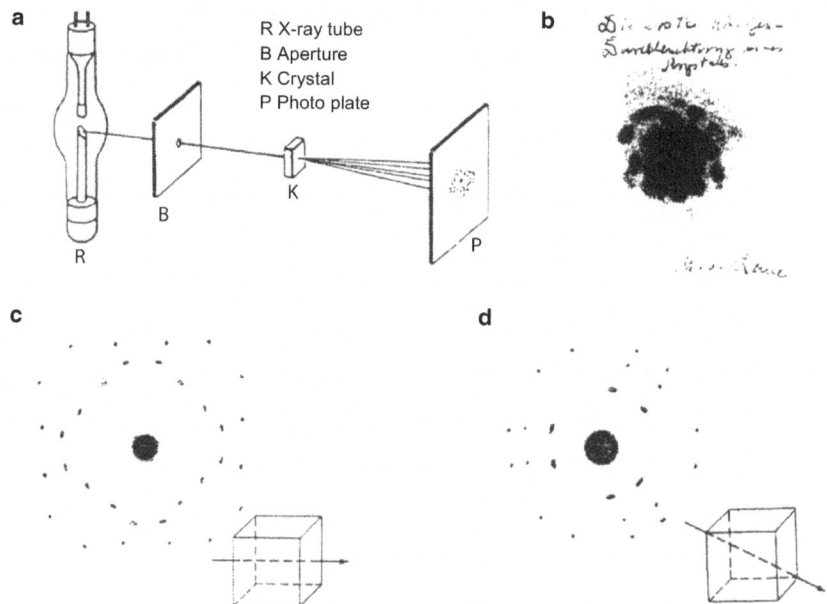

Fig. 6.4 Laue's notes on the hypothesis of using a crystal as a specific lattice for the interference of X-rays [5]

A corresponding description for *chemical processes* could be achieved by determining mass and volume ratios in reactions, by the stoichiometry of converted quantities of atoms, ions or molecules, by corresponding sum or structure symbols derived from them, which finally led to the common reaction symbols used today. Also solubility products, equilibrium and rate constants or reaction orders are characteristic parameters for chemical processes.

In today's laboratories, the description of substances and processes is predominantly carried out by *instrumental analytical methods*: such as paper, thin layer or column chromatography, gas chromatography (GC), mass spectrometry (MS), combinations of GC and MS, atomic absorption spectrometry (AAS), infrared spectroscopy (IR), nuclear magnetic resonance spectroscopy (NMR), X-ray structure analysis. The corresponding large-scale equipment is not only very expensive, but also requires professional supervision by specially trained personnel. Chemistry students get to know the large-scale equipment in the experimental laboratory, teachers and students can have them explained in chemical institutes of a university or during tours of industrial research laboratories.

However, school equipment has also been developed that conveys some of the mentioned methods in a clear and didactically reduced way: for example by

Wiederholt [6] for *gas chromatography* and for *differential thermal analysis*, by Brockmeyer [7] for *X-ray structure analysis* (see also corresponding model experiments for gas chromatography (Sect. 6.6: V6.1) and for X-ray structure analysis (Sect. 6.6: V6.2).

6.1.3 Synthesis of New Substances

The oldest function of "experimenting" was probably the more or less pronounced "art of trying" to produce substances that were separated from natural mixtures or presented as completely new. Thus, in 1708, Johann Friedrich Böttger succeeded in reinventing porcelain in Meissen near Dresden, which was already known in China in the 6th century. According to Schwedt [8], some substances are listed in Table 6.1 with the approximate time when they were first produced.

Currently, we read almost weekly about new substances such as novel drugs, memory metals, electrically conductive polymers, or nanotechnology materials. The enormously large number of future syntheses of new substances in chemical laboratories is neither limited nor predictable.

6.1.4 Experimental Skills and Abilities

Future chemistry teachers must not only be able to experiment themselves, but do so embedded in the classroom discussion with a group of students, furthermore, they should also convey the experimentation to the learners as much as possible. They therefore have *skills* to plan, carry out and evaluate experiments, they also possess the manual *abilities* to handle appropriate apparatus, devices and chemicals of school chemistry properly and to consider safety and disposal aspects.

Beyond the basic experimental training, the student teachers should therefore develop a repertoire of school experiments on teaching-relevant topics in subject-didactic experimental internships. They must gain experience in dealing with experiments that require special safety requirements. These include, among other things, dealing with

- Hydrogen, steel cylinders and assembly of the valves, production of hydrogen in the gas generator and pneumatic collection in the standing cylinder, explosive gas mixtures and explosive gas samples,
- Methane, butane, portable butane gas burner and changing gas cartridges, methane-air and butane-air mixtures and their rapid reactions in standing cylinder,
- Alkali metals, their storage under petrol and cutting with a knife, reactions with oxygen, water or halogens, their disposal in alcohol,
- Halogens, the production of chlorine in the gas generator and its metal reactions, bromine and its storage, iodine and its sublimation,

6.1 Scientific focus: Experiment, Experimental Skills, Safety

Table 6.1 Extraction of some substances in the history of mankind [8]

8000 B.C.	Ceramics
3700 B.C.	Jewelry made of copper, silver, and gold
3000 B.C.	Antimony and lead in Babylon, bronze in Egypt
2900 B.C.	Glass in Egypt
2400 B.C.	Indigo colors in Egypt
2000 B.C.	Sulfur from hot springs
1200 B.C.	Tin and zinc in India
1000 B.C.	Iron through furnace
500 B.C.	Purple and madder colors, soda, potash, gypsum, mortar, alum, caustic potash in Rome
400 B.C.	Mercury in Greece
20–80	Soap, mineral colors, alloys in Rome
500	Borax, saltpeter, heavy metal compounds in India
600	Porcelain in China
850	Ammonium chloride, acetic acid, and lead white (lead hydroxycarbonate)
900	Paper in Cairo
1100	Inks and paints
1227	Ethanol as a remedy
1230	Gunpowder in China
1300	Sulfuric acid (distillation)
1565	Zinc vitriol (zinc sulfate)
1580	Benzoic acid
1648	Hydrochloric acid, nitric acid
1669	Phosphorus
1671	Litmus as an indicator
1708	White porcelain in Meissen
1727	Silver nitrate for first photos
1746	Sulfuric acid (lead chamber process)
1747	Sugar from beets
1766	Hydrogen
1773	Oxygen, Nitrogen
1808	Sodium, Potassium, Magnesium, Calcium, Strontium, Barium
1810	Chlorine
1827	Aluminium
1828	Urea
1855	Lithium
1867	Dynamite

(continued)

Table 6.1 (continued)

1870	Indigo
1884	Artificial silk
1894	Argon, Noble gases
1898	Polonium, Radium
1909	Bakelite
1913	Ammonia
1924	Insulin
1928	Penicillin etc.

- toxic and self-igniting white phosphorus, with red phosphorus, and the proper disposal of residues,
- Measuring devices for quantitative experiments such as beam balance and digital balance, syringe, volumetric pipette, measuring pipette, burette, conductivity tester, pH meter etc.
- Batteries and transformers, electrical circuits and corresponding accessories such as cables, crocodile clips, light bulbs, voltage and current meters.

In the *experiments V6.3 to V6.10* (Sect. 6.6), the handling of these metals is demonstrated using the example of "alkali metals", reactions with air, water and halogens are described. Important school experiments on essential topics can be found in the experimental literature (e.g., [9–12]). If students or learners in schools are also to carry out demonstration experiments with special effects independently, these should be carefully tried out under supervision due to the specific hazard potentials.

During the practical implementation of school experiments, participants learn about a variety of *laboratory-specific devices*. To provide them with an initial knowledge of laboratory devices and their specific names, they can be given a compilation of laboratory devices (Fig. 6.5). In particular, handling of the gas burner for future experimentation should be practiced (Fig. 6.6)—in this context, it has become interesting to offer learners a "laboratory driving license". Knowledge of the devices is particularly important in order to be able to plan possible student experiments for the lessons or to carry out spontaneous teacher experiments in addition to the classroom discussion, on the other hand, it is advantageous to be able to organize the chemistry collection in the school properly as a Chemistry teacher.

To record measurement series and simultaneously obtain graphical representations, the connection of measuring devices to the *computer* is suitable. With suitable hardware and software, measuring devices and computers are connected, measurement values are recorded in a suitable manner and evaluated in a variety of ways. Since hardware and software develop quite quickly, it is refrained from listing the currently relevant literature or programs at this point.

Selected laboratory equipment

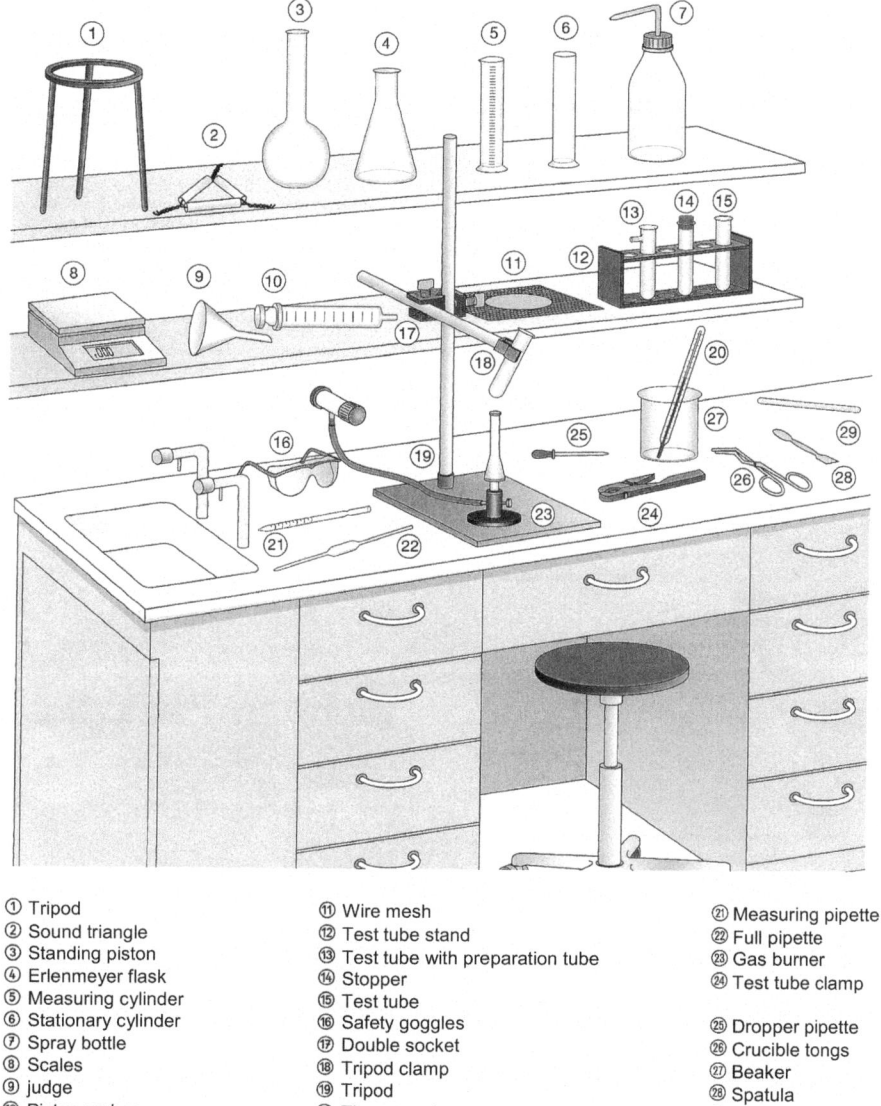

① Tripod
② Sound triangle
③ Standing piston
④ Erlenmeyer flask
⑤ Measuring cylinder
⑥ Stationary cylinder
⑦ Spray bottle
⑧ Scales
⑨ judge
⑩ Piston prober

⑪ Wire mesh
⑫ Test tube stand
⑬ Test tube with preparation tube
⑭ Stopper
⑮ Test tube
⑯ Safety goggles
⑰ Double socket
⑱ Tripod clamp
⑲ Tripod
⑳ Thermometer

㉑ Measuring pipette
㉒ Full pipette
㉓ Gas burner
㉔ Test tube clamp

㉕ Dropper pipette
㉖ Crucible tongs
㉗ Beaker
㉘ Spatula
㉙ Glass rod

Fig. 6.5 Selected laboratory devices. (With kind permission from Schroedel Publishing [13])

6.1.5 Safety and Disposal

It goes without saying that there must be no accidents during experiments for educational purposes, that negligence or at least gross negligence must be ruled out:

| Internship | **Handling the gas burner** |

During experiments in chemistry lessons, substances often have to be heated. Gas burners for natural gas or propane gas are used for this purpose. Only if you know how to use them can you handle them safely.

Experiment 1: "Driver's license" for the gas burner

Materials: Gas burner.

Implementation:
Preparation
1. Place the gas burner on a fireproof surface so that it cannot tip over.
2. Connect the gas hose of the burner to the gas supply line of the table.
3. Close the gas supply and the air supply to the burner.

Caution: Tie up long hair and always wear safety goggles. The gas burner must not be left unattended during work.

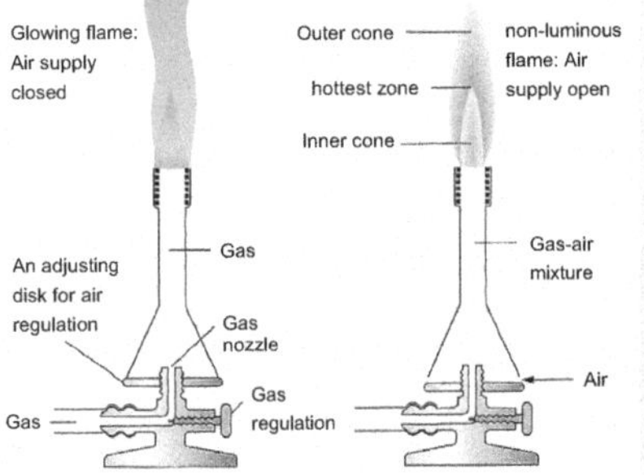

Glowing flame:
Air supply closed

Outer cone — hottest zone
Inner cone

non-luminous flame: Air supply open

An adjusting disk for air regulation
Gas
Gas nozzle
Gas regulation
Gas

Gas-air mixture
Air

Commissioning and setting the gas burner
1. Open the valve on the gas supply line and then the gas supply to the burner. Ignite the escaping gas. Work quickly, but without rushing!
2. Change the height of the gas flame with the gas regulator screw on the burner until the flame glows bright yellow.
3. Then open the air supply until you get a blue, non-luminous flame.
4. The height of the burner flame should correspond to the width of your hand. Only work with the blue, but not yet roaring, burner flame.
5. To extinguish the burner, close the valve on the gas supply to the burner.
6. Repeat the process until you have it under control. Then you will have your gas burner license.

Attempt 2:
Examination of the burner flame

Materials: gas burner, magnesia sticks.

Implementation:
1. Light the gas burner and set a blue, slightly roaring flame. The flame consists of an inner light blue cone and an outer dark blue cone.
2. Use a magnesia stick to examine the two flame cones. First hold the rod still in the inner cone for a while. Then slowly pull it through the flame from bottom to top.

Tasks:
a) Describe your observations.
b) Explain where the hottest zone of the flame is. Pay particular attention to the edges of the flame and the transition from the inner to the outer flame cone.

Attempt 3:
Melting glass

Materials: Gas burner, glass tube (approx. 30 cm).

Implementation:
1. Hold the glass tube firmly at both ends and heat it in the middle above the inner flame cone. The glass tube should be rotated evenly.
2. As soon as the glass begins to soften, quickly pull both ends apart outside the flame.
3. Melt another glass tube at one end. Turn it and heat the end until the glass is red-hot.

Task:
Describe your observations.

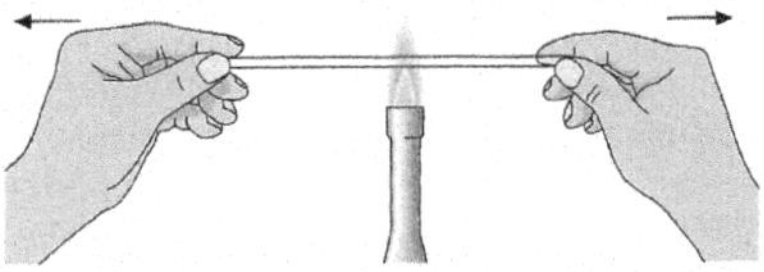

Fig. 6.6 Practical course on handling the gas burner. (With kind permission from Schroedel Publishing [13])

6.1 Scientific focus: Experiment, Experimental Skills, Safety

Laboratory leaders and teachers can avoid the accusation of gross negligence by clearly warning all interns of dangers or ordering specific safety measures. A good technical and practical-experimental training of chemistry teachers at universities and seminars is the best guarantee for proper and accident-free experimenting. In addition, they must be familiar with the collection and *safety facilities* of their school or laboratory, know the location and use of fire extinguishers, fire blankets, extinguishing sand, emergency showers and first aid kits. They also have to ensure that proper labeling of standing bottles and the proper *disposal* of waste and residual chemicals are guaranteed—otherwise they must create the conditions for this. Furthermore, it must be checked whether suitable substitutes can be found for dangerous substances and how these actually react in the experiment.

Uniform for all is the *"Ordinance on Dangerous Substances"* (Hazardous Substances Ordinance, GefStoffV): It particularly stipulates the labeling of chemical containers with the known hazard symbols, with hazard warnings and safety advice (R- and S-phrases). The TRGS 450 also regulates the *"Handling of Hazardous Substances in the School Sector"*. From their annexes it becomes apparent that the following substances may be used in teacher experiments, but not for student experiments:

- Very toxic substances such as carbon disulfide, nitrobenzene, tetrachloromethane (carbon tetrachloride), white phosphorus, potassium cyanide etc.
- Carcinogenic or teratogenic substances such as benzene, nickel, cobalt, chromium(VI) compounds in the form of dusts, 1,2-dibromomethane etc.
- Explosive substances such as explosives, black powder, mixtures of oxidizable substances with potassium or sodium chlorate, picric acid etc.
- Substance mixtures containing pathogens, such as pathogenic bacterial and fungal cultures, fecal wastewater etc.

As part of a UN conference in 1992, the new *Global Harmonized System of Classification and Labeling of Chemicals* was agreed upon (see examples by the practical in chap. 6.6). These new GHS hazard symbols apply worldwide and replace the symbols commonly used in individual countries. In universities and schools, chemical containers and storage bottles must be labeled accordingly, work instructions for experiments must contain appropriate safety instructions. The experimental instructions of this book also each contain a safety bar with hazard symbols for brief information about possible dangers.

6.1.6 Disposal

There are specific guidelines for lecturers and teachers based on the new symbols—teachers must familiarize themselves with these regulations on hazardous substances and their disposal before starting work. They begin with considerations on how to reduce or even avoid dangerous or environmentally harmful waste when experimenting, and how to conduct experiments in such a way that dangerous waste does not arise in the first place. These considerations can be worked out together with the students and be part of active environmental education. If an alternative implementation cannot be realized, the question arises as to how to properly dispose of problematic chemical residues that occur.

For disposal, *three paths* can be distinguished:

1. Conversion of hazardous substances into harmless substances and their disposal into the wastewater: For example, alkali metal residues are to be converted in spirits, these solutions are to be diluted with water and the diluted solutions are to be discharged into the wastewater.
2. Conversion of hazardous substances and disposal of the products in collection containers: For example, chromate waste can be reduced with sodium sulfite solution and green chromium(III) salt solutions can be placed in the heavy metal salt container.
3. Disposal in specific collection containers: For example, hydrocarbons are stored separately from halogenated hydrocarbons in appropriate containers and later handed over to a disposal company.

There have proven to be *four collection containers* for laboratories in schools:

1. Acidic and alkaline solutions of high concentration,
2. toxic inorganic substances, such as solutions of heavy metal salts,
3. halogen-free organic substances, such as gasoline or toluene,
4. halogenated organic substances, such as dibromethane or bromotoluene.

Collection container 1 can be disposed of in the wastewater after neutralization of the mixture and strong dilution. Filled collection containers 2-4 are to be taken to a disposal site from time to time, which must be proven by the school authority.

6.2 Teaching Processes: Functions, Selection Criteria and Forms of Experiments

The described scientific functions of the experiment also correspond to didactic functions, such as gaining knowledge by setting up and testing hypotheses. Thus, facts that are unknown to students can be "rediscovered" by them—they learn the scientific method of knowledge. This *"discovery learning"* requires the setting up of hypotheses, the derivation of individual cases and their testing by planning, carrying out and evaluating experiments. A detailed description of the "research-developing teaching method" can be worked out at Schmidkunz-Lindemann [14]. However, there are also other functions of the experiment in chemistry lessons.

6.2.1 Functions of the Experiment

In chemistry lessons, functions of the experiment are discussed that have no correspondence in the scientific discipline, such as for motivation, for illustration, for repetition, for simulation of technical procedures, for understanding historical experiments. They are exemplified by *experiments on the topic "Alkali metals and metal hydroxides."*

6.2.1.1 Introduction and Subject-Related Motivation

V6.11 Sodium Hydroxide on the Scale
Students are familiar with the scale from everyday life: They weigh themselves on the personal scale, weigh a postal shipment with the letter scale or two kilograms of oranges on the market with the beam scale. However, the mass of the portion to be weighed always remains constant during the weighing process.

However, if you put 20–30 sodium hydroxide beads on a watch glass and weigh them on the digital scale, you strangely find that the mass is constantly increasing over time. The young people are not familiar with this phenomenon and are motivated by the cognitive conflict to discuss and explain this phenomenon (Chap. 3).

Since only the air from outside is added and the beads begin to shine strangely moist, the observer can conclude the water vapor from the air and clarify his assumption with an experiment: He dissolves sodium hydroxide in water and recognizes that this salt is not only well soluble, but also reacts strongly exothermic with water: $Na^+OH^- \rightarrow aq \rightarrow Na^+(aq) + OH^-(aq)$. However, there are other gases in the air that could be responsible for the increase in mass of the beads.

6.2.1.2 Raising a Question

V6.12 Reactions of Sodium Hydroxide with Components of the Air
Students are informed that even dry air reacts with sodium hydroxide. Which gas from the dry air is responsible for a mass increase that was observed in V6.11 (Sect. 6.6)? Is it nitrogen, oxygen, or carbon dioxide—or all three main components of the air? To test all three gases, an experiment is designed, possibly filling a syringe with the pure gases and bringing them into reaction with solid sodium hydroxide in the attached test tube. Surprisingly, only the last mentioned gas reacts, in the experiment a clear volume decrease can be seen:

$$2NaOH + CO_2 \rightarrow Na_2CO_3 + H_2O$$

or

$$2OH^-(aq) + H_2CO_3(aq) \rightarrow CO_3^{2-}(aq) + 2H_2O$$

6.2.1.3 Testing Hypotheses

V6.13 Reactions of Sodium Hydroxide Solution with Carbon Dioxide
If both pure water and carbon dioxide react with sodium hydroxide, then the sodium hydroxide solution should also react with the gas. In particular, the hypothesis arises that a highly concentrated sodium hydroxide solution dissolves a larger volume of carbon dioxide than the same volume of a less concentrated solution. An apparatus is again designed using the syringe, the hypothesis is tested. From collecting initial experiences with solid sodium hydroxide on the scale (V6.11) to the finding that the gas carbon dioxide also noticeably reacts with solid sodium hydroxide (V6.12), these findings are put into context, from which the hypothesis is derived and experimentally tested.

The result after the execution is that 5 mL of a concentrated, approximately 10-molar solution absorbs the entire gas volume of 50 mL carbon dioxide in the syringe, that 5 mL of the 1:10 diluted 1-molar solution reacts with absorption of a smaller volume, that 5 mL of the 0.1-molar solution only takes up a very small amount of gas. The hypothesis is thus confirmed.

6.2.1.4 Collecting Data

V6.14 pH Values and Concentrations of Sodium Hydroxide Solutions
To measure, for example, which pH values apply to the solutions used in V6.13 (Sects. 6.2.1.3, 6.6), one can initially use universal indicator paper for the first rough estimate, then calibrate a pH meter for the alkaline range and measure exact values. Since the highly concentrated solution is unsuitable for measurements, one limits oneself to the 1-molar and 0.1-molar solution: One measures the pH values 14 and 13 and possibly relates them to the concentrations of hydroxide ions.

Further data of solutions from the kitchen, bathroom, and garage can be examined or pH values of solutions from everyday life can be researched. They start at pH 0 of the acid of a car battery and end with pH 13 of a drain cleaner (sodium hydroxide solution).

6.2.1.5 Illustrating a Theoretical Relationship

V6.15 Dilution Series to Illustrate pH Values
The pH values obtained in V6.14 (Sects. 6.2.1.4, 6.6) cannot be easily classified by the learners and correlated with the concentration of the $OH^-(aq)$ ions in the solution. To make this relationship clear, a 1-molar sodium hydroxide solution is given, the concentration of the $OH^-(aq)$ ions is set at $c = 1$ mol/L or 10^0 mol/L and the pH value 14 is assigned via the ion product. This solution is diluted by the factors 1:10, 1:100 and 1:1000, pH values 13, 12 and 11 are measured and the concentrations of the $OH^-(aq)$ ions are derived with $c = 10^{-1}$ or 10^{-2} or 10^{-3} mol/L. This makes it clear that the pH value increases or decreases by one unit when the concentration of the solutions changes by a factor of 10.

Fig. 6.7 **a** Downs cell for the production of sodium [15], **b** School experiment for the extraction of lithium

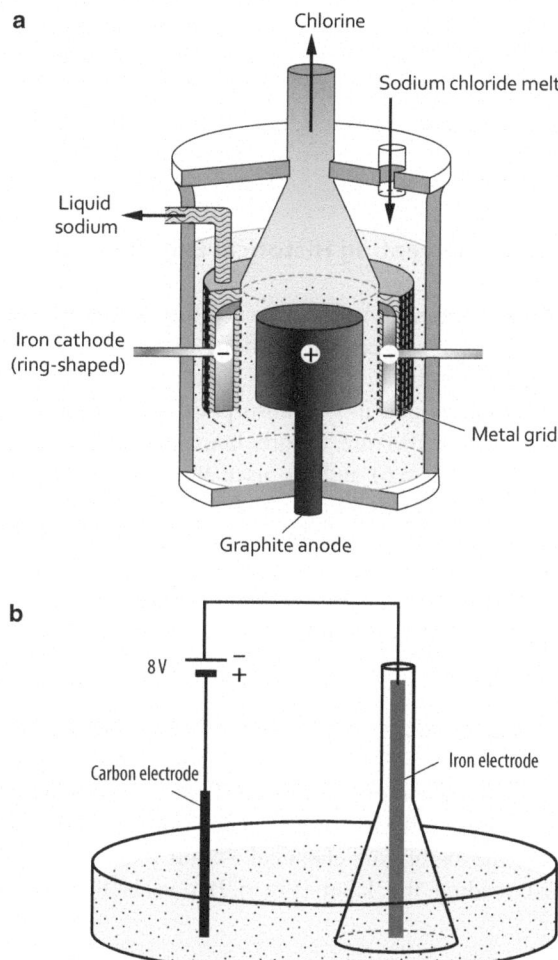

The same series of measurements can be carried out with a 1-molar hydrochloric acid solution: The pH values 0, 1, 2 and 3 are to be measured, concentrations of the H⁺(aq) ions with c = $10^0 = 1$ or 10^{-1} or 10^{-2} or 10^{-3} mol/L are to be derived.

6.2.1.6 Simulating Technical Processes

V6.16 The production of Sodium
In a special molten salt electrolysis apparatus (Downs cell), pure alkali metals are industrially produced, especially sodium from pure sodium chloride: $2Na^+Cl^- \rightarrow 2Na + Cl_2$. Cylindrical metal grids are necessary to prevent mixing and reaction of the resulting sodium and chlorine (Fig. 6.7a).

In school experiments, alkali metals can also be obtained from salt melts. In the present experiment, lithium is produced from lithium chloride, to which about 30% potassium chloride is added to lower the melting temperature. The mixing and reaction of both resulting elements, lithium and chlorine, must be prevented by separating the positive and negative pole of the electrolysis arrangement with a funnel (Fig. 6.7b).

6.2.1.7 Recreating Historical Experiments

V6.17 Elemental Analysis According to Liebig
In this analysis, certain portions of organic substances react with oxygen, the gaseous combustion products carbon dioxide and water vapor are quantitatively determined in two absorption containers (Fig. 6.8). Sodium hydroxide and potassium hydroxide absorb large amounts of the gas carbon dioxide (V6.13, Sects. 6.2.1.3, 6.6) and are used in the combustion analysis. In particular, Justus von Liebig successfully introduced the "potash apparatus" he developed, which completely absorbs the formed carbon dioxide from the gas stream in five sequentially connected and half-filled glass balls (Fig. 6.8). A similar apparatus can be schematically designed and demonstrated to recreate this analysis method. In addition, a version is possible in which the formed gas carbon dioxide collects quantitatively in the syringe (Sects. 6.6: V6.17).

6.2.1.8 Revising and Deepening Understanding of Facts

V6.18 Reaction of "NaOH-Al-Type" Drain Cleaners
To meaningfully revise and deepen understanding of alkali metal hydroxides and lyes, the bathroom chemical "drain cleaner" can be introduced and examined. Just reading the label (Fig. 6.9) reveals the ingredients sodium hydroxide and

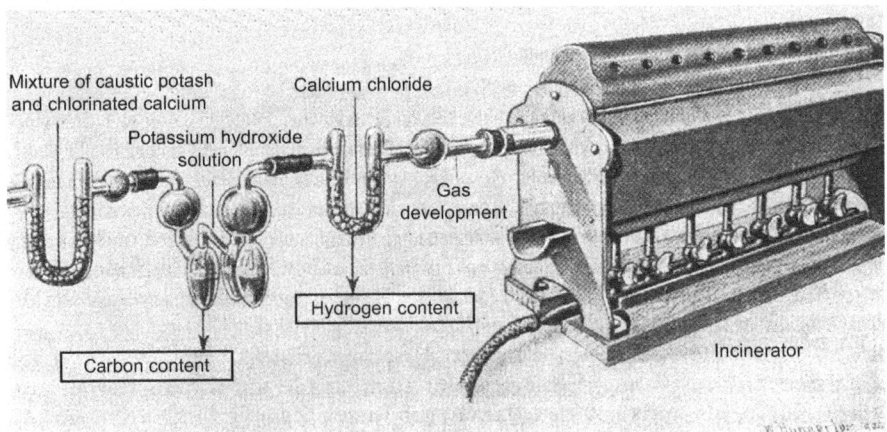

Fig. 6.8 Schematic of the historical elemental analysis according to Liebig [16]

6.2 Teaching Processes: Functions, Selection Criteria ...

Fig. 6.9 Label of a NaOH-Al-type drain cleaner [17]

aluminum, and the examination indeed shows the presence of silver-colored metal splinters next to the known white substance in granular form. Mixing with water leads to a strongly exothermic and alkaline reaction, the addition and dissolution of small paper snippets illustrates the decomposition effect of hot, concentrated soda lye. Aluminum splinters are added to create a "swirling effect" by forming the gases hydrogen and ammonia, which accelerates the cleaning (Chap. 9).

6.2.1.9 Checking Learning Success

V6.19 Reaction of Alkaline Earth Metals with Water
The learning success can be checked to see to what extent the students are able to transfer the known reactions of alkali metals with water to similar reactions of magnesium and calcium with water. The formation of hydroxide solutions and hydrogen is known, but solid hydroxides also precipitate in the form of suspensions. To formulate the reaction symbols, the composition of the alkaline earth metal hydroxides must be conveyed: $Ca^{2+}(OH^-)_2$ or $Mg^{2+}(OH^-)_2$. If the prerequisites are present, the corresponding redox reactions can be derived and interpreted with electron transfers : $Ca + 2H_2O \rightarrow Ca^{2+}(aq) + 2OH^-(aq) + H_2$.

6.2.1.10 Practicing Experimental Skills

V6.20 Student Experiments on the Lithium-Water Reaction
In order to be able to realize student experiments also with regard to the dangerous alkali metals and caustic alkalis, experimenting with lithium is possible. It can be handled more safely than sodium or even potassium: reactions with water are safely possible in a beaker and usual indicator tests can be carried out. Hydrogen can also be shown in a closed apparatus such as a water-filled stand cylinder and used to demonstrate hydrogen (sodium should not be reacted in this way!). By measuring the volume of hydrogen, even a quantitative stoichiometric calculation of the reaction can be carried out: 1 mol Li atoms + 1 mol H_2O molecules $\rightarrow$ 1 mol Li^+(aq) ions + 1 mol OH^-(aq) ions + 1/2 mol H_2 molecules.

6.2.2 Selection Criteria for Experiments

After the various functions of experiments have become clear, further criteria for selecting experiments in chemistry lessons are offered and reflected: The respective experiment should preferably

- be suitable for the age group, i.e., motivating and interpretable,
- build on existing knowledge of the students,
- be didactically "rich", i.e., represent a significant fact of chemistry,
- be tailored to the conditions of the school collection,
- have a high probability of success,
- not pose a safety risk during execution,
- be feasible within a reasonable time span,
- end with a clearly visible effect,
- be usable as a student experiment.

For the **specific teaching matter** of the chemical reaction of two elements to a compound, selection criteria can also be found, for example for a first experiment common in schoolbook literature (Table 6.2):

$$\text{element A} + \text{element B} \rightarrow \text{compound AB; energy conversion}$$

1. A, B, AB should be as familiar as possible to the students from everyday life.
2. The reaction should be feasible as a student experiment.
3. The reaction should take place in a reasonable time.

Table 6.2 Fulfillment of criteria 1–12 for the selection of a first experiment for the chemical reaction of two elements to a chemical compound

Examples from textbook literature	1	2	3	4	5	6	7	8	9	10	11	12
Formation of												
Iron sulfide		×	×	×	×	×	×		×	×	×	
Copper sulfide		×	×	×	×	×	×		×	×	×	
Copper sulfate hydrate		×	×	×	×	×	×	×	×	×		
Magnesium oxide	×	×	×	×		×	×		×	×	×	×
Calcium oxide	×	×	×	×		×	×		×	×	×	×
Sodium chloride			×	×		×		×	×	×	×	×
Potassium bromide			×	×	×	×			×	×	×	×
Aluminium bromide			×	×	×	×	×		×	×	×	
Silver iodide		×	×	×	×	×	×	×	×	×	×	×
Water			×	×			×	×	×	×	×	×
Carbon dioxide	×	×	×	×			×		×	×	×	×

4. The equipment effort should be low.
5. A, B, AB should be visibly solid or liquid substances.
6. A, B, AB should differ significantly in their properties.
7. The mixture of A and B should be easily separable.
8. AB should be decomposable into the starting materials.
9. The reaction should show a clear energy conversion.
10. Safety requirements should be met.
11. A and B should be illustrated by structure models.
12. AB should be illustrated by structure models.

6.2.3 Implementations of the Experiment

The commonly used form for conducting experiments is the *demonstration experiment*, which the teacher demonstrates to the students and evaluates with them. The *student experiment* is considered didactically important and valuable, but often neglected due to the argument of lack of teaching time or lack of material equipment. Nevertheless, student experiments should not be missing in any modern chemistry class. Many experiments can be successfully used in a simplified and optimized form in class, especially when the experimental skills of the students are systematically promoted. The initial "loss of time" is a necessary investment in competence acquisition and pays off later. A hybrid form is the *student demonstration experiment*, which the teacher arranges with a group of students and carries out with them. The experimental forms for chemistry lessons are summarized in Table 6.3.

Especially with demonstration experiments, *principles of Gestalt psychology* [18] should be observed to draw the viewer's attention to the essentials and thus

Table 6.3 Some implementations of experiments in class	
	Teacher demonstration
	Student demonstration
	Student experiment
	• equal work
	• division of labor
	• independent
	Real experiment
	Thought experiment
	Media-mediated experiment
	Qualitative, quantitative
	Varying by substance quantity:
	• Macro scale
	• Semi-micro scale
	• Micro scale

optimize perception. Intended observations are thus facilitated and evaluations of the experiments are made more successful:

- Parts of an apparatus are to be arranged for the viewer so that involved substances flow from left to right.
- Stand material is covered as much as possible and should not interfere with the observation of planned effects. Stands are arranged behind the apparatus from the viewer's perspective.
- As simple apparatus as possible are selected—unnecessarily complicated and difficult to understand devices or vessels are to be avoided.
- Connecting hoses or pipes run as smoothly as possible in a horizontal arrangement. Substance flows are guided as straight as possible.
- Objects and chemicals not directly related to the experiment should not interfere on the experimental table; they should be removed from the field of view as much as possible.
- The planned effect is to be visually, acoustically, olfactorily or tactilely enhanced if it is insufficiently recognized from a distance.

It is particularly possible to use the *overhead projector* to enhance planned effects: both through a projection and enlargement of petri dishes and solutions contained therein, as well as through the brighter light of the projector when experimenting on the illuminated glass plate and thus better illuminating. By specifically aligning the mirror, the apparatus on the experimental table can also be directly illuminated with the light of the overhead projector, or only the part of the apparatus that delivers the effect can be illuminated.

If a small *television camera* (gooseneck camera) is available, it is also possible to transfer an insufficiently observable effect enlarged to the monitor or to project it with the video data projector. This experimental method is also suitable for working with small amounts of chemicals: "Chemistry en miniature" [19]. The semi-micro technique of Häusler [20] or the cuvette technique of Kometz [21] also make experimenting with small amounts of substance possible.

6.2.4 Organizational Procedure of Experimental Teaching

The preparation of experimental teaching in chemistry should not be limited to singular lessons. It is usually a series of experiments from a teaching unit to be distributed over several lessons, individual experiments are to be tried out the day before the lesson, slides for the overhead projector or worksheets for student experiments are to be prepared and duplicated for the lesson. The individual steps of the preparation can be structured as in Table 6.4.

Table 6.4 Preparation and Implementation of Experimental Teaching

Preparation (the day before):	Perform the planned student experiment Design the worksheet for the experiment Try out planned demonstration experiments Draw a slide for the used apparatus Provide devices and chemicals Prepare safety measures
Implementation:	Present problem, let students discuss Explain and assemble devices (by students) Draw apparatus (by students) (board or slide) Allow students time to copy diagrams Explain, implement safety measures Realize experiment (by students) Make observable effects clear Allow reading of measuring devices Clearly end experiment (safety)
Observation, Measured Values:	Collect observations, formulate by students Note observations (board, chronological order) If necessary, repeat implementation partially Represent measured values in tables and/or graphically Discuss measurement errors and sources of error
Explanation (Evaluation):	Evaluate individual observations with students If necessary, repeat parts of the experiment (by students) Offer and discuss aids for explanation Refer to prior knowledge/known experiments Discuss observations first individually, then integratively Develop model conceptions for the structure of substances Derive formulas and reaction symbols from the models Design experiment protocol by students Have all students transfer the protocol into their notebook If necessary, assign protocol as homework *Structure of the protocol*: • Topic • Problem statement • Implementation • Observation • Evaluation (Error discussion)
Follow-up (next lesson):	Answer questions about the experiment/protocol Transfer additions to the protocol Return to the problem statement and conclude the problem Derive the new topic from the result

6.3 Learners: Playfulness and Curiosity, Experimental Skills

For students aged 10–15, experiments in chemistry lessons are of particular importance, as their playfulness and curiosity suggest a great openness to experiments. Therefore, simple phenomena and measurements such as density, solubility, melting and boiling temperatures of substances are highly welcome by these young people and recommended for implementation as student experiments. Often, very interested students already own a purchasable *experiment kit* and know many simple experiments: These students stand out particularly because of their knowledge and experimental skills. The following criteria in particular need to be reflected upon.

First Manual Skills in Handling Experimental Equipment
In primary school science lessons, first student experiments on topics such as air and combustion or water and solutions have been carried out, but usually not with glass equipment from the chemistry lab, but rather with familiar objects from the kitchen. Therefore, the proper handling of experimental equipment and chemicals, taking into account initial safety regulations, needs to be practiced in order to gradually achieve the skills for problem-oriented, independent experimentation (see Figs. 6.5 and 6.6 in Sect. 6.1.4). One particularity should be noted: Boys often like to push themselves to the forefront, take over the experimentation and assign the role of observer or note-taker to the girls. Teachers should try to balance this or let them experiment in separate groups.

Getting Used to Precise Observation
So far, casual glances need to be gradually changed in favor of targeted observation, which should already be determined during the planning of the experiment. In the case of quantitative measurements, measurement sizes and units need to be made clear, measurement devices need to be presented particularly vividly and demonstrated carefully before they are used by students in independent experiments: the proper handling of digital scales, for example, needs to be explained first before they are used independently by learners during experimentation.

Recording Thought Processes
The ability to experiment is optimized when students first create simple protocols in class or as homework. Not only is the logical sequence of individual steps in the experiment reflected and learned until it becomes routine, but measurement values can also be represented in table form or through graphical representations. The latter task is particularly difficult for beginners and should be practiced on initial measurement series, such as recording temperatures of a melting or boiling process depending on the energy supply. The pre-structuring of data through compact tables and the further transfer of data into graphical representations need to be worked out and practiced on examples of increasing complexity. They can finally be created and printed on screen with computers and suitable programs.

6.4 Social Reference Fields: Experiments on Everyday Life and Environment

Students are familiar with substances and their transformations from everyday life and their living environment, they know—for example through questions about sorting and treatment of household waste—about the endangerment of the

6.4 Societal Reference Fields: Experiments on Everyday Life and Environment

environment by problematic substances. Therefore, the following reflections of a content-related and action-oriented nature are necessary and sensible with regard to social reference fields:

- *Experiments on environmental protection issues:* As soon as the treatment of a certain topic (such as air, water, soil, ecology) touches on the issue of environmental protection, possible experiments in this regard should be demonstrated and discussed. For example, Eleni Daoutsali [22] was able to positively evaluate a planned lesson on the car catalytic converter, which included both real experiments with an exhaust gas catalytic converter and model experiments on the catalysis of carbon monoxide on the model platinum catalyst [22].
- Also with regard to practical *environmental education*, ways can be reflected together with the learners for planning and carrying out experiments in which as few pollutants as possible are produced (when choosing a precipitation reaction, do not precipitate lead salts such as yellow lead iodide) or the disposal of pollutants is unproblematic (neutralizing and diluting acidic or alkaline solutions before their disposal into the wastewater).
- *Experiments and models on the greenhouse effect and ozone issues:* The current discussion on climate change has also reached students. It therefore seems urgently necessary to discuss this threat to humanity and, if possible, to demonstrate the known experiments on the greenhouse effect or to show with oscillating molecule models how the energy absorption by the molecules of the greenhouse gases takes place. In particular, Nina Harsch has shown through her teaching concept that successful teaching on the greenhouse effect and ozone issues is possible [23, 24].
- *Experiments for Application in Everyday Life and Technology*: As soon as a connection to everyday life can be made from an experiment, this should be established or demonstrated for subject-related motivation. If the industrial production of certain substances takes place at the school location (sugar from beets, fertilizing salts from the mine, metals by electrolysis, food such as fish or vegetable preserves, etc.) or another chemical industry is present, suitable experiments should also refer to these techniques and thus show close connections with chemistry in everyday life and technology on site. This also includes company tours, for example for drinking water extraction, wastewater clarification or waste recycling.
- *Excursions and Company Explorations*: They give students insights into problems of the mutual influence of laboratory experiments and technical implementation on a large scale (for example, wine distillation in class and large-scale production of spirits, galvanic cell in class and large-scale production of various battery types, electrolysis in class and industrial aluminum extraction).
- *Historical Experimental Developments for Society*: Students should learn about the great importance of the production methods of many substances on social life (Stone Age, Bronze, Iron, Silicon Ages) or to what extent certain groups of substances have changed social life (Liebig's research and the importance of mineral fertilizers for agriculture and food production, fuels and distillation of petroleum, development of plastics, textiles, medicines, dyes, building materials etc.).

In a certain way, war was also the "father of many things" and accelerated the development or production of some substances: invention of gunpowder in the Middle Ages by the Chinese and resulting firearms, sugar from beets after the continental blockade by Napoleon and thus the import ban on sugar cane from overseas, ammonia and nitrate synthesis for the production of large quantities of explosives for the First World War, synthetic gasoline and rubber for warfare in the Second World War, development of nuclear-weapon-capable uranium and dropping of the first atomic bombs at the end of the Second World War. Such connections should not be omitted in class, rather their knowledge can contribute to the young people recognizing, discussing and aligning their political actions with other war-related, threatening developments.

- *Group Dynamic Processes in Solving Experimental Tasks*: Through the division of labor in student groups during experiments, the social behavior of group members, common interests or coordination within the group and consideration of the individual for the benefit of group goals are promoted. In particular, the joint coordination between boys and girls in an experimental group can positively influence their cooperation and reduce any reservations. Finally, it is also the safety during experiments that demands discussion in the group, promotes environmentally conscious handling of chemicals and thus expands environmental awareness and skills for environmentally friendly action.
- *Exhibitions of Experimental Results by Student Groups*: If, for example, own investigation results of air, soil or river or drinking water qualities in the place of residence are published by posters, subject-related discussions of the "experts" with the classmates can follow, contacts to visitors not belonging to the school and to other social groups can be initiated. The program "Youth researches" also leads to the opening of chemistry lessons to our society and to a great recognition of astonishing achievements of individual students. These forms of education for competence, ability to criticize and democratic openness of our young people is a very rewarding goal of chemistry lessons!

6.5 Exercise Tasks

A6.1
Experiments fulfill certain functions in chemistry and in chemistry lessons. Explain three functions each for the subject of chemistry and for chemistry lessons, each with an example. Sketch the function of an experiment for a chemistry lesson in a teaching context of your choice.

A6.2
Hypothesis testing serves both for knowledge in science and for knowledge in chemistry lessons. Describe a historical hypothesis of the natural sciences and its

testing. Design a situation in chemistry lessons that leads to a hypothesis, sketch possible experiments for its testing.

A6.3
For the execution of student experiments in chemistry lessons, both identical and divisional forms of experimentation are common. Explain three teaching examples of your choice. Design a worksheet for the learners for one form of experimentation.

A6.4
Experiments for quantitative measurements are of particular importance. Name five measuring devices suitable for chemistry lessons and the corresponding quantities for data acquisition with examples. Describe or draw the corresponding experimental setups.

A6.5
Safety measures and disposal play a major role in experimental teaching. Name five important safety activities each using the example of a school experiment. Describe important disposal activities and list essential collection containers for chemical residues.

6.6 Examples of Experiments

V6.1 Gas Chromatography in School Experiment

Problem
Common chromatography devices of research laboratories are closed apparatuses that one cannot look into (Black Boxes). Therefore, for the illustration of the analysis procedure, it is didactically sensible to demonstrate an apparatus that shows all functions openly by using a glass cooler (see picture). With the presented apparatus, for example, propane and the two isomers of butane can be separated from the camping gas mixture and displayed one after the other by glowing flames.

Material
Gas chromatograph, glass tip, gas syringe (10 ml), butane gas burner; Hydrogen

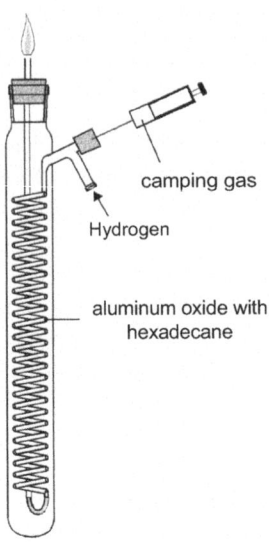

Procedure
The cooler contains aluminum oxide as a white carrier substance, whose surface has been provided with low-volatile hexadecane ($C_{16}H_{34}$) and is therefore suitable for the separation of hydrocarbons. Hydrogen is connected and passed through as carrier gas until the oxyhydrogen test is negative. Hydrogen is ignited at the glass tip. The gas syringe is filled with gas from the butane burner, 5 ml are injected through the septum into the stream of hydrogen. The experiment is ended by turning off the hydrogen stream. Caution: After switching off, hydrogen and air mix and even some time afterwards there is still an oxyhydrogen mixture in the cooler!

Observation
The hydrogen flame burns invisibly (if necessary, prove with a piece of paper). About 30 s after injecting the butane, the colorless flame lights up brightly and goes out, after a short time it lights up twice more.

V6.2 School X-ray Device for Crystal Analysis

Problem
Most devices for instrumental analysis are very expensive and can be viewed in research institutes. For X-ray structure analysis, the teaching material industry has developed a school device with which—in addition to the known shadow images—the interference patterns of single crystals (Laue diagrams) or crystal powders (Debye-Scherrer images) and gloss angles of the crystals (Bragg's angles) can be demonstrated.

Material
School X-ray device, accessories such as X-ray film or Polaroid X-ray diffraction cassette, film developer and fixer; Lithium fluoride single crystal, sodium chloride powder

Implementation
First, shadow images can be demonstrated, as the learners usually know them: For example, a chicken leg is irradiated and the bone is made clearly visible.

To record Laue diagrams, the device is converted: The single crystal is adjusted, behind it the photographic plate or a Polaroid X-ray cassette is arranged. A fine X-ray beam is to be masked out and directed onto the single crystal. After sufficient exposure time, the photographic plate is developed or the Polaroid image is pulled out of the cassette. The crystal powder is treated similarly.

Observation
In the first case, the learners recognize a shadow image, as they have seen it during a doctor's visit. In the second case, in addition to the primary beam, a symmetrical pattern of diffraction points can be seen, which contains information about the structure of the salt crystal. Finally, diffraction rings can be seen, which can also be evaluated by experts.

Precautions for the following experiments with alkali metals
- Put on safety glasses.
- Do not touch pieces of alkali metal with bare hands, but with tweezers.
- Before debarking, carefully dab off the adhering petroleum with filter paper.
- Carefully cut off barks with knife and tweezers.
- Put barks and metal residues in spirits, dilute after complete reaction and dispose of.
- Bring pea-sized pieces of the metals to react.
- Extinguish alkali metal fires not with water, but with sand.
- Dilute alkaline solutions with water and rinse away.

V6.3 Cut surfaces of alkali metals

Problem
Alkali metals are stored under petroleum because both the water vapor of the air and oxygen and carbon dioxide react spontaneously with the metals. To illustrate this storage measure to the students, pieces of metal are to be cut and the cut surfaces observed in the air.

Material
3 watch glasses, tweezers and knife, filter paper; Lithium, Sodium and Potassium

Implementation
A piece of each metal is freed from petroleum with filter paper and cut through. The cut surfaces are observed.

Observation
In the order of Lithium, Sodium and Potassium, the silver-shining cut surface tarnishes more and more quickly and turns dark.

V6.4 Reactions of alkali metals with oxygen

Problem
The tarnishing of the alkali metals is interpreted by the learners as a reaction with air or with oxygen of the air, so that a burning can be predicted and the assumption can be checked by igniting. It should be noted that the metals also react with water vapor and carbon dioxide in the air.

Material
Tripod with wire net, burner, pea-sized pieces of the three alkali metals

Implementation
A piece of each metal is placed under the hood on the wire net, ignited with the roaring flame of the burner, and the color of the flames is observed.

Observation
Lithium burns with a reddish flame, Sodium with a yellow and Potassium with a violet color, white smoke rises, which can sting very corrosively in the nose. In all three cases, a white combustion product remains (in the case of contamination with petroleum, black reaction products can also occur).

V6.5 Reactions of alkali metals with water

Problem
As discussed for V6.4, students may also suspect the water vapor in the air as the reason why the cut surfaces of the metals react in air. A reaction with water can therefore be formulated as a hypothesis and tested in the experiment.

Materials
3 beakers, 3 watch glasses, test tubes, universal indicator paper; pieces of the three alkali metals

Procedure
The beakers are filled halfway with water. One piece of metal is placed in each glass and covered with a watch glass. After the reactions, the solutions are tested with universal indicator paper. A small portion of the solutions is evaporated in the test tube, the resulting white solids are tested with moist indicator paper.

Observation
All three metals react with water, the intensity increases from lithium to sodium and potassium. Sodium melts during the reaction, potassium melts, ignites, and burns with a violet flame. All three solutions react strongly alkaline. After the water has evaporated, a white solid remains, which again shows an alkaline reaction with moist indicator paper.

▶ **Note** Covering the beakers with a watch glass each serves for safety: The sodium or potassium ball possibly sticking to the edge of the glass during the reaction can burst and be thrown out of the glass. This can be avoided by adding a few drops of dish soap to the water before the reaction.

The metal pieces can also be placed on moist filter paper under the fume hood: In this case, they react directly to the white solid. Caution, wear safety goggles! Sodium and potassium ignite in the process, and there is also the risk of corrosive beads of the metals or metal hydroxides splashing around.

V6.6 Detection of hydrogen in the sodium-water reaction

Problem
The hissing sound already indicates a gas that forms at the water surface during the reaction. If the reaction is carried out in a closed apparatus, the gas can be collected and identified as hydrogen.

Materials
Large test tube with side tube and stopper, small test tubes, syringe, wood chip, universal indicator paper; pieces of sodium, gasoline

Procedure
The test tube with side tube—clamped in the stand—is filled halfway with water, a little gasoline is layered on top to prevent the reacting sodium from overheating. The well-moving syringe is to be connected airtight. A lentil-sized piece (about 100 mg) is placed in the test tube, it is closed with the stopper and the syringe is observed. After the reaction has ended, the gas from the piston sampler is led into a small test tube (opening downwards!), the burning wood chip is held in: whistling sound! The resulting solution is tested with universal indicator paper: blue coloration.

Observation
The piece of sodium sinks in the gasoline and develops a colorless gas on the water surface. It briefly rises again in the gasoline, then sinks again and reacts again with water. The process repeats until the piece of sodium has completely reacted and the syringe is about half filled. The gas-air mixture can be ignited in the small test tube and reacts violently with a whistling sound: the gas hydrogen was mixed with air. The briefly visible hydrogen flame is colored yellow. The remaining solution reacts alkaline.

▶ **Note** The observations are also to be evaluated quantitatively: According to stoichiometry, at room temperature 2 · 23 mg of sodium develop the hydrogen volume of 24 mL, a 100-mg piece of sodium thus reacts to about 50 mL of hydrogen. The hydrogen is mixed with air both in the syringe and in the test tube: the spark test must therefore not be carried out with the syringe, but only in the open test tube.

V6.7 Reaction of sodium with chlorine

Problem
The great reactivity of the alkali metals towards oxygen and water suggests that the halogens also react violently. In addition, the reaction of sodium with chlorine is particularly interesting because the reaction product of two toxic starting materials suddenly becomes a food: table salt or sodium chloride. Therefore, it may exceptionally also be tasted carefully.

Materials
Gas generator, syringe, 20-cm-long glass tube, test tube; potassium permanganate, concentrated hydrochloric acid, sodium pieces

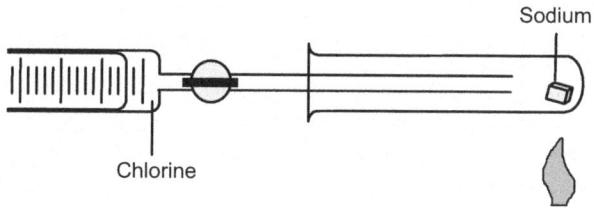

Procedure
Under the fume hood, chlorine is generated from potassium permanganate and hydrochloric acid in the gas generator and the syringe is filled with chlorine. Sodium is heated in the horizontally clamped test tube until it glows yellow (see picture). The chlorine is slowly passed over the glowing sodium using the syringe and attached glass tube, so that the bright yellow flame is maintained. The white substance on the pulled-out glass tube can be carefully tasted (not the contents of the test tube!).

Observation
The sodium reacts with a bright yellow flame, white crystal powder settles on the glass tube. It tastes distinctly like table salt. For further confirmation, the properties of the white powder are compared with authentic table salt.

▶ **Note** The often damaged test tube may contain residues of sodium—it should be carefully treated with ethanol and rinsed with water, then disposed of in the glass waste.

V6.8 Reaction of Potassium with Bromine

Problem
A second alkali metal-halogen reaction is demonstrated to exemplify further properties of two element families: those of the alkali metals and the halogens. Depending on the state of knowledge, these redox reactions can also be interpreted as electron transfers.

Materials
Goblet glass, filter paper, adhesive tape, spatula; lens-sized potassium pieces, bromine (storage bottle)

Procedure
The opening of the goblet glass is half covered with filter paper, it is to be fixed with adhesive tape. In the fume hood, about 2 mL of liquid bromine are added to the goblet glass. A piece of potassium is prepared, placed on the filter paper, transported into the goblet glass with a spatula and the hand is withdrawn. Caution—violent reaction!

Observation
It immediately bangs loudly, a white dust cloud is instantly formed during the violent reaction: crystals of white potassium bromide. The cloud can also be brown colored by bromine vapor.

V6.9 Flame Colors of Alkali Metal Salts

Problem
The flames during the combustion of alkali metals in air already show colorations (see V6.4), which are characteristic for these salts. These colors are better observed when the salts are evaporated in the hot burner flame: they can thus serve analytical purposes and indicate the respective metal in its compounds.

Materials
3 watch glasses, cobalt glass; magnesium oxide rods, lithium-, sodium- and potassium chloride

Procedure
A little lithium chloride is placed in a watch glass, sodium chloride in another, and potassium chloride in a third. These samples are moistened with water. The end of a magnesium oxide rod is heated in the non-luminous gas burner flame until the flame no longer lights up. With the rod, one now dips into the respective salt and holds it in the flame, for each new test the rod is broken off and re-glowed. The potassium flame is viewed through a cobalt glass. A new, re-glowed rod is touched at one end with wet fingers, it is to be held in the non-luminous flame.

Observation

The burner flame is colored deep red, yellow, and violet by the salts in the given order. The last color is better recognized when observed through the blue cobalt glass. Traces of sodium salts in the sample can also color the flame of potassium salt yellow, the yellow color is absorbed by the blue glass. In the last test, the flame lights up yellow: The moist sweat on the fingers contains sodium chloride.

▶ **Note** In this attractive student experiment, the experimenters often let the salt crystals fall into the burner tube, accepting that the flame colors of the following salts will be altered. To avoid this, the burner is clamped horizontally over a base in a stand: The salt residues fall onto this base and can be easily disposed of.

V6.10 Electrolysis of Sodium Hydroxide

Problem
In the electrolysis of the melts of alkali metal salts, the respective alkali metal is obtained at the negative pole, the halogen at the positive pole (see V6.16). In the electrolysis of sodium hydroxide solution, not the metal, but the gas hydrogen is formed at the negative pole, and the gas oxygen is formed at the positive pole. The gases hydrogen and oxygen develop in a volume ratio of 2 : 1.

In this context, we also speak of the "decomposition of water" and of a special "decomposition apparatus". It can also be used for the electrolysis of many other solutions to investigate the resulting gases.

Material
Hofmann's decomposition apparatus with platinum electrodes, DC power source with cable, wood chip; diluted sodium hydroxide solution

Procedure
The decomposition apparatus is filled with sodium hydroxide solution, connected to the DC power source, and the electrolysis is carried out at a voltage of about 5–10 V. Electrolysis is carried out until one of the two measuring tubes is filled with gas. The volumes in both measuring tubes are to be read. A test tube is placed over the tap at the negative pole, the tap is opened, the escaping gas is collected and tested for flammability. The tap over the positive pole is opened and a glowing wood chip is held to the opening. Both taps must be closed after gas extraction.

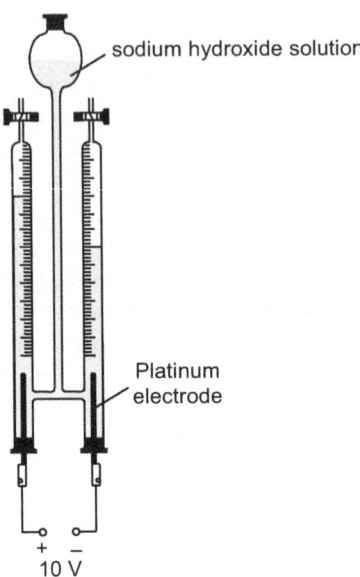

Observation
Gases develop in a volume ratio of 2 : 1. The gas of the larger volume burns starting with a bang: hydrogen. The other gas causes a very bright flame on the glowing wood chip: oxygen.

V6.11 Sodium Hydroxide on the Scale

Problem
A sample of solid dry sodium hydroxide does not show a constant mass on the scale: It becomes a few hundred milligrams heavier within a few minutes. For young students, this is a cognitive conflict: Why do the white beads on the scale get heavier? Upon close observation, one can see that the beads shine moistly, that they become watery after hours and melt. The learners therefore suspect that the water content of the air is responsible for the reaction. A look at the properties of sodium hydroxide shows a very good solubility: 100 g of the salt dissolve in 100 g of water.

Material
Digital scale, watch glass, beaker; universal indicator paper, sodium hydroxide (dry)

Procedure
The watch glass is provided with 20–30 sodium hydroxide beads and placed on the switched-on scale. After 10 min, the reaction product is put into the beaker and dissolved in a little water. The solution is tested with indicator paper.

Observation
The portion of sodium hydroxide becomes continuously heavier, by about 1 g after 10 min. The beads shine moistly, they dissolve in water with the development of heat. The solution turns indicator paper deep blue.

V6.12 Reactions of Sodium Hydroxide with Components of the Air

Problem
Students may ask whether other components of the air such as nitrogen, oxygen or carbon dioxide also react with sodium hydroxide and contribute to the increase in mass. At this point, they can independently plan an experiment to clarify these questions. They will think of a syringe, fill it gradually with the mentioned gases and let each react with a few beads of sodium hydroxide: Observable volume changes indicate the reaction of the respective gas. Other experimental apparatuses are also possible.

Materials
Test tubes with perforated stoppers, piston samplers; sodium hydroxide, nitrogen, oxygen, carbon dioxide

Procedure
A test tube is filled with some beads of sodium hydroxide, the syringe is filled with one of the mentioned gases. A gas sample is transferred into the open test tube, then the piston sampler is connected gas-tight to the test tube with the tap open. The experiment is repeated with each of the other two gases.

Observation
Only the gas carbon dioxide reacts noticeably with sodium hydroxide: The gas volume is constantly decreasing. The other two gases do not react.

V6.13 Reactions of Sodium Hydroxide Solution with Carbon Dioxide

Problem

In the preliminary experiments, it was observed that both water and carbon dioxide react with sodium hydroxide. From these initial experiences, it can be deduced that a sodium hydroxide solution should also react with the gas—and the higher the concentration of the solution, the better. The hypothesis might be: The more concentrated a sodium hydroxide solution is, the more vigorously it reacts with carbon dioxide, the faster a volume decrease occurs in a closed apparatus. The students are again asked to plan an experiment to test this hypothesis and to develop an apparatus for it.

Possible Materials

Test tubes with perforated stoppers, three syringes with tap, concentrated (about 10-molar) sodium hydroxide solution, carbon dioxide

Possible Procedure

The concentrated sodium hydroxide solution is diluted twice by a factor of 1:10, resulting in an approximately 1-molar and a 0.1-molar solution. 5 mL of each of these three solutions are placed in three test tubes, the three syringes are filled with 100 mL of carbon dioxide. The glasses are each flushed with 50 mL of the gas from the syringes, so that the gas replaces the existing air. After the flushing process, the stoppers are firmly placed on the test tubes, these are shaken and the syringe is observed in each case.

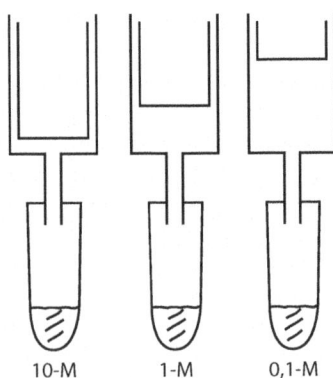

Observation
The concentrated sodium hydroxide solution reacts immediately with the entire gas volume of 50 mL, the other two solutions react more slowly: With the 1-molar solution, about 20 mL remain, the 0.1-molar solution reacts with only a few mL of the carbon dioxide (see picture).

V6.14 pH Values and Concentrations of Sodium Hydroxide Solutions

Problem
Important data when working with acidic and alkaline solutions are the pH values. To know which pH values are present for the solutions used in V6.13, one can first estimate the values by dipping universal indicator paper and reading the pH value from the color scale. With a calibrated pH meter, the 1-molar and the 0.1-molar solution are tested: pH values of about 14 and 13 result. The corresponding concentrations of OH^-(aq) ions are: $c = 10^0$ and 10^{-1} mol/L (10^0 is mathematically identical to the numerical value 1). To understand these numbers, the relationship between concentrations of hydronium ions and hydroxide ions via the ion product for aqueous solutions must be established.

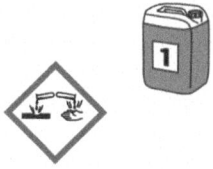

Materials
pH meter; Sodium hydroxide solutions from V6.13, universal indicator paper

Procedure
The pH values of the sodium hydroxide solutions are first tested with indicator paper. The two diluted solutions are to be tested with a calibrated pH meter, the corresponding concentrations of OH^-(aq) ions should be estimated.

Observation
The indicator paper turns deep blue. The pH values of the two diluted solutions are approximately 14 and 13, the concentrations are calculated to be $c(OH^-) = 1$ mol/L and 0.1 mol/L, respectively.

V6.15 Dilution Series to Illustrate pH Values

Problem
The learners should be made even more aware that diluting an acidic or alkaline solution by a factor of 10 leads to a pH value jump of one unit. For this purpose, a 1-molar sodium hydroxide solution is given and assigned a pH value of 14. By

diluting this solution three times by the factors 1:10, 1:100, and 1:1000 and after measuring the pH values 13, 12, and 11, it is illustrated that diluting by a factor of 1:10 reduces the pH value by one unit. The pH values measured in V6.14 are now also more illustrative. A similar experiment can follow with the dilution series of a 1-molar hydrochloric acid.

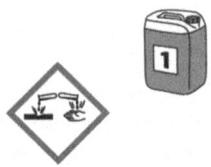

Materials
pH meter, three measuring cylinders (100 mL), volumetric pipette (10 mL); special indicator paper (for estimating pH values in the range 11-14), sodium hydroxide solution ($c = 1$ mol/L)

Procedure
10 mL of the solution are placed in the first measuring cylinder, diluted with water to exactly 100 mL and mixed well by shaking. This solution is then diluted 1:10 again and the resulting solution is diluted 1:10 once more. All solutions are tested with a pH meter and indicator paper, pH values and concentrations are compared.

Observation
The pH values are 14, 13, 12, and 11. The pH value thus changes by one unit when diluted by a factor of 1:10. The corresponding concentrations of OH^-(aq) ions are $c = 1$ or 0.1 or 0.001 or 0.0001 mol/L.

V6.16 Electrolysis of a Lithium Chloride Melt

Problem
Although solutions of alkali metal salts can be electrolyzed without any problems, hydrogen is deposited at the negative pole instead of the expected metal. To obtain the metal, the alkali metal halide must therefore be melted and this melt electrolyzed.

However, melt electrolysis involves the technical problems of working at high temperatures and high current densities: To lower the melting temperature, a salt mixture is therefore used, the DC power source must be suitable for a high current flow.

Materials
Evaporating dish (Pyrex glass), watch glass, funnel, iron and carbon electrode, burner, tripod with clay triangle, DC power source with high ampere fuse; lithium chloride, potassium chloride

Procedure (see picture of the apparatus in Fig. 6.7)
A mixture of 21 g anhydrous lithium chloride and 7 g potassium chloride is prepared, filled into the evaporating dish and provided with electrodes. The iron electrode is to be connected as the negative pole, the carbon electrode as the positive pole, a direct voltage of 8 V is to be set. The mixture is melted over the clay triangle with a roaring burner flame and electrolyzed. After about 10 min of electrolysis, the process is interrupted and the molten lithium is placed on a watch glass—it is surrounded by a thin layer of chloride. After cooling, the metal is prepared, in particular the reaction with water and the resulting hydrogen is demonstrated.

▶ **Note** If the lithium should ignite during electrolysis, do not extinguish with water!! Use sand or table salt!

Observation
The desired metal lithium is formed at the negative pole, and green, pungent smelling chlorine gas at the positive pole. Lithium reacts with water, resulting hydrogen burns with a red flame.

V6.17 Elemental Analysis according to Liebig

Problem
Sodium hydroxide and potassium hydroxide played a major role in the historical combustion analysis—especially Liebig successfully used the "potash apparatus" he developed, which completely absorbs formed carbon dioxide from the gas stream (Fig. 6.8). A similar apparatus can be demonstrated schematically with a combustion tube and two connected U-tubes for the absorption of water vapor and carbon dioxide (Fig. 6.8).

The experiment described here on the reaction of butane shows a simplified version of elemental analysis, in which only the formed carbon dioxide is quantitatively captured in a syringe. Since four CO_2 molecules form from one C_4H_{10} molecule, the gas volume increases fourfold, from 20 mL of butane, 80 mL of carbon dioxide exist. The water vapor condenses at room temperature, the water droplets remain in the apparatus.

Materials
Two syringes, combustion tube with copper oxide (wire form) enclosed in quartz wool tufts, connecting hoses; butane gas, lime water

Procedure
A syringe is filled with 20 mL of butane gas, the apparatus is assembled and clamped (see picture). The copper oxide is heated vigorously with the roaring flame, slowly guiding the butane gas back and forth over it until the volume is constant. The volume of the resulting gas is read off and the gas is passed through a small amount of lime water.

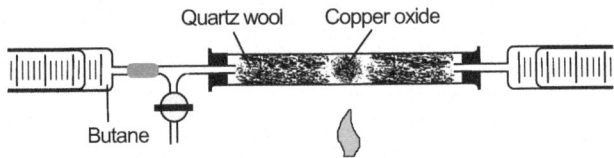

Observation
The copper oxide turns into shiny red copper. About 80 mL of a colorless gas is formed, the test with lime water indicates carbon dioxide.

V6.18 Reaction of "NaOH-Al-Type" drain cleaners

Problem
To meaningfully repeat and deepen treated facts about substances such as alkali metal hydroxides and lyes, the bathroom chemical "drain cleaner" can be introduced and examined. Just reading the label (Fig. 6.9) reveals the ingredients aluminum and sodium hydroxide, the examination confirms the presence of silver-colored metal splinters next to the known white substance in granular form. The combination with water shows the strongly exothermic and alkaline reaction, by adding little pieces of paper the decomposition effect of hot, concentrated soda lye can be shown.

Materials
Test tubes, spatula; drain cleaner "Drain Free" or similar, universal indicator paper, dil. hydrochloric acid, wooden splints, newspaper

Procedure
The substance mixture is visually examined, white powder separated from metal splinters. The powder is dissolved in water, the solution tested with indicator paper. The metal is given in hydrochloric acid, the resulting gas is collected with a second test tube and tested for hydrogen with a burning wooden splint. A few spatula tips of the mixture are mixed with a little water and observed. Newspaper snippets are added to the concentrated solution.

Observation
The solution turns indicator paper deep blue, the metal dissolves with the formation of a colorless gas that burns in the air with a whistling sound: hydrogen. The substance mixture reacts strongly exothermic with water with gas formation: smell of ammonia. The solution decomposes newspaper.

V6.19 Reaction of alkaline earth metals with water

Problem
It can be checked for learning success to what extent the students are able to transfer the known reactions of alkali metals with water to the similar reactions of magnesium and calcium with water. The formation of hydroxides and hydrogen is known, however, the solid hydroxides precipitate in the form of suspensions. When formulating the reaction symbols, the composition $Me(OH)_2$ must be assumed according to the ion charges Ca^{2+} and Mg^{2+}.

Materials
Glass trough, graduated cylinder with cover glass, discharge tube, refractory test tube, magnesium oxide trough, perforated stopper, glass tube, test tubes and clamp; universal indicator paper, magnesium shavings, calcium shavings, sand

Implementation
The refractory test tube is filled with a spoonful of wet sand and clamped horizontally. In the middle part of the test tube, the magnesium oxide channel is placed with magnesium chips, the glass is closed with a stopper and a discharge tube, the water-filled stand cylinder is prepared (see picture). The metal is heated vigorously until it glows, then steam is passed over the metal by heating the sand. The gas that forms is collected pneumatically, then ignited. At the end of the reaction, the stopper is immediately removed to prevent the water from rising back from the tub.

A test tube—held with a wooden clamp—is filled with a few spatula tips of calcium and then added with a little water, while a second test tube is held over it with the opening facing downwards during the reaction. The milky suspension of the first glass is tested with indicator paper, the opening of the second glass is brought close to a flame: caution explosion!

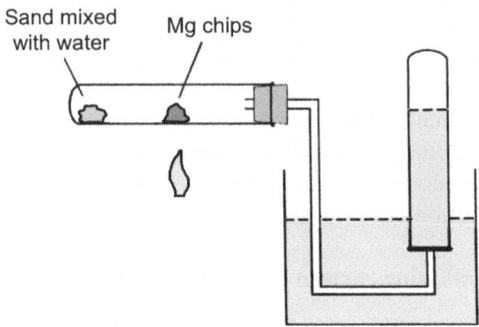

Observation
The magnesium reacts in the heat with a bright flash, a rapid gas development begins, the stand cylinder fills up, the gas burns with a whistling sound: hydrogen.

Calcium reacts strongly exothermic already at room temperature to a solution, which proves to be alkaline: calcium hydroxide solution ("lime water"). The solution is mixed with solid, white calcium hydroxide, a milky suspension can be seen. The gas formed burns with the known whistling sound: hydrogen.

V6.20 Student experiments on the lithium-water reaction

Problem
In order to be able to realize student experiments also with regard to the dangerous alkali metals and caustic alkalis, experimenting with lithium is possible. It can be handled more safely than sodium or even potassium: The reaction with water is not only possible without danger in a beaker, but also quantitatively in a closed apparatus, such as in a pneumatically filled stand cylinder with water (sodium and potassium should not be taken!).

Working material per group
Beakers, glass dish, small stand cylinder with cover glass, tweezers; three lentil-sized pieces of lithium, universal indicator paper, phenolphthalein solution, matches

Experiment 1
Fill the glass dish half full of water, take a piece of lithium with the tweezers and throw it onto the water surface. Record your observations!

Experiment 2
You have noticed a gas development in experiment 1: It is suspected that the gas produced is hydrogen. Try to collect and detect the hydrogen. To do this, rinse out the glass dish and fill it half with water. Then fill the stand cylinder to the brim with water, close it with the cover glass, hold the plate firmly on it, turn it over and immerse it in the dish in such a way that the opening is below the water surface. Now pull the cover glass away.

Take the second piece of lithium with the tweezers, hold it in the water under the opening of the stand cylinder and let it slide into the stand cylinder. After the reaction has finished, cover the stand cylinder under water with the glass plate (press firmly!), take it out of the dish and turn it over again. Remove the cover glass, immediately ignite the gas with the match. Record your observations.

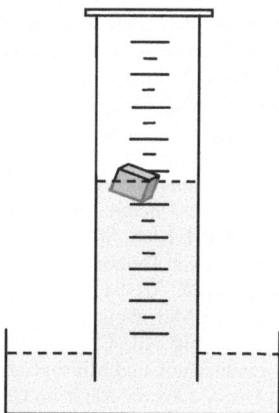

Experiment 3
The solution formed during the reaction is called lithium lye, alkalis can be detected using indicator paper. Test pure water and the lithium lye formed in experiment 2 with indicator paper by briefly dipping a strip of paper and comparing it with the color scale. Estimate the two pH values by comparing them with the color scale.

Experiment 4
Another indicator that can indicate lithium lye is phenolphthalein solution. Drop a few drops into pure water and into the lithium lye. Record the observations.

Experiment 5
Rinse out the dish, fill it half full with water and add a few drops of phenolphthalein solution. Throw the third piece of lithium into it. Record your observations!

Task
Explain all observations. Record reaction symbols in words and—as far as possible—also with formulas. Draw your model idea of the involved atoms, ions or molecules before and after the reaction.

References

1. Liebig J (1840) Der Zustand der Chemie in Preußen. Ann Chem Pharm 34:97
2. Ströker E (1967) Denkwege der Chemie. Alber, Freiburg
3. Kekulé A (1866) Lehrbuch der Organischen Chemie. Bände 1–3. Chemie, Erlangen
4. Von Laue M (1961) Mein physikalischer Werdegang. Eine Selbstdarstellung. Wiley, Braunschweig
5. Von Laue M (1923) Die Interferenz der Röntgenstrahlen. In: Ostwalds Klassiker Nr. 204. Engelmann, Leipzig
6. Wiederholt E (1999) Gaschromatographie – Nachweis von in Wasser gelöstem Sauerstoff und Stickstoff. MNU 52:92
7. Brockmeyer H (1973) Röntgenstrahlen im naturwissenschaftlichen Unterricht. Aulis, Köln
8. Schwedt G (1991) Chemie zwischen Magie und Wissenschaft. VCI, Weinheim
9. Glöckner W et al (1994) Handbuch der Experimentellen Chemie. Aulis, Köln
10. Häusler K et al (1991) Experimente für den Chemieunterricht. Oldenbourg, München
11. Gilbert L et al (1994) Tested demonstrations in chemistry. ACS, Granville
12. Shakhashiri B (1983) Chemical demonstrations. University of Wisconsin Press, Madison
13. Asselborn W et al (2016) Chemie heute 7. NRW. Schroedel, Braunschweig
14. Schmidkunz H, Lindemann H (1992) Das forschend-entwickelnde Unterrichtsverfahren. Westarp, Essen
15. Jäckel M et al (1988) Chemie heute Sekundarbereich II. Schroedel, Hannover
16. Strube W (1981) Der historische Weg der Chemie. Deutscher Verlag, Leipzig
17. Eisner K et al (1992) Elemente Chemie I. Klett, Stuttgart
18. Schmidkunz H (1983) Die Gestaltung von Demonstrationsexperimente nach wahrnehmungspsychologischen Erkenntnissen. NiU-P/C 31:131
19. Roesky H (1997) Chemie en miniature. VCI, Weinheim
20. Häusler KG (1993) Die Halbmikrotechnik. NiU-Chemie 41(2):10
21. Kometz A, Krech K (1998) Küvettentechnik und Mikroglasbaukasten. Chem Sch 45:348
22. Daoutsali E, Barke H-D (2011) Der Abgaskatalysator im Chemieunterricht. Pdn-chemie Sch 1(60):33
23. Harsch N, Estay C, Barke H-D (2011) Treibhauseffekt, Ozon und Saurer Regen. Pdn-chemie Sch 3(60):20
24. Harsch N, Barke H-D (2014) Treibhauseffekt, Ozon und Saurer Regen – eine Soll-Ist-Zustandserhebung und ein darauf aufbauendes Unterrichtskonzept. MNU 67:408

Further Reading

25. Barke H-D, Harsch G (2011) Chemiedidaktik kompakt. Lernprozesse in Theorie und Praxis, 1. Aufl. Springer, Berlin, Heidelberg

Models and Model Representations 7

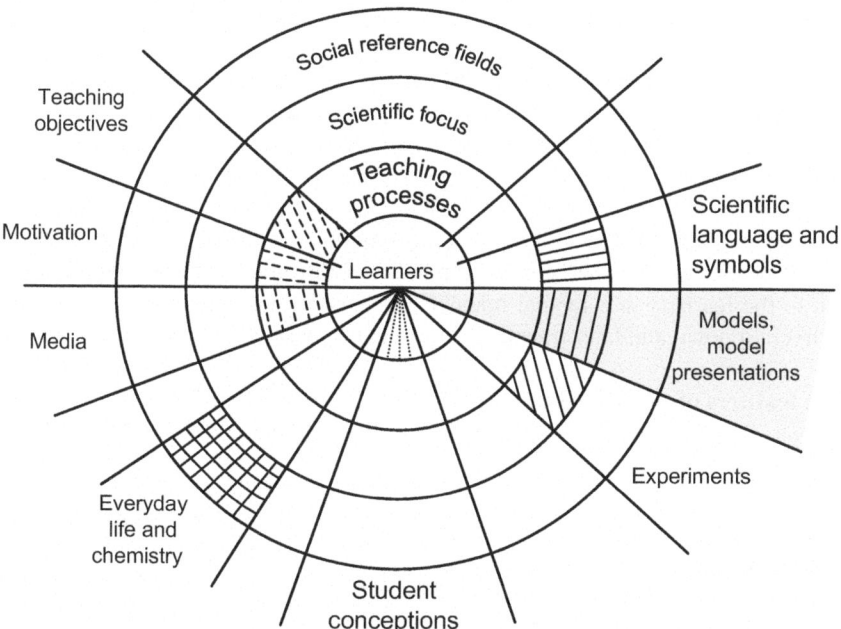

In San Diego/California, one of the authors (H.-D. Barke) was invited to give a slide presentation about Germany. In a picture book *Germany*, he found a lot of beautiful views, including the famous Neuschwanstein Castle near Füssen in Bavaria. Since Americans like it, he projected this picture first. As soon as it was visible, a lady loudly interrupted: "Beautiful — like our castle in Disneyland." The speaker wanted to correct: "In Disneyland there is a copy of the castle, this photo shows the original near Füssen in Bavaria" — but the lady did not listen at all. The difference between the original and the model was not that important to her!

In the mentioned example, the original and both models can basically be compared: Both the photo of the castle and the model castle in Disneyland can be examined in detail for detail with the Bavarian original. In the natural sciences, models of chemical structures such as crystal lattices or molecular models cannot be compared with the submicroscopic originals, because these are fundamentally not visible — neither with the magnifying glass nor with the microscope.

Since the smallest particles of matter are not visible, attempts have always been made to design suitable model conceptions. Thus, Lémery had created the following particle conception in the 17th century for the effect of acids:

> One cannot better explain a hidden thing than by attributing to its particles, from which it consists, such figures that correspond with their effects: so I want to say, the acidic sharpness of a liquid thing consists in the pointed particles, which are in motion: and hopefully no one will want to persuade me that the acid has no points, because all experience testifies to this: for it causes on the tongue such stings, which either come quite equal or very close to those, which one receives from very sharply pointed materials [1].

Lémery had never been able to see acid particles, but succumbed to the temptation to formulate the quoted model conception by speculative transfer of macroscopic properties to the smallest particles of matter. Young learners also often transfer properties of substances to the submicroscopic level: "Sulfur particles are yellow, sugar particles taste sweet".

However, since models and model conceptions are of great importance for the understanding of chemistry, it is an important and difficult didactic task to convey them to the learners in a factual manner. For this purpose, general characteristics are first laid down and later transferred to scientific models.

Main features of the model concept
After an empirical analysis of the generally used model concept, Stachowiak [2] distinguishes three basic features:

- *Imaging feature*: "Models are always models of something, namely images and thus representations of certain natural or artificial originals".
- *Reduction (or idealization) feature*: "Models do not take into account all properties of the original system they represent, but only those that seem relevant to the respective model creators and users".
- *Subjectification feature*: "Models always fulfill their representation and replacement function only for certain subjects, limited to certain mental or actual operations of their users, and within certain periods of time".

If we take the photo of Neuschwanstein Castle as an example, it depicts the building and the surrounding landscape with fields, trees, paths, and mountains in the background in a reduced size: The *imaging feature* is fulfilled. Some of many reductions of the original are the missing spatial dimension, the non-existent play of light and shadow on the walls and windows of the castle, or the absence of the movements of trees and branches in the wind: *Reduction feature*. The specific view of the castle in the photo or the section of the landscape are subjectively selected by the photographer according to his purposes: *Subjectification feature*.

If we base on the model conception of Mr. Lémery, we see the intention to *depict* the effect of acids with the "sharp particles". In his conception, he *reduces* his model arbitrarily to the subjective sensation of acid solutions on human tongues. The many other properties of acids known experimentally at that time are completely ignored by Lémery — his model conception has hardly any general relevance, it was completely *subjectively* chosen and only applied to him, "within a certain period of time".

7.1 Scientific Focus: Models and their Functions

Chemistry gained recognition and success as a science when it overcame the stage of trial and error in the alchemy of the Middle Ages and developed *first model concepts for the structure of substances* in the 18th and 19th centuries beyond mere laboratory experiments and descriptions of substances. Exemplary stages of knowledge are the following:

- **Dalton** postulated in 1808 that there are as many types of atoms as there are elements, and proposed the first atomic mass table, which was corrected and expanded mainly by **Berzelius** in the following decade. By comparing experimentally determined mass ratios in substances and atomic masses, empirical analysis became possible, leading to the knowledge of the compositions of many substances, which were increasingly differentiated and described with empirical formulas. Details on the history can be found in chapters 14, 17 and 18 (electronic appendix).
- **Kekulé** derived the valence theory in 1865 from his experiences: With the tetravalency or the tetrahedral model of the carbon atom, he created, together with **van't Hoff**, the first models of spatial molecular structures. With these model concepts, it became possible to predict the structure of many molecules, confirm them experimentally, and plan successfully syntheses of novel substances (Chapter 20 in the electronic appendix).
- **Laue** recognized in 1912 the three-dimensional structure of crystal lattices by diffraction and interference of a masked X-ray beam on salt crystals, thus creating the first valid model concepts of solids. All subsequent structure investigations provided models of the structure of many crystalline substances, enabling the synthesis of new substances (Chapter 24 in the electronic appendix).

7.1.1 Model concept and knowledge in the natural sciences

The model concept and associated cognitive processes can be understood using the scheme by Steinbuch [3] (Fig. 7.1).

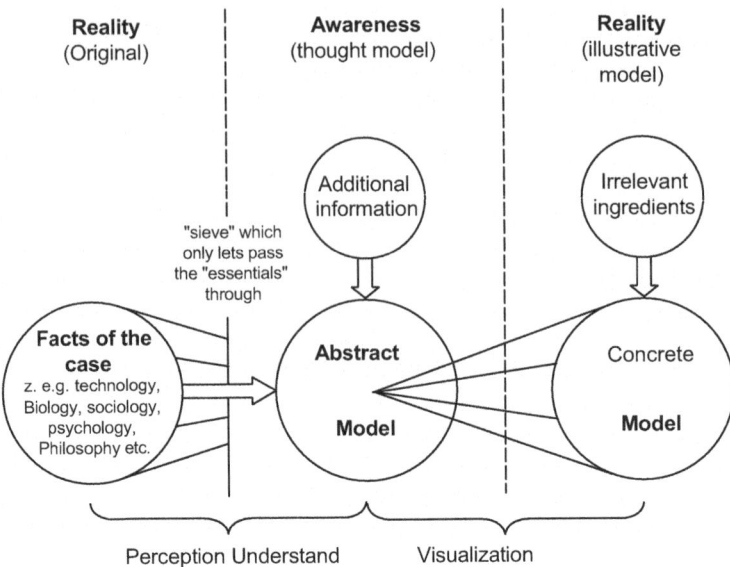

Fig. 7.1 Schema "Thinking in Models" according to Steinbuch [3]

Some complex fact of *reality*, an original, is mapped through perception into an abstract model, a thought model, in which only the "essential" is used, the relevant in the given context. To this end, certain information, such as generally accepted laws of logic or physics, is added. This provides a thought model for future thought processes. This abstract thought model can be projected back into reality for illustration by building a concrete perceptual model or also by graphic representation. However, these inevitably contain irrelevant accessories, i.e., those that the thought model to be represented does not contain [3].

The sequence proposed by Steinbuch "Original → Thought model → Perceptual model" is not absolutely necessary in every case; e. g. the thought model does not necessarily have to be logically preceded by the perceptual model — scientists occasionally use concrete models before they move on to abstract thought models.

Laue's path of knowledge however, went from the facts via the conceptual model to the visual model (Fig. 7.2): The *interference pattern* is the original, which was created by the Laue experiment on the salt crystal (Chap. 24), symmetries of the interference pattern are passed through the "sieve" as the essential and are combined with calculations of known interferences of light on two-dimensional physical grids: *Additional information*. Laue based them on the new calculations regarding three-dimensional diffraction grids — the result was a model conception for the spatially symmetrical structure of the crystal from smallest particles, which act as diffraction centers for the X-ray beam: *Conceptual model* in the researcher's consciousness. For the reader who cannot follow the mathematical

7.1 Scientific Focus: Models and their Scientific Functions

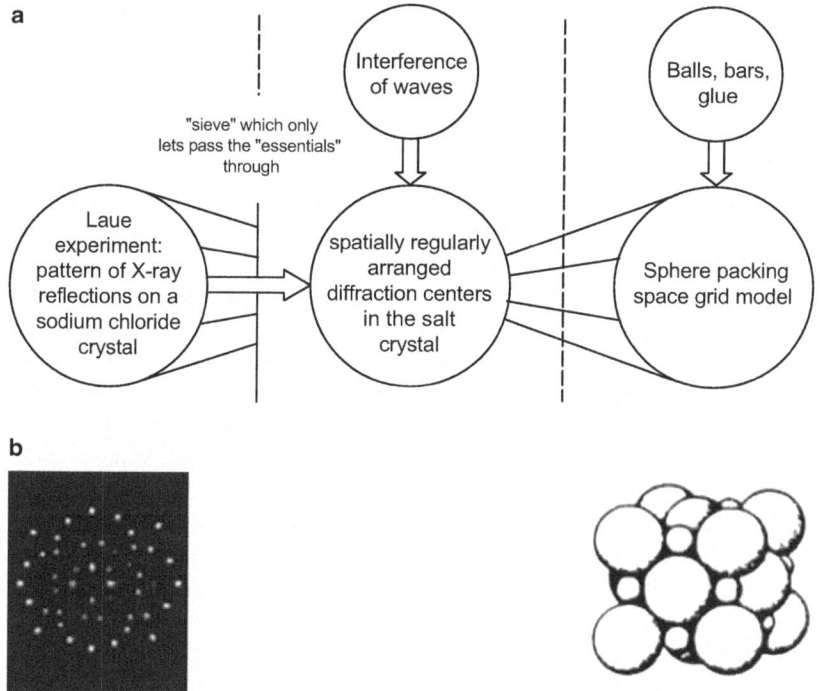

Fig. 7.2 "Thinking in models" and application to Laue's path of knowledge (Chap. 24)

path, Laue drew a cubic *grid model* at the end of his 30-page publication, which vividly shows the positions of the ions in a salt crystal: *Visual model*.

Later, irrelevant ingredients such as balls of different sizes and colors, rods and glue were used to illustrate the conceptual model and to construct concrete *sphere packing or space lattice models*. The resulting visual models of the structure of crystals can be, in addition to space lattice models and sphere packings, also representations of crystallographic elementary cells. Using the example of the well-known sodium chloride structure, the two types of models are illustrated (Fig. 7.3).

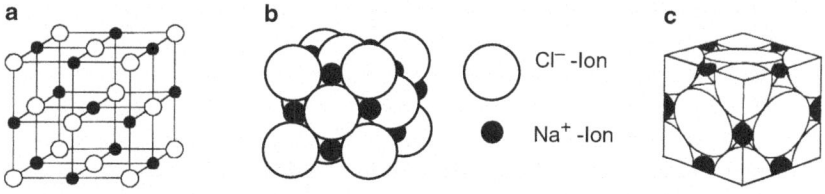

Fig. 7.3 Space lattice, sphere packing and elementary cell of the sodium chloride structure

The *process of understanding through models* is also formalized by Kircher [4]; his schema is explained using the sodium chloride crystal and the models related to it (Fig. 7.3 and 7.4):

- The original **O** is a natural rock salt crystal with a well-developed cubic shape, with smooth surfaces, straight edges, and right angles.
- The model **M** chosen is the sphere packing (Fig. 7.3b), in which chloride ions are represented by large spheres and sodium ions by small spheres.
- The student or the subject **S** can now understand the original O by using the model M for knowledge: The sphere packing model thus becomes the mediator between the subject S and the original O:
  1. There are *properties x and y* of the crystal, which have corresponding model properties x' and y'. For example, if x is chosen for the known spatial arrangement of the sodium and chloride ions in the crystal, then x' represents a corresponding arrangement of the large and small spheres with the coordination number 6 in the model. If y represents the known radius ratio of the two types of ions in the sodium chloride crystal, then y' refers to the corresponding ratio of the diameters of the chosen spheres in the sphere packing model. x and y represent the parameters depicted by the model, which according to Stachowiak are the *mapping features*, or according to Steinbuch constitute the *essence* that is let through by the "sieve".
  2. There is the *property z* in the original, which finds no correspondence in the model. For example, the salty taste of the crystal cannot be reproduced in the model, the model is shortened by this property: *shortening feature* according to Stachowiak. Further such shortenings are density or melting temperature of the crystal — however, the model builder never intended to transfer such original properties to the model.
  3. There can be the *property w'* in the model, which finds no correspondence in the original. For example, the choice of colors — for instance, white for the large spheres and black for the small spheres — is a model property that

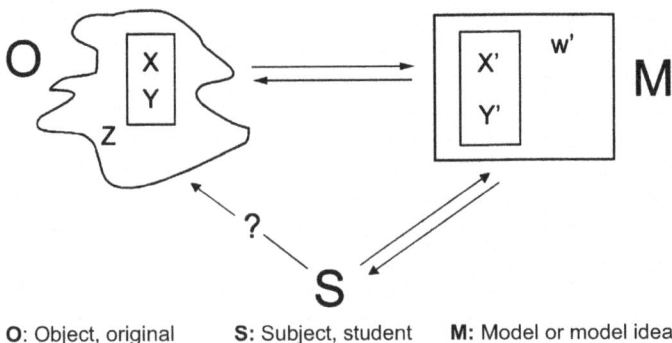

O: Object, original   S: Subject, student   M: Model or model idea

Fig. 7.4 Schema for the process of understanding through models according to Kircher [4]

is completely irrelevant and is subjectively decided by the model builder: *Irrelevant ingredients* according to Steinbuch. Further such irrelevant ingredients are model materials such as wood, pulp or styrofoam, are adhesives between the spheres like glue or velcro.

The *space lattice model* (Fig. 7.3a) has as its mapping feature exclusively the cubic arrangement of the ions in the crystal, however, it shows both coordination numbers 6 in a vivid way through the possibility to look into the model: A sodium ion is surrounded by 6 chloride ions, the chloride ion also by 6 sodium ions. The striking connecting rods between the intersection points are only necessary for model construction, with respect to the original they are completely irrelevant ingredients and have no mapping function. Unfortunately, the connecting rods are often interpreted by learners as "bonds" and compared with the covalent bond — therefore the sphere packing is the didactically most advantageous model.

Both the sphere packing model and the space lattice model do not have equal numbers of the two types of spheres, but the ratio 14 : 13. Therefore, the *model of the elementary cell* (Fig. 7.3c) is added: it shows three mapping features: The cubic structure of the ion grid, the size ratio of the ions and the correct number ratio of the ions 1 : 1. If you count all the pieces of the large and small spheres of the model together, you get 4 large and 4 small spheres. Transferred to the original, an aggregate of 4 sodium ions and 4 chloride ions can be represented by the special symbol $\{(Na^+)_4(Cl^-)_4\}$.

The unit cell with this symbol can be considered as the *smallest unit of the NaCl structure* — just as the C_2H_5OH molecule is considered the smallest unit of ethanol (Chap. 8). From $\{(Na^+)_4(Cl^-)_4\}$ symbols such as $\{(Na^+)_1(Cl^-)_1\}$ or $(Na^+)(Cl^-)$ or even NaCl can be derived, however, the latter — and almost universally used — symbol is so abbreviated that the ions are no longer displayed: novices therefore often see the NaCl symbol as a molecular model and develop corresponding misconceptions about the structure of table salt (Chap. 2).

The sphere packing model also illustrates the *ionic bond* of the sodium and chloride ions in the salt crystal: The electrostatic attraction forces of oppositely charged ions and the repulsion forces of like-charged ions in the ionic lattice are in equilibrium, in which the attraction forces predominate and ensure the solid cohesion of the ions. This results in the high melting temperature of 800°C. In the ionic lattice, the direct attraction of a cation applies to six anions in octahedral arrangement and that of an anion to six cations in the same arrangement — both coordination numbers are therefore 6. They also indicate that the ratio of ions is 1:1, as required by the ion charges of both types of ions.

7.1.2 Thought Models in Chemistry

In the scientific discipline, *thought models* are constantly changed by new experiences and empirical findings, so it is hardly possible for a longer period of time to specify *the* current atomic model or *the* current model for chemical bonding.

Quantum Mechanical Atomic Model
The structure of the electron shell of the atom or ion is described by the principal quantum number n ("K-, L-shell"), by the azimuthal quantum number l (s-, p-, d- and f-"subshells"), by the magnetic quantum number m and the spin quantum number s. A maximum of two electrons of different spins can form a common electron cloud, an orbital (Pauli principle). Starting from wave-particle duality, wave functions were developed that make statements about energy conditions and electron probability densities (Schrödinger equation). The combination of wave functions leads to the description of atoms by atomic orbitals, of molecules by molecular orbitals — accordingly, electron pair bonds in molecules can be mathematically captured, structures of newly postulated molecules can even be predicted today by means of quantum calculations (molecule design).

Historical Models of Atomic Structure
In teaching and instruction of the subject chemistry, historical models for atomic structure are often used for didactic reasons, e.g.

- Mass Model (Dalton 1808)
- Mass-Charge Model (Thomson 1897)
- Nucleus-Shell Model (Rutherford 1911)
- Shell Model of the Electron Shell (Bohr 1913)
- Electron Cloud Repulsion Model (Gillespie, Kimball 1966)

Models for Chemical Bonding
The corresponding models for two types of bonds are to be viewed from two perspectives:

1. The *effects* of forces in space can be made tangible for learners by
   a. *directed* binding forces, which emanate from the atom and act in specific directions of space, are represented by snap buttons or rods: In the CH_4 molecule, for example, one imagines four bonds to four H atoms, which are arranged like the corners of a tetrahedron.
   b. *undirected* binding forces, which act evenly around an atom or ion, are symbolized by "naked" spheres that do not dictate any direction of bonding. In the copper crystal, undirected attractive forces emanate from each Cu atom to 12 other Cu atoms in the neighborhood, hence the coordination number is 12.
2. The *nature* of the electrical forces is usually described by mathematical models based on the distribution of electron densities. It is useful to distinguish the following *limit cases of chemical bonding*:
   - Electron pair bonding (covalent bonding, atomic bonding)
   - Ionic bonding (ionic bond, ionic relationships)
   - Metal bonding (metallic bonding)
   - Hydrogen bridge bonding (hydrogen bonding)

7.1 Scientific Focus: Models and their Scientific Functions

- Dipole forces between permanent and induced dipole moments in molecules
- Van der Waals forces or intermolecular forces between molecular surfaces

Models for Chemical Structure

The mathematical capture of atomic structure and chemical bonding is often only a means to an end, to make statements about the chemical structure. On this basis, it is the goal of many instrumental analytical methods to trace and describe the arrangement of atoms or ions in given substances. From the determined structures, the *structure symbols* are derived as their abbreviations.

Limit cases of chemical structures can be sketched as follows:

- Molecule structure (specification of atom types and number of involved atoms, bond lengths and bond angles)
- Atomic lattice structure (specification of atom types and lattice constants, unit cell, ratio of atoms in crystals)
- Metal lattice structure (specification of atom types and lattice constants, unit cell, ratio of atoms in atomic lattices of alloy crystals)
- Ion lattice structure (specification of ion types and lattice constants, unit cell, ratio of ions in ion lattices of salt crystals)
- Molecule lattice structure (specification of molecule types and lattice constants, unit cell)
- Cluster structures (for example in complexes or nano-particles)

Models for Chemical Reaction

Particle rearrangements in chemical reactions are described both by *model concepts* and by *reaction symbols*:

- Atom rearrangements in reactions of metals to alloys
- Ion rearrangements in hydration and precipitation reactions
- Proton transfers in acid-base reactions
- Electron transfers in redox reactions
- Ligand transfers in complex reactions
- Addition, substitution, elimination and rearrangement reactions in organic chemistry

7.1.3 Visual Models in Chemistry

In chemistry, one predominantly works with abstract conceptual models. As soon as it is desired, suitable visual models are developed for didactic reasons (Fig. 7.3): With regard to many conceptual models of *chemical structures*, concrete, easily understandable visual models can be built, such as for molecule and crystal lattice structures.

Models of Molecular Structures

The spatial arrangement of atoms in a molecule is indicated using spatial coordinates, bond lengths, and bond angles, which are experimentally determined in the laboratory. These data can be illustrated using the following models (Fig. 7.5):

- *Space-filling model* (Fig. 7.5a): The space occupied by the atoms is taken into account, atom spheres are assembled into a specific molecular model according to the bond lengths and angles. The volumes of the atom spheres are dimensioned so that they represent a defined percentage of the entire electron shell, usually it is 90%.
- *Ball-and-stick model* (Fig. 7.5b): The relative atomic volumes are not taken into account, only bond distances and bond angles. For this purpose, balls are used, which are the same size for all types of atoms and are usually color-coded, the balls are held together by connecting rods or push buttons,
- *Stick model* (Fig. 7.5c): To emphasize bond distances and bond angles, no balls are used, only sticks of appropriate length and with correct angles to each other.

Models of Crystal Lattice Structures

Technically, the unit cell would be sufficient as a conceptual model for each structure and the expert could derive the entire crystal structure from it. To make it clear for learners, a specific concrete lattice section is chosen, such as in models like space lattices or sphere packing (Fig. 7.3):

- *Space lattice model*: Equal-sized spheres — possibly of different colors — are connected by connecting rods with neighboring spheres until the desired section from the structure is reached, such as the section corresponding to the unit cell (Fig. 7.3a),
- *Sphere packing model*: The known size ratio of the involved atom or ion types is taken into account, selected spheres are firmly glued or stacked according to known structural parameters until the desired section is shown, such as a unit cube (Fig. 7.3b),
- *Unit cell*: It is derived from the unit cube model for the present example and ensures an accurate numerical ratio of the atoms or ions (Fig. 7.3c), by translation in all three spatial directions, arbitrarily large structure sections can be obtained,
- *3D drawing*: Red-green drawings of chemical structures are fixed with the corresponding red-green glasses and spatially interpreted [5].

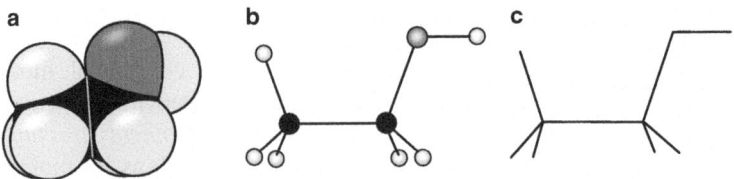

Fig. 7.5 Space-filling model, ball-and-stick, and stick model of the structure of the C_2H_5OH molecule (ethanol)

7.2 Teaching Processes: Models and their Didactic Functions

Fig. 7.6 Kit "Structures of Metals" by GEOMIX [6]

In addition to the well-known molecular model kits, there are various kits in the teaching materials industry that allow the construction and comparison of sphere packing:

- "Structures of Metals" (Fig. 7.6): This kit contains square and triangular fields in the lid, with which all three basic structures of metals can be recreated in the form of sphere packing with precisely shaped wooden balls (d = 3 cm).
- "Model Kit for Crystal Lattice Structure" (Fig. 7.7): A plastic plate with depressions in a triangular pattern allows the laying of a base layer with colored plastic balls (d = 1 cm) and the stacking of the desired sphere packing.
- "Solid-State Model Kit" (Fig. 7.8): Base plates with holes of various patterns are suitable for holding pins, onto which drilled glass balls of different diameters and colors can be arranged according to the chemical structure.

7.2 Teaching Processes: Models and their Didactic Functions

In the fields of biology and geography, students have already gained some experience, for example with models of human organs such as the heart or kidney, or with models like maps or globes — these areas of models are illustrative for them. Therefore, they usually also like the introductory lessons in the subjects of biology and geography: They stay in the familiar area of direct comparison of original and

Fig. 7.7 "Model Kit for Metal Lattice Structure" by LEYBOLD [7]

Fig. 7.8 "Solid-State Model Kit", Institute for Chemistry Education, University of Madison, Wisconsin, USA [8]

model. Even in chemistry lessons, the first models of states of aggregation and dissolution processes are still relatively illustrative.

As soon as the inevitable formulas and reaction symbols are dealt with in chemistry, the illustrative nature ends. Chemistry therefore becomes difficult to understand because chemical formulas and reaction equations belong to the *abstract thought models* and are only understood at the earliest in the developmental stage of formal operations (Sec. 4.3).

Accordingly, for the teaching process, the question is which illustrative models such as molecule models, sphere packing, and space lattices can be used for the chemical structure, so that formulas and equations are less abstract for the learners on this basis. All bonding models or models for the structure of the individual atom are abstract mathematical constructs — their treatment must be postponed until a first understanding of the structure of substances has been achieved with the illustrative structure models: first the chemical structure, then the chemical bond!

However, some chemistry and physics educators view the illustrative nature of particle images and sphere packing for chemical structures very critically. For example, Buck [9] refers to the particle concept as an "unmodel" and opposes the usual representations of sphere arrangements for illustrating states of aggregation:

> We should not draw the spheres at all, we are actually dealing with centers of force. Teachers and authors who should know better easily accept that atomic and molecular orbitals are asymptotic, extend in principle into the entire universe, and that the boundary lines are drawn arbitrarily, usually at 85%. The illustrative nature of the images is — to put it briefly — the decisive error [9].

Elsewhere, Buck suggests triggering the "jump to the atoms" through a certain sequence of slides of increasing and then decreasing complexity and discussing system properties at each stage:

> Egg → chicken coop → farm → village → country → earth → universe → earth → city → school → student → hair → hair fiber → ?. The next slide? There isn't one, because such slides do not exist [10].

This discussion is certainly very stimulating and could indeed take place before the introduction of smallest particles in class. However, the conclusion from the "non-existence of the next slide" cannot be to immediately move on to the "force centers" or "infinitely extended core-shell systems" with the students. For developmental psychological reasons, tangible circles and spheres, or even cubes or Lego bricks, must first be chosen as models for smallest particles. The discussion of the shape, color, or material of the models as "irrelevant ingredients" opens up the opportunity to convey the scientific model concept at this level. If initially the simple particle model is applied as a preliminary model idea and developed further in the course of the following lessons via the Dalton atomic model to the core-shell model, then at the end of this lesson the "arbitrarily drawn boundary lines" can be meaningfully discussed and the views on circles and spheres can be relativized.

Parchmann and colleagues [11, 12] have highlighted the special importance of models and model ideas for the **basic concepts** "substance-particle relationships" and "structure-property relationships" and described the ways in which initial models can be developed with learners [12]. Regarding the "structure-property concept", Pfeiffer [13] has made it clear that not only the structures of molecules are important, but also solid-state structures. Examples include alloys of copper and gold, such as CuAu, and the ionic lattice of sodium chloride. Herdt [14] also calls for the consideration of basic concepts and bases the aforementioned two relationships on a "structure orientation as a key concept": "A key concept enables — according to the literal meaning — the unlocking of a previously blocked path of knowledge, ideally the independent exploration of the learning object. This is exactly what applies to the principle of structure orientation: It combines the substance-particle concept and the structure-property concept in terms of content" [14].

7.2.1 Conveying Chemical Facts through Model Ideas

Figure 7.1 shows the scheme "Thinking in Models" and thus the process of knowledge acquisition in chemistry "from left to right": The chemist develops a "thought model" through suitable "additional information" and transfers it into a "concrete model" for visualization purposes. The learner naturally cannot take this path, but it is possible for him to pave the way in the scheme "from right to left" by dealing with visual models: He works with the concrete models and develops increasingly advanced thought models of corresponding facts in his consciousness (see also a related conception [15]). It is accepted that these first thought models initially "interfere" with the visual models and the irrelevant ingredients are not immediately recognized as such — but gradually a factually appropriate abstraction can take place more and more.

Johnstone [16] has designed a chemical triangle regarding this problem, which assigns substances and chemical reactions (chemical transformation of substances) to the "macro level", at least meaning everything that can be seen, felt, and smelled (Fig. 7.9). The "submicro level" is assigned to atoms, ions, and molecules and corresponding chemical structures and structure models; finally, all

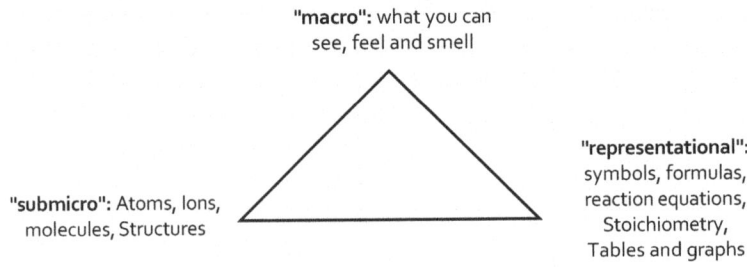

Fig. 7.9 Johnstone's "Chemical Triangle" [16]

7.2 Teaching Processes: Models and their Didactic Functions

abbreviations of chemical structures such as symbols, formulas, reaction equations, and stoichiometric calculations are summarized in the "representational level".

Johnstone notes that it is difficult for chemistry learners to grasp all three levels of the triangle at the same time: "It is psychological folly to introduce learners to ideas at all three levels simultaneously. Herein lies the origin of many misconceptions. Trained chemists can keep these three levels in balance — but not the learner" [12].

Gabel [17] makes it clear that the big mistake in chemistry teaching is to immediately switch from the "macro level" to the "representational level" or "symbolic level" without using the "submicro level" for visualization: "The primary barrier to understanding chemistry, however, is not the existence of the three levels of representing matter. It is that chemistry introduction occurs predominantly on the most abstract level, the symbolic level" [17].

The constant oscillation between the three levels of representation is also a central concern for the design of learning processes in the PIN concept [18] of organic chemistry, and the START concept [19]of general chemistry for beginners

According to the findings of Johnstone and Gabel, it also seems sensible to interpret initial experimental observations of students using concrete models and to initially omit the chemical symbols. After the introduction of initial model concepts, such as the particle or Dalton model, facts should be interpreted with this model as far as possible at this level. Chemistry teaching should be *"dual-track"*, it should be *structure-oriented* and at the same time *experiment-oriented*

1. Track: Phenomena and laboratory experiences
2. Track: Structural models and model concepts (cf. [15])

When planning lessons, it is necessary to decide early on which phenomena to deal with in the first weeks of chemistry lessons without any model interpretation and to initially explain in everyday language. After the introduction of the first model concept — such as the particle model — phenomena and experiments should then be consistently selected that are suitable for interpretation with the *particle model*. If the *Dalton model* is then used to introduce the concept of the atom, from this point on this model should consistently underlie the explanations (Fig. 7.10).

For example, dissolution processes are explained at the level of the *particle model* with sugar particles and water particles, without going into the structure of sugar or water molecules (Fig. 7.11). The chemical reaction of carbon with oxygen to form carbon dioxide can only be meaningfully interpreted with the rearrangement of atoms at the level of the *Dalton atomic model* (Fig. 7.12): The C atoms of the carbon crystal and the O atoms of the O_2 molecules of oxygen rearrange to form CO_2 molecules of carbon dioxide. The rearrangements of the atoms should be discussed for as many further reactions as possible — for example in hydrogen-oxygen reactions or in sulfur-oxygen reactions. Many such and further reactions are illustrated by Haupt and Moritz [20].

1. Rail: *Phenomena and Laboratory experience*	Solution process, Diffusion, Distillation	Chemical reactions of gases, Gas laws	Redox reactions of metals and Salt solutions
↓	↓	↓	↓
2. Rail: *Structural models and Model concepts*	Particle conception: Arrangement of the small particles before and after	Dalton model: Models for the Molecules before and after	Core-shell model: Electron transition from metal atom to Ion and vice versa

Fig. 7.10 Examples of "dual-track" approach in chemistry teaching

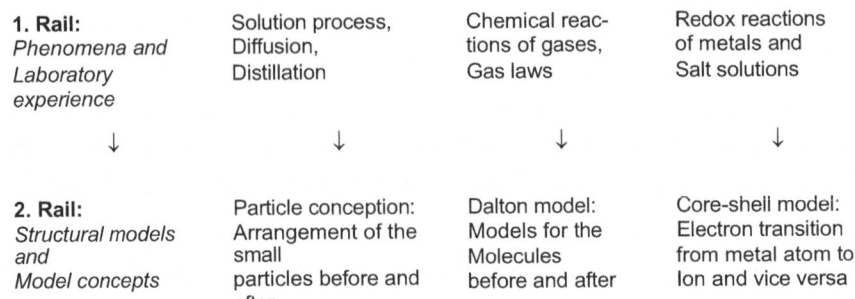

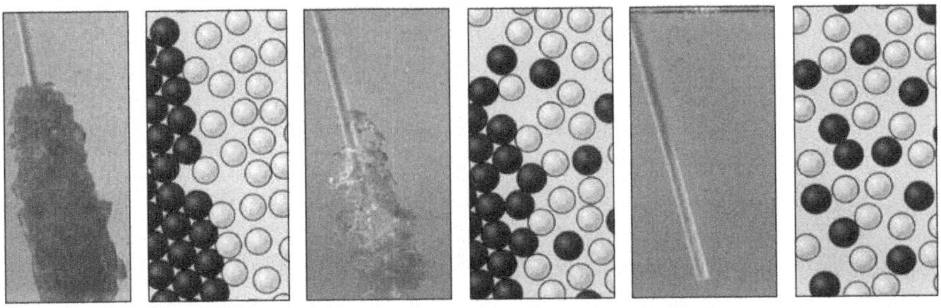

Fig. 7.11 The dissolution of rock candy interpreted by the particle model [21]

Carbon (s) + oxygen (g) → Carbon dioxide (g); exothermic

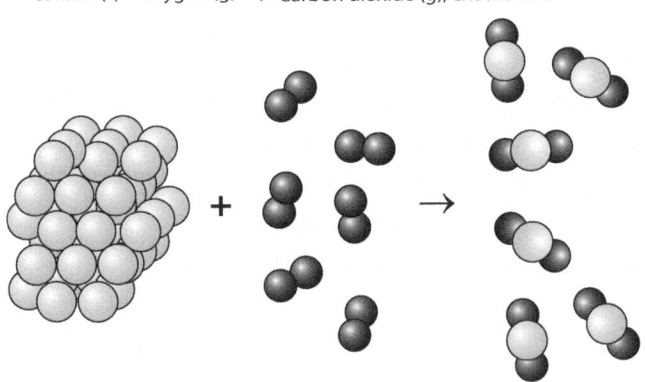

Fig. 7.12 The carbon-oxygen reaction interpreted by the Dalton model [21]

It should be noted that, according to surveys by Marohn [22], confusion can arise if molecules are initially represented as spheres, but later the sphere is supposed to represent a single atom. This demonstrably leads to misconceptions in the interpretation of dissolution processes and changes of state: In the students' imagination, molecules split into individual atoms. When boiling, H and O atoms form from water molecules, ethanol molecules decompose into H_2, O_2, H_2O or C_2H_6 molecules [22].

To prevent this misunderstanding, in the context of the START concept [19], the spherical shape is reserved exclusively for atoms. Particles in the sense of undifferentiated molecules are symbolized by squares or rectangles, spatially by cubes or cuboids.

7.2.1.1 Sphere Packing

Since the introductory phase of chemistry teaching often focuses on working with metals, the *structures of metals* in the form of sphere packing should play an important role (Fig. 7.13): Students accept sphere packing "playfully" when they have the opportunity to create these models themselves by stacking spheres and discovering the coordination number 12 — the number of spheres that touch a sphere inside the packing. Therefore, in the practical course, building the sphere packings corresponding to the three metal structures and important structural sections is suggested with M7.1 to M7.7 (Sec. 7.6). Further information can be found elsewhere [23].

If the gaps in the cubic closest sphere packing are taken into account, filling all octahedral gaps results in the sodium chloride structure (Fig. 7.14). If all tetrahedral gaps are also filled, the lithium oxide structure results; if they are only half occupied, it is the zinc blende structure (Fig. 7.14). In the practical course, building the *structures of these and other salt crystals* is suggested with M7.8 to M7.15 (Sec. 7.6). Additional information on salt structures can be found elsewhere [24].

Sphere packings are excellent models for metal crystals because the metal atoms actually arrange themselves like spheres in a closest sphere packing. They

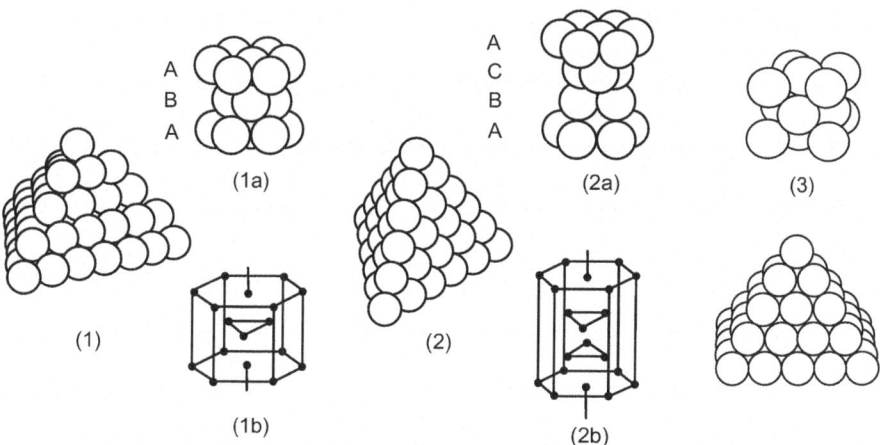

Fig. 7.13 Hexagonal and cubic closest packings as models for metal structures

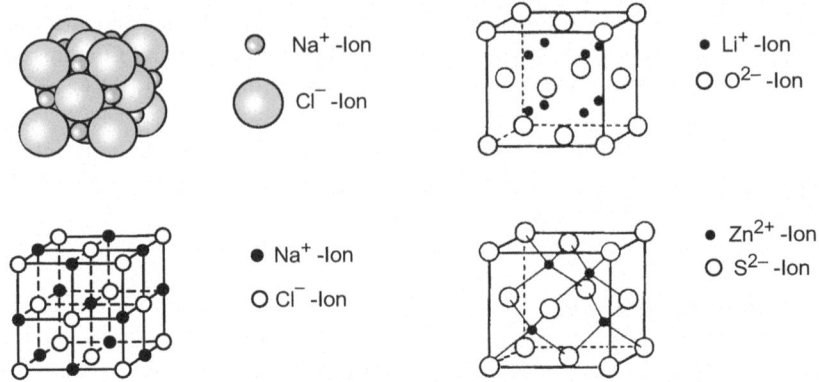

Fig. 7.14 The cubic closest sphere packing as a basic structure for salt structures

are also an important prerequisite for the models of the salt structure. This structure is significant because without it, students often develop "homemade misconceptions": The students talk about "NaCl molecules", electron transitions from the Na atom to the Cl atom "Na → e → Cl", Na^+Cl^- ion pairs, etc. [25]. If they reflect on corresponding structural models and discover the coordination numbers 6 for both types of ions, or even build them themselves, then the scientifically accurate model ideas can successfully prevail. Barke and Harsch [26] suggested in the article "Building Structural Models — Understanding Ionic Lattices" the construction of sphere packings in which the cubic face-centered elementary cube is built in.

7.2.1.2 Beaker Models

Once the concept of ions has been introduced on the basis of the extended Dalton's atomic model, possibly with the periodic system of "atoms and ions as basic building blocks of matter" (Fig. 8.7 in Chap. 8), beaker models can also help to illustrate and thus better understand the structure of salt crystals and salt solutions (Fig. 7.15).

Precipitation reactions can be represented in this way (Fig. 7.16), to show that only one insoluble salt precipitates, while the other salt remains dissolved, meaning that these ions remain unchanged in solution and do not combine: In the USA, they are therefore also called "spectator ions" — they have nothing to do

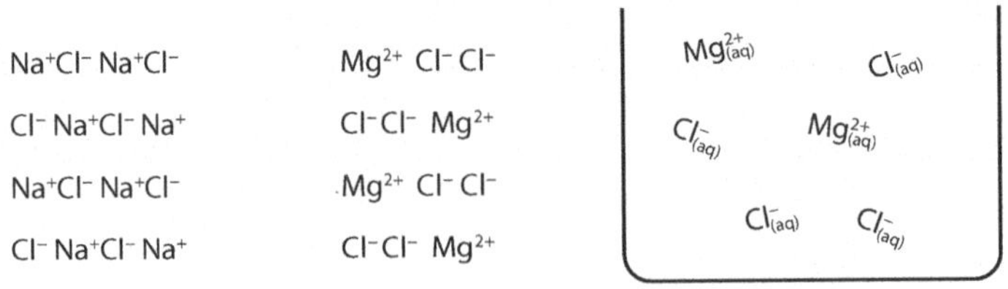

Fig. 7.15 Ion symbols for the schematic construction of salt crystals and salt solutions

7.2 Teaching Processes: Models and their Didactic Functions

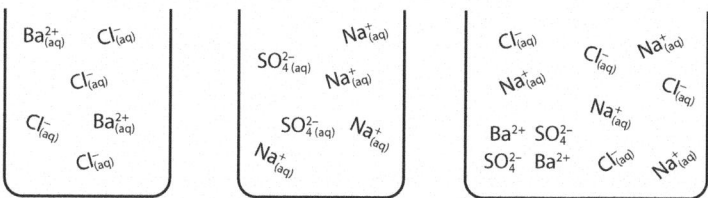

Fig. 7.16 The precipitation reaction of barium sulfate as a beaker model. (Note: The use of the symbol (aq) for hydrated ions is a matter of convention. For beginners it is absolutely necessary; later on it may be tiresome and eventually omitted if they know the meaning)

with the actual precipitation reaction. If you write the usual overall equations, a combination of these ions is often also associated. Moreover, the beaker models of Fig. 7.16 and 7.17 show the ion symbols with (aq) symbols, to make clear the hydration shell of 4, 5 or 6 H_2O molecules and to represent them separately in the solution. The H_2O molecules also present in the solutions are not illustrated in these images — they can be added optionally.

Weak acids can be visualized by clearly showing in the drawing that the acid molecules are present in very large numbers and the associated ions are present in equilibrium only to a small fraction corresponding to the low degree of protolysis of about 1% (Fig. 7.17). If it is only abstractly mentioned in class that the weak acids form an equilibrium of molecules and ions and that the degree of protolysis is very small, then only a few students understand the matter. With the beaker model, they can imagine something; they develop a *mental model* of strong and weak acids in this way.

Neutralization is also an abstract matter for many students if it is argued exclusively with reaction equations: They associate the reaction symbol HCl + NaOH → NaCl + H_2O with the formation of the salt sodium chloride. With the beaker model (Fig. 7.18), they not only see the central reaction of hydronium ions with hydroxide ions to water molecules, but also that the sodium and chloride ions remain unchanged and have nothing to do with neutralization: They are "spectator ions". They also understand that the number of ions remains the same, that formally hydronium ions are replaced by sodium ions. This helps in discussions of electrical conductivity during neutralization: The well-conducting hydronium ions are replaced by the sodium ions, which contribute less to conductivity, the commonly depicted conductivity curve drops until the equivalence point, before it rises again.

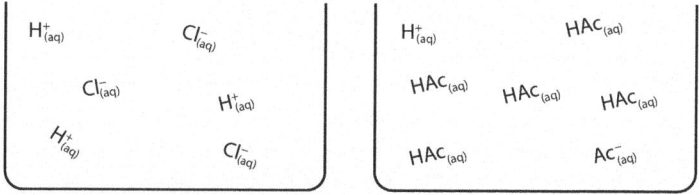

Fig. 7.17 Strong and weak acids in the beaker model

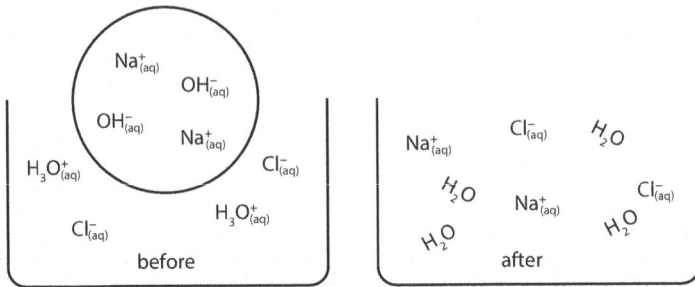

Fig. 7.18 The neutralization of hydrochloric acid and sodium hydroxide in the beaker model

Hanna Klitzke [27] found in an empirical study with some tasks on beaker models that a large part of chemistry education students still use molecule symbols instead of the expected ion symbols. With a "squiggly line", the surface of a solution is also marked — and thus the water of the solution is not symbolized with H_2O molecules, but understood as a continuum at the macro level (Fig. 7.9). The usual neutralization reaction has been symbolized by a third of the subjects with the formation of H_2O molecules, but erroneously also with the formation of "NaCl" [27]. In general, there are difficulties with solid salts: They are described with molecule symbols like NaCl and $BaSO_4$ and not correctly with an ion lattice.

For these reasons, the beaker models should be incorporated into both university teaching and school chemistry lessons: The sole reference to "ion lattices" and "hydrated ions in solutions" seems not to be understood — only the drawing of these models develops an understanding of the existence of ions in solid salt crystals as well as of hydrated ions in solutions of acids, bases and salts.

7.2.1.3 Molecular Models

As soon as gases are to be interpreted structurally, the structure of corresponding molecules is to be illustrated by molecular models. Since the valences of nonmetal atoms in kits are usually predetermined by snap buttons or rods of the model spheres, it is relatively easy for learners to design such molecular models independently and to become familiar with the valences of the different types of atoms over time (see also Fig. 7.5 and [28]).

The rods from sphere to sphere in the molecular models can be seen by the learners as a bond between the respective atoms and can be referred to as a *directed bond* between the atoms. Subsequently, it is also possible to interpret the cohesion of metal atoms in metals or ions in salts as an *undirected bond* and to distinguish it as a different type of bond from that in molecules [15]. One can also introduce the terms *finite* atomic associations for molecules, and speak of *infinite* associations of atoms or ions in metal and salt crystals.

If at least two different molecular construction kits are available and two models are built for the same molecule, one avoids the *imprinting on a single model* with all irrelevant ingredients such as material and colors — the learners thus only have to find out the structure of the molecule as the common feature of both models and recognize it as a characteristic of the image. The same applies retrospectively also for sphere packing: If two models with spheres of different materials and colors are used for the same metal structure, there is no one-sided imprinting on materials or colors. Finally, it should be pointed out that dealing with these spatial structure models trains the *spatial imagination* of young people [29] — an important cognitive ability that is needed far beyond the understanding of chemistry in other subjects as well as in profession and everyday life! Stereo pictures [5] may also be helpful.

Furthermore, for a good model understanding, it is appropriate to point out the ability of atoms in molecules to vibrate (Fig. 7.19): The CO_2 molecules have become so prominent because they cause the increased greenhouse effect of our earth, which goes beyond the natural one, due to too high concentrations in the air. Due to the specific molecular structure, the molecule absorbs infrared radiation that is reflected back into space by the sunlit earth. The valence and deformation vibrations of the molecules store this energy and lead to the warming of the atmosphere. This may then lead to drastic climate changes on earth that are no longer reversible — this should also be discussed with young people, because they are the ones who may suffer from climate change later on (Chap. 9).

7.2.1.4 Volume Models

Avogadro's law states that for gases, a specific volume results for a certain number of molecules — regardless of their type — under standard conditions, at 0 °C and standard pressure the volume V = 22.4 L applies for 1 mol of molecules, for room temperature and normal pressure one can approximately calculate with V = 24 L. If this relationship is not to remain so abstract and misunderstood, volume models are also common, into which a concrete, albeit very small constant number of molecular models is drawn (Fig. 7.20).

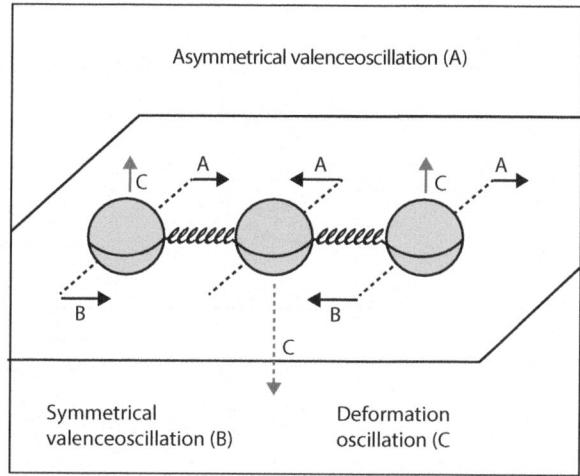

Fig. 7.19 Model ideas of the ability of atoms in molecules to vibrate

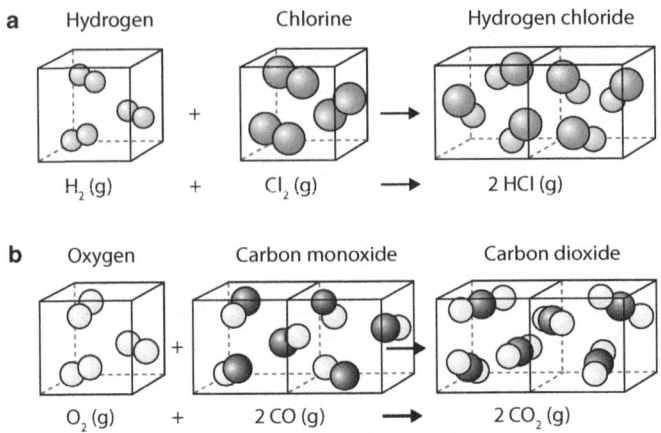

Fig. 7.20 Volume models for gases and their interpretation with Avogadro's law [21]

7.2.2 Adaptation and Extension of Models in Chemistry Education

Chemistry education usually begins by conveying the simple particle model, with which a smallest particle is agreed upon for each pure substance: for the substance copper the copper particle, for water the water particle, for sugar the sugar particle.

The assignment of carbon particles poses difficulties: as substances, there are both *diamond* and *graphite*. However, there are no specific graphite particles and different diamond particles. Carbon particles are present in both substances, they construct the diamond lattice in a specific way, and the graphite lattice in a different spatial arrangement (Fig. 7.21). Accordingly, carbon particles are neither colorless nor black: colors are properties of substances, colors are *not* particle properties — they are *irrelevant ingredients*!

It is shown in Table 7.1 that the concept of the model in the mediation process is not defined once and for all time, but changes *spiralcurricularly* depending on purpose and level of knowledge. This conveys to the learners in chemistry lessons

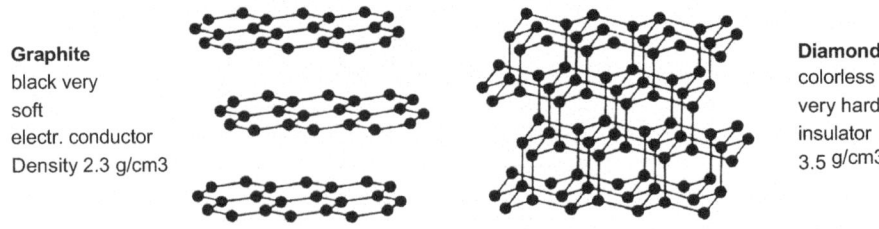

Fig. 7.21 Arrangement of carbon atoms in the diamond lattice and in the graphite lattice [21]

7.2 Teaching Processes: Models and their Didactic Functions

that models and model ideas are to be expanded according to new insights and knowledge — as was historically the case.

Not all models listed in Tab. 7.1 need to be addressed and their limits discussed, but for each introductory lesson, it must be decided anew when, after discussing some phenomena in chemistry lessons, the first model, possibly the *simple particle model*, should be introduced with which examples and experimental demonstrations and consistently applied for a certain period. Afterwards, the decision must be made when the transition to the *Dalton's atomic model* takes place and models of atoms and molecules are conveyed, and whether the introduction of atoms and molecules also involves an *extended Dalton's atomic model* with the treatment of ions planned to develop appropriate model ideas for salts and salt solutions (Chap. 8). Finally, the *nuclear-shell model* of the atom and the *shell model of the electron shell* must be conveyed.

It can also be decided to ignore the simple particle model and start immediately with the Daltonian atomic model — in this case, the learners do not have to work out a model change after a short teaching time. Conversely, the Dalton model cannot be ignored.

Table 7.1 Models and model ideas, their possibilities of use and limits

Particle model (Particle model)	**Interpretation:** States of aggregation, state changes, kinetic gas theory, diffusion, dissolution processes, chemical reaction without change of particle type (for example in the formation of alloys) **Limits:** ↓
Dalton model (Mass model, mass-charge model)	Atom, atomic mass, element, compound, periodic table, chemical reaction, law of conservation of mass in chemical reactions, rearrangement of atoms and ions, ion, ion symbol, ionic lattice, non-directional bond Molecule, molecule symbol, molecule structure, directional bond, sum and structure symbol, reaction symbol etc. **Limits:** ↓
Nucleus-shell model (Rutherford model)	Nucleus, protons, neutrons, radioactivity, atomic shell, fast moving electrons, electron clouds, electrolysis, metal-nonmetal reaction and electron transfer **Limits:** ↓
Shell model of the atomic shell (Bohr's atomic model)	Periodic table and octet rule, chemical bond, ionic bond, electron pair bond, acid-base reaction and proton transfer, redox reaction and electron transfer, complex reaction and ligand transfer **Limits:** ↓
Electron pair repulsion model (VSEPR)	Electron cloud, bond angle, bond length, spatial structure of molecules etc. **Limits:** ↓
Orbital model (wave mechanical model)	Orbital, hybridization, structure of the benzene molecule, prediction of lattice or molecule structures etc. **Limits:** Insights in the future

7.2.3 Further Functions of Models and Model Ideas

Beyond the mediation functions of the models for understanding the structure of matter and thus for understanding chemical facts, there are further functions.

7.2.3.1 Reduction of Anthropomorphic Ideas in Learners

In Chap. 2, preliminary ideas such as "sun rays remove the puddle" or "acids eat up metals" have been described on the topic of "student ideas". These statements should be discussed and questioned through the use of models and model ideas: The puddle disappears because the water evaporates and the water particles mix with the air particles, acids react with metals and Mg^{2+}(aq) ions and H_2 molecules are formed from Mg atoms and H^+(aq) ions through electron transfer, a salt solution and hydrogen are produced. Despite initial difficulties with the model concept, learners will gradually be able to replace their original formulations with newly acquired model ideas. For this, illustrative models of the structure of matter must be used and corresponding facts made understandable. In this way, increasingly accurate mental models develop in consciousness.

7.2.3.2 Reduction of Complex Relationships

The didactic reduction of difficult relationships is an essential task of teachers: Wherever possible, complex relationships are tried to be reduced so that the intended statement remains factually appropriate. If the element and compound definition — as happened in the 1950s — is reduced to the model idea that elements are made up of atoms, compounds of molecules, this reduction is factually incorrect and later to be corrected. If the complex structure of atoms is reduced to the model idea of valencies and one speaks of the valency 4 of the C atom or the valency 1 of the H atom, this reduction is justifiable and applicable to the composition and structure of the methane molecule. This idea is not wrong and later not to be corrected — but it can be extended in later lessons to the shell model of the electron shell and thus to bonding electron pairs.

7.2.3.3 Generalization of Facts

Starting from the use of a model, students can often generalize an overarching fact. If, for example, the CH_4 tetrahedron is presented as a model for the methane molecule and the extension to the ethane molecule is demonstrated with the molecule construction kit, the students can develop the molecule models of all further homologues and isomers of the alkanes through a generalization. They train spatial ideas of the structure of these molecules and are able to derive common structure symbols or semi-structure symbols or to see these symbols spatially and to build the spatially correct structures with the help of the molecule construction kit.

7.2.3.4 Illustration of Reactions

If a chemical reaction is to be made clear, it is best to demonstrate structure models of the substances before and after the reaction to the learners (Chap. 8). For example, the formation of an ester can be made clear by presenting the structures

7.2 Teaching Processes: Models and their Didactic Functions

of the involved carboxylic acid and alcohol molecules before the reaction through molecular models. After the synthesis and detection of the ester, the splitting off of water molecules can then be made clear on the model, thus making the reaction symbol with semi-structural formulas understandable. In other words: No chemist today will describe the formation of ethyl acetate with pure sum symbols:

$$C_2H_6O_1 + C_2H_4O_2 \rightarrow C_2H_8O_2 + H_2O$$

For reactions in organic chemistry, it is quickly agreed: The molecular structures before and after the reaction are to be taught and possibly semi-structural symbols are to be derived from them — otherwise, sum formulas of the specified type would have to be memorized and reproduced without understanding the chemistry. But note: In history (Liebig) they have been very important!

For solid-state reactions of inorganic chemistry, it is often thought that symbols like Na, Zn or Al for corresponding metals, with NaCl, ZnS or Al_2O_3 for these salts are sufficient — also in this respect, the information-poor sum symbols are not suitable for understanding corresponding reactions. Model ideas of the structure of metals from metal atoms and of the structure of salts from ions must also be added; for salts, it would generally be advantageous to always also indicate the ion symbols with the ion charges in formulas: Na^+Cl^-, $Zn^{2+}S^{2-}$, $(Al^{3+})_2(O^{2-})_3$ (see also [30] and [31] and Chap. 8).

7.2.3.5 Correlation of Empirical Facts and Mathematical Terms

Descriptions of many facts are based on non-illustrative mathematical terms. For example, the student in the exam conversation often gives the lexical definition of the negative logarithm of the H^+ ion concentration (sometimes also the "H^+ ion constellation" [32]), but can rarely apply the definition spontaneously. If you ask about the ion concentrations in the lime water of pH 12, it takes a while until the correct value $c(H^+) = 10^{-12}$ mol/L is derived, from this the value $c(OH^-) = 10^{-2}$ mol/L is concluded and possibly also the concentration $c(Ca^{2+}) = 0.5 \cdot 10^{-2}$ mol/L is given.

So it initially seems more important to set up a dilution series of a 10^{-1} molar hydrochloric acid with pH 1 and to show that each dilution 1 : 10 reduces the concentration of the H^+ ions by one power of ten, namely to the concentrations 10^{-2}, 10^{-3} and 10^{-4} mol/L, and accordingly the pH values 2, 3 and 4 result. If you repeat the dilution series of a 10^{-1} molar sodium hydroxide solution with the pH value 13 and measure the pH values of the diluted solutions, you get the values 12, 11 and 10 (see V6.15, Chap. 6). In this way, the learner acquires a better understanding of pH values than through the pure logarithm definition alone.

7.2.3.6 Illustration and Simulation of Chemical Engineering Processes

For example, if a process for the production of plastics is to be discussed exemplarily, an experiment can take over the function of a model. For example, the experimental production of a nylon thread from the basic substances adipic acid and hexamethylenediamine is shown by pulling the forming polyamide out of the beaker in the form of a thread after overlaying both substances. This experiment then serves as a model for the production of nylon in chemical engineering.

7.2.3.7 Formulating Forecasts and Hypotheses

Model concepts allow the prediction of properties and reactions. For example, if the structures of alkane and alkanol homologues are known in class, predictions about the solubility of these homologues can be made using corresponding structural concepts: alcohols, whose molecules have short C-chains, are not soluble in gasoline, but in water; alcohols with molecules of very long C-chains are not soluble in water, but in gasoline. Corresponding experiments are planned and used to test these hypotheses.

The history of natural sciences also offers many examples. For instance, Watson had initially cut out the shape of the base molecules from cardboard and later from zinc sheet metal for the prediction of the base combinations in the suspected DNA double helix [33]:

> The metallic purine and pyrimidine models I needed were not ready in time. So I spent the rest of the afternoon cutting out accurate models of the bases from thick cardboard. I began to slide the bases back and forth and arrange them in pairs in a different, also possible way. Suddenly, I realized that an adenine-thymine pair held together by two hydrogen bonds had the same shape as a guanine-cytosine pair.

These sentences hint at the great importance the very simple cardboard models had for the insights into the structure and function of nucleic acids. Based on these model concepts, in combination with X-Ray diffraction experiments by other scientists, Watson and Crick were able to build a complete structural model of DNA and formulate forecasts or hypotheses that were in line with all known facts and provided new imaginations, such as accurate model concepts for the reduplication of DNA and the processes leading to the origin of life. The essential steps of these insights are presented in detail in Chapter 25.

7.3 Learners: Experiences with Models

Students come to chemistry class in several ways with their model experiences: They own toy models such as Barbie dolls, car or ship models, they have fun with these models due to their toy character, and they also know models and model concepts from other school subjects.

7.3.1 Toys

Children are interested in comparing their Barbie doll with themselves or another person. They find out that many properties match between the original and the model: for example, the position and shape of the mouth, nose, eyes, and ears. However, they also recognize many functions that the doll model does not show, such as for eating or inhaling and exhaling air.

These experiences are contrary to the understanding of scientific-chemical models. If the young student takes a molecule model in hand, he naturally cannot

7.3 Learners: Experiences with Models

compare it with the original, he is not able to count the number of atoms bound in the molecule on the original molecule. Therefore, it should be made clear to the learners that all information about the structure of molecules comes from laboratory experiences. Historically, elemental analysis was developed to determine the composition from the masses or volumes of the reactants and products (see V6.17, Chap. 6), today the devices of instrumental analysis are available to measure even more accurately bond distances and bond angles in molecules, constants in metal lattices or ion lattices. If you present the results of such analyses to the learners by building or having them build molecule or lattice models, they can develop accurate mental models in consciousness based on the models and understand chemistry.

7.3.2 Fun with Models

The affective component in dealing with models, which is always positively developed in children, can be effectively transferred to working with models of the structure of matter if it is linked to the action-oriented introduction to the scientific model world — for example, with the construction of dense sphere packings or some molecule models with the molecule construction kit.

If you show students the sphere packing for the sodium chloride structure and ask them to rebuild the model with white cellulose balls (Ø 30 mm) and red balls (Ø 12 mm), they do this with great fun and proudly show this model at home. They may even discuss and explain the model with parents and siblings, explaining what it is a model of: With the consciousness of "experts", they pass on chemical facts to other people! The mentioned cellulose balls can be purchased cheaply from Richter [34], instructions for a model construction practical can be found in section 7.6.

Even older students and teachers or even teachers at teacher training conferences show a very positive affection for structure models of all kinds: They like to convince themselves of the number and type of isomers in alkane molecules by assembling corresponding molecule models using construction kits. If seasoned teachers are provided with the building material in the form of foam or plexiglass balls, they also like to assemble the complex models of elementary cells with all the half, quarter, and eighth balls [35]. The fun of building structure models seems to have no age limit!

7.3.3 Models from Other School Subjects

Students often bring rich experiences on the topic of "models" from other school subjects. Some school subjects are listed.

7.3.3.1 Biology
School collections usually contain models for the eye, for the ear, or for the human skeleton (the skeleton in biology collections is rarely an original). In these cases, the model character, abbreviations, and irrelevant ingredients of the models are

very evident. They are very vivid and — because of their relevance to the individual — very motivating models that are well suited for discussing the concept of a model. However, since this model concept is based on the direct comparison of the original and the model, it does not apply to chemical facts. As soon as the DNA structure comes into play and the function of the base molecules for the construction of the double helix is to be explained, one is also in the field of abstract chemical model ideas in biology.

From biology, students may also know the model of population dynamics, within which they can discuss "predator-prey relationships" using the well-known example of periodically changing numbers of foxes and mice in the forest. With such models, they learn — in contrast to *static* models — also examples of *dynamic* models and can possibly relate them to the dynamic model of chemical equilibrium.

7.3.3.2 Geography
Maps are also easily understandable models for learners, for example those for their hometown or for well-known hiking trails. The globe is also comparable to the original as a model in the age of space travel when the earth is shown as a photo or film from the space capsule (before the age of space travel, the globe was a model that could not be directly and completely compared to the original).

Models for the nickel-iron core in the earth's interior, on the other hand, are those that cannot be obtained directly through sensory comparison. They are derived from experiments on the earth's surface with measurements of the electric fields around the earth. This approach corresponds to the scientific process of knowledge and provides desired model ideas of the earth's interior based on this.

For several decades, dynamic models for the development of climate change have been added. For example, from information about ice cores from the polar ice, the climate conditions centuries ago can be traced back and predictions for the future can be derived. These model ideas are currently the basis for politicians around the world to reduce the combustion of fossil fuels worldwide in order to stabilize the carbon dioxide content in the atmosphere and not let it increase further.

7.3.3.3 Mathematics
Geometric drawings can be understood as models. For example, there are models for right-angled, other models for isosceles triangles: They represent the imagined triangles. If the students even build spatial models of cubes, cuboids, octahedrons or tetrahedrons using provided nets made of cardboard and describe them mathematically, good conditions are provided to effectively use these models also for chemistry lessons and to successfully convey the structure models of cubic symmetry based on them.

If the students finally draw the spatial models in perspective in mathematics lessons, these skills can not only be taken up and further developed in chemistry lessons when drawing structure models, but the spatial imagination of the students is also trained and improved [29]. The promotion of this ability seems to be important not only in mathematics or chemistry lessons for a good understanding, a trained spatial imagination is also an important prerequisite for learning many professions!

7.4 Social Reference Fields: Interdisciplinary Model Ideas

Working with models has a pronounced interdisciplinary character and represents a contribution to general education that is hardly to be underestimated: In addition to the insight into the high importance of the model concept specifically in chemistry and beyond in mathematics and all natural sciences, the significance of models in many other areas can also be reflected, for example in the

Industry:	Material cycles or composite systems
Economy:	Models for money or goods cycles
Sociology:	Behavior patterns of certain groups of people
Politics:	Voting behavior of interest groups
Ecology:	Cycles of substances in nature and ecological systems
Psychology:	Models of human information processing

Accordingly, many experiences with concrete and mental models should be collected in class, so that the different meanings and uses of models for many other school subjects and areas of one's own life world can be understood — an interdisciplinary contribution of chemistry lessons!

7.5 Exercise Tasks

A7.1
The closest cubic sphere packing can also be described by octahedra and tetrahedra in a ratio of 1:2 or by arranging elementary cells in all three spatial directions. Create corresponding solids using the nets (Fig. 7.22) and demonstrate the space filling arrangement in each case.

A7.2
In the school collection, you will find a NaCl sphere packing, a NaCl lattice, and a NaCl elementary cell. Identify the features of the models and list the main irrelevant ingredients for these models.

A7.3
The particle model is of particular importance for introductory teaching. Explain five observations from nature and the laboratory that can be accurately explained with this model. What chemical phenomena show the limits of this model?

A7.4
In typical chemistry lessons, the particle model, Dalton's atomic model, and the shell model of the atom are introduced one after the other as model concepts. Choose (a) a substance and (b) a chemical reaction and create model sketches based on all three models. Discuss the differences.

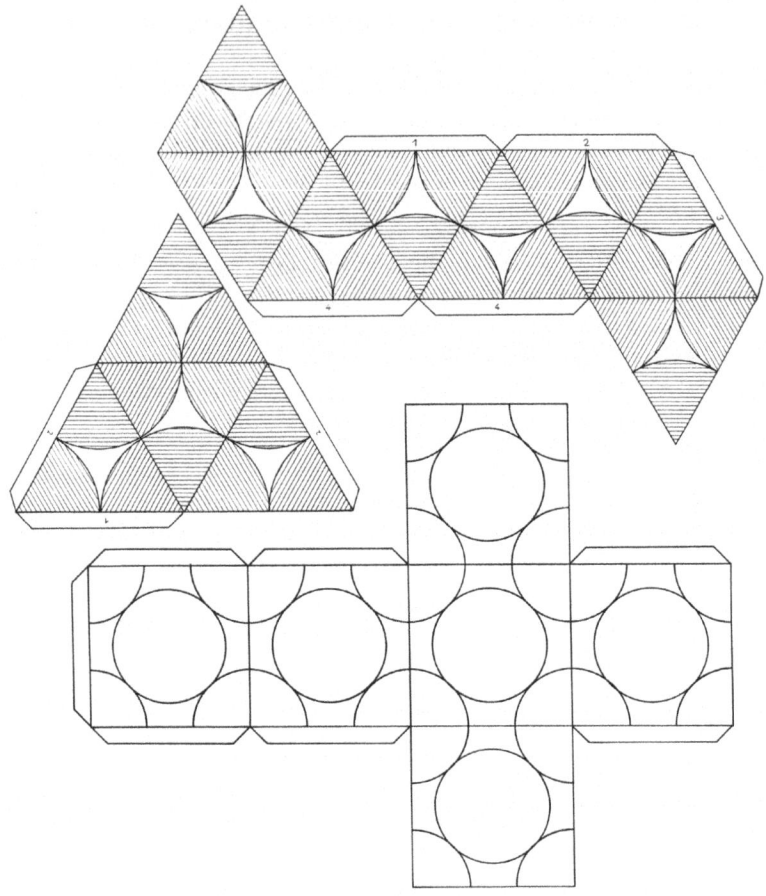

Fig. 7.22 Nets of octahedra, tetrahedra, and elementary cells as templates for constructing spatial sections from the closest cubic sphere packing

A7.5
Chemical equilibrium can be illustrated by model experiments, but also by experiences from everyday life. Give an example in each case and establish connections to chemical equilibrium.

7.6 Model Building Workshop: Structures of Metals and Salts

The workshop is aimed at both students in introductory classes who are familiar with the simple particle model and want to illustrate the structure of metals using sphere packing models with the help of their teachers, where one sphere corresponds to one metal particle. On the other hand, in advanced teaching, the concept of ions based on the extended Dalton's atomic model is necessary to model the structure of various salt crystals from ions. In section 7.6.1, the workshop tasks are set, and in section 7.6.2, solutions to the tasks are offered.

7.6.1 Tasks and Construction Instructions

Material
100 balls $d = 30$ mm [34], 50 balls $d = 12$ mm [34], triangular wooden frame ($a = 17.5$ cm), square wooden frame ($a = 15$ cm), modeling clay, adhesive, two equilateral ball triangles, each glued together from six balls with $d = 30$ mm

Structures of Metals
To describe the structure of metal crystals from smallest particles with structure models, closest sphere packings are suitable (1 metal particle $\equiv$ 1 sphere):

M7.1
Fill the triangular wooden frame tightly with a layer of balls in a triangular pattern. Pack as many ball layers on top as possible. Draw the ball layers.

M7.2
The coordination number refers to the number of balls that touch a ball inside the packing. Determine the coordination number in the closest sphere packing! Draw three ball layers in such a way that this number is recognizable.

M7.3
There are two types of closest sphere packings with the coordination number 12 possible:

a. in the layer sequence ABCABC …
b. in the layer sequence ABAB …

Build both packings! Draw the ball layers with a triangular pattern in such a way that (a) and (b) become clear.

▶ **Definition** A *layer sequence ABCA* … is present when the 4th layer of spheres is congruent with the 1st layer when viewed from above. The *layer sequence ABA* … is present when the 3rd layer of spheres is already congruent with the 1st layer (the layers in the triangular pattern are always meant!).

▶ **Information** In the ABCA sphere packing, a cubic section can be found as the elementary body. Therefore, this packing is also called cubic closest sphere packing (lat. *cubus*: cube).

M7.4
Next to the depicted sphere packing, which consists of 14 spheres, draw the corresponding lattice model by representing a cube in perspective, indicating only the centers of the spheres instead of the whole spheres, and connecting these points.

M7.5
Glue the depicted *elementary cube*, which consists of 14 spheres, together using the two sphere triangles and two additional spheres. Try to incorporate it into the cubic closest packing starting from the triangle pattern (see M7.1).

Draw two possibilities for the construction of the elementary cube from 14 spheres:

a. Linking of layers in the triangle pattern (1 + 6 + 6 + 1),
b. from layers in the square pattern (5 + 4 + 5).

M7.6
Take the square wooden frame, create the cubic closest packing starting from the square pattern, and also build the elementary cube into it. Determine the coordination number. Draw sphere layers in such a way that this number is recognizable.

> **Information**
> The following models show the construction of metal crystals from smallest particles:
>
> 1. The hexagonal closest sphere packing with the layer sequence ABA: It represents the way crystals of magnesium, zinc, etc. are built from their smallest particles. It is also said that they form crystals of hexagonal symmetry or crystals of the Mg type.
> 2. The cubic closest sphere packing with the layer sequence ABCA: It represents the way crystals of copper, silver, gold, etc. are built from their particles. It is also said that they form crystals of cubic symmetry or crystals of the Cu type. The elementary cube has a sphere in each face center, so it is called *cubic face-centered*.
> 3. The name "cubic face-centered" is intended to highlight the difference to the *cubic body-centered* sphere packing: It is no longer the closest packing, the coordination number is 8. Metal crystals of tungsten and the alkali metals realize this structure: W type.

M7.7
The sphere packing of nine spheres shows the elementary cube of the cubic body-centered structure. Next to the packing, draw the lattice model by sketching a cube

7.6 Model Building Workshop: Structures of Metals and Salts

in perspective, indicating only the centers of the spheres instead of the spheres themselves, and connecting these points.

Structures of Salts

The construction of many metal crystals from atoms can be modeled by packings of spheres *of one kind*, the construction of salt crystals from ions requires at least *two kinds* of spheres. The following will first build models for the construction of the *sodium chloride crystal* (table salt), and finally the models of three other salts.

M7.8

The Na^+ ions of sodium chloride are represented by spheres with $d = 12$ mm, the Cl^- ions by spheres with $d = 30$ mm. Using the triangle frame, create a sphere packing as dense as possible with both types of spheres. Draw the sphere layers.

M7.9

Determine the coordination numbers for both types of spheres! Draw sphere layers in such a way that the coordination numbers for both types of spheres are recognizable.

M7.10

In the densest sphere packing, there are two different sizes of gaps. Determine the number of spheres that form the gaps, and draw the gap-forming spheres for both types of gaps (perspectively or in the form of sphere layers): a) large gap, b) small gap.

Information

In the densest sphere packings, two different types of gaps can be found, convince yourself using the model M7.8:

1. The large gaps are formed by 6 spheres in an octahedral arrangement: Octahedral gaps (OL)
2. The small gaps are formed by 4 spheres in a tetrahedral arrangement: Tetrahedral gaps (TL)
3. In the densest sphere packings, the numbers of spheres, OL, and TL are present in the *ratio 1 : 1 : 2*.

The structure of the salt crystal from smallest particles can therefore be described as follows: The Cl^- ions form the cubic densest packing, all octahedral gaps are occupied by smaller Na^+ ions. The coordination is 6/6, the ratio of ions is 1 : 1, the formula is $(Na^+)_1(Cl^-)_1$. or shortened NaCl.

How is the *cubic shape* of the salt crystals explained?

M7.11
Take the *elementary cube* from M7.5, fill the octahedral gaps with smaller spheres.

a. Complete the model drawing (see picture).
b. Draw the space lattice model next to it (see M7.4).

M7.12
Create the cubic densest packing using the square frame and both types of spheres. Draw the sequence of layers.

M7.13
Convince yourself that the elementary cube can be incorporated into the sphere packing starting from the triangle pattern (see M7.8) as well as from the square pattern (see M7.12). What position does it take in each case?

Draw the elementary cube using a) sphere layers in the triangle pattern, b) sphere layers in the square pattern.

M7.14
A model for the aluminum oxide structure:

a. Glue together 3 layers of 15 spheres each (see picture).
b. Attach 10 small spheres to each according to the specified pattern (see picture).
c. Stack the three layers so that the sequence of layers is ABA.

Note that the coordination number of small spheres is 6, that of large spheres is 4. What is the ratio of spheres?

▷ **Information** In *aluminum oxide*, the O^{2-} ions form a hexagonal densest packing, the octahedral gaps are only 2/3 occupied by Al^{3+} ions. The coordination is 6/4, the ratio of ions is 2 : 3. The following formulas are derived: $\{(Al^{3+})_2(O^{2-})_3\}$ or shortened Al_2O_3.

M7.15
Form some small spheres from modeling clay that fit into the tetrahedral gaps of large spheres. Build the corresponding elementary cube with large and small spheres

a. for the zinc blende structure,
b. for the lithium oxide structure.

▷ **Information** *Zinc blende* can be described as a cubic close packing of S^{2-} ions, half of whose tetrahedral voids are occupied by Zn^{2+} ions. The coordination is 4/4, the formula for the unit cell is $\{(Zn^{2+})_4(S^{2-})_4\}$, the empirical formula is ZnS.

7.6 Model Building Workshop: Structures of Metals and Salts

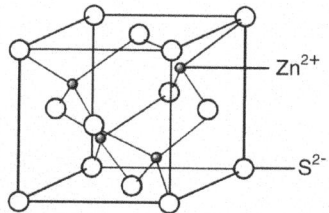

▷ **Information** *Lithium oxide* can be described as a cubic close packing of O^{2-} ions, all of whose tetrahedral voids are occupied by Li^+ ions. The coordination is 4/8, the formula for the unit cell is $\{(Li^+)_8(O^{2-})_4\}$, the empirical formula is Li_2O.

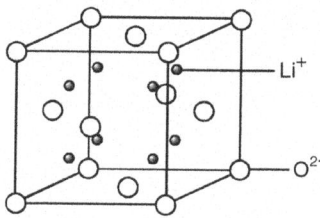

7.6.2 Solutions and Drawings for the Tasks

M7.1

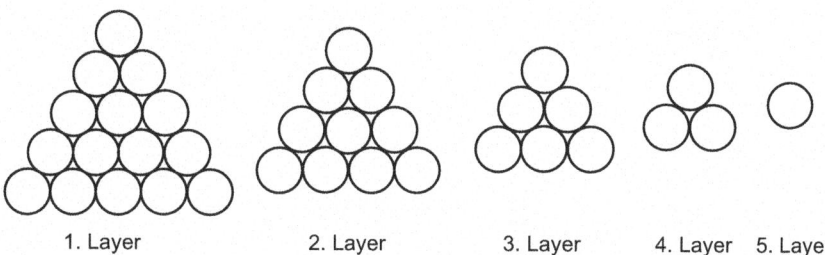

M7.2

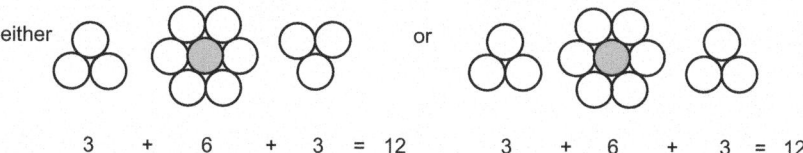

M7.3

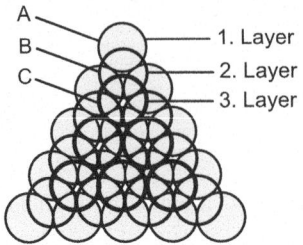 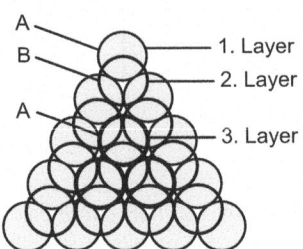

M7.4

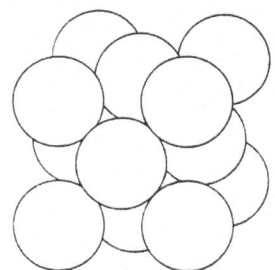 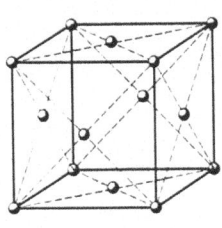

M7.5

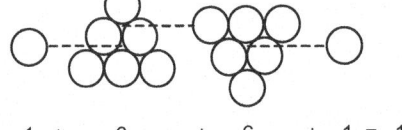 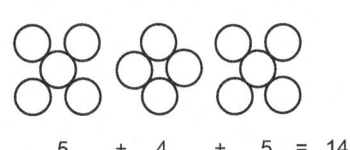

1 + 6 + 6 + 1 = 14   5 + 4 + 5 = 14

M7.6

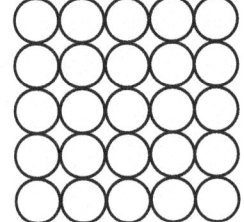 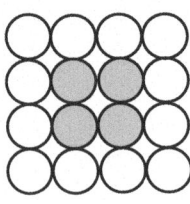 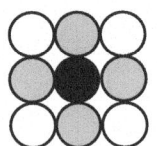 

7.6 Model Building Workshop: Structures of Metals and Salts

M7.7

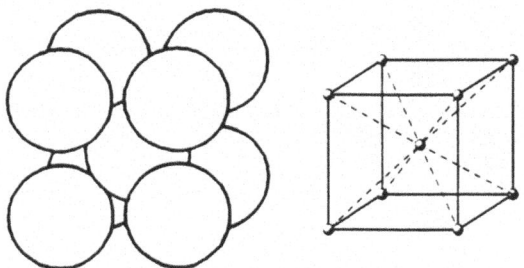

M7.8

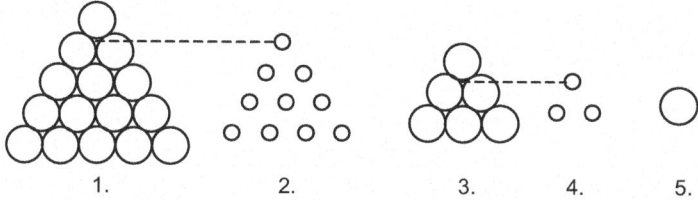

1.   2.   3.   4.   5.

M7.9

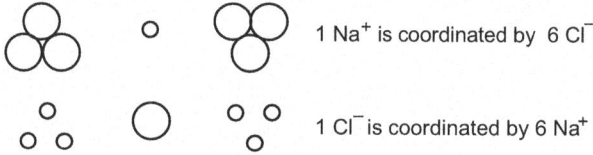

1 Na$^+$ is coordinated by 6 Cl$^-$

1 Cl$^-$ is coordinated by 6 Na$^+$

M7.10

a
6 Balls:
Octahedron
gap

b
4 balls:
Tetrahedron
gap

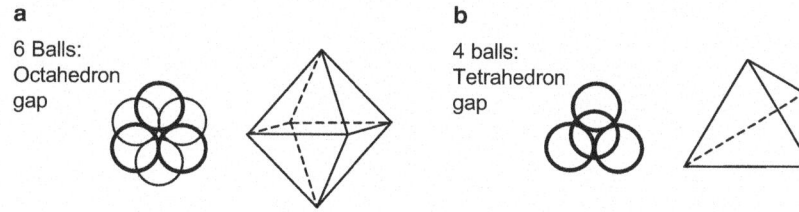

M7.11

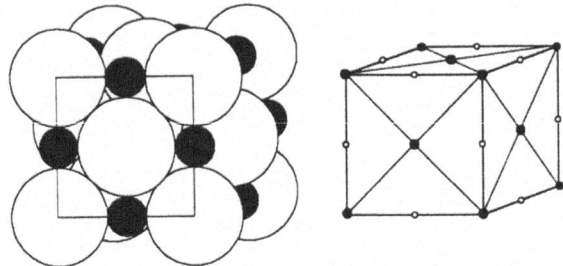

M7.12

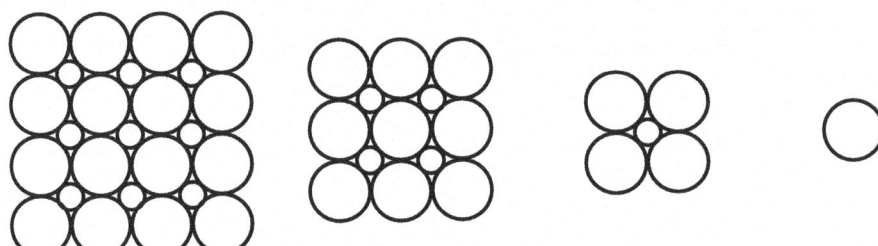

7.13

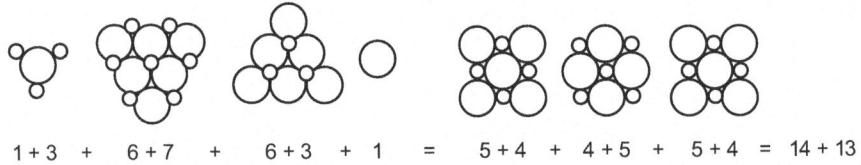

1 + 3   +   6 + 7   +   6 + 3 + 1   =   5 + 4   +   4 + 5   +   5 + 4 = 14 + 13

References

1. Häusler K (1998) Highlights in der Chemie. Aulis, Köln
2. Stachowiak H (1965) Gedanken zu einer allgemeinen Theorie der Modelle. Studium Gen 18:432
3. Steinbuch K (1977) Denken in Modellen. In: Schäfer G et al (Hrsg) Denken in Modellen. Westermann, Braunschweig
4. Kircher E (1977) Einige erkenntnistheoretische und wissenschaftstheoretische Auffassungen zur Fachdidaktik. Chimdid 3:61
5. Harsch G. u. Schmidt R. (1981) Kristallgeometrie. Packungen und Symmetrie in Stereodarstellungen. Diesterweg, Frankfurt
6. Geomix Modelle: Ratec, Körberstr. 15, 60433 Frankfurt, Germany

7. Leybold Didactic: Postfach 1365, 50330 Hürth, Leybold, Germany
8. Institute for Chemical Education of the University of Madison: 1101 University Av., Madison WI, 53707, USA, University of Madison, USA
9. Buck P (1994) Die Teilchenvorstellung – ein Unmodell. ChemSch 41:412
10. Buck P (1994) Wie kann man die „Andersartigkeit der Atome" lehren? ChemSch 41:11
11. Bünder W, Demuth R, Parchmann I (2003) Themenheft Basiskonzepte. PdN-ChiS 52(1):1
12. Parchmann I (2007) Basiskonzepte – ein geeignetes Strukturierungselement für den Chemieunterricht? NiU-Chemie 18(100):6
13. Pfeiffer P (2007) Struktur-Eigenschafts-Konzept – chemische Zusammenhänge erschließen, verstehen und anwenden. NiU-Chemie 18:36
14. Herdt C (2016) Strukturorientierung – ein Schlüsselkonzept. PdN-ChiS 65(5):44
15. Sauermann D, Barke H-D (1998) Strukturchemie und Teilchensystematik. Chemie für Quereinsteiger, Bd. 1. Schüling, Münster (Siehe auch im Internet: wikichemie. Zugegriffen: Mai 2018)
16. Johnstone AH (2000) Teaching of chemistry – logical or psychological? CERAPIE 1:9
17. Gabel D (1999) Improving teaching and learning through chemistry education research: a look to the future. J Chem Ed 76:548
18. Harsch G, Heimann R (1998) Didaktik der Organischen Chemie nach dem PIN-Konzept. Vom Ordnen der Phänomene zum vernetzten Denken. Vieweg, Braunschweig
19. Harsch G, Heimann R, Benmokhtar S, Wagner A (2014) Das START-Konzept. Teilchenmodelle und Formelsprache im Chemieanfangsunterricht. Aulis, Stark, Hallbergmoos
20. Haupt P, Moritz P (2008) Modelle chemischer Substanzen für den Anfangsunterricht. Aulis, Köln
21. Asselborn W et al (2010) Chemie heute. Schroedel, Braunschweig
22. Marohn A (2008) Schülervorstellungen zum Lösen und Sieden. MNU 61:451
23. Sauermann D, Barke H-D (1998) Struktur der Metalle und Legierungen. Chemie für Quereinsteiger, Bd. 2. Schüling, Münster (Siehe auch im Internet: wikichemie. Zugegriffen: Mai 2018)
24. Sauermann D, Barke H-D (1998) Ionenkristalle mit einfachen Gitterbausteinen. Chemie für Quereinsteiger, Bd. 4 (Siehe auch im Internet: wikichemie. Zugegriffen: Mai 2018)
25. Barke H-D (2006) Chemiedidaktik – Diagnose und Korrektur von Schülervorstellungen. Springer, Heidelberg
26. Barke H-D, Harsch N (2015) Strukturmodelle bauen – Ionengitter verstehen. PdN-ChiS 5(64):28
27. Klitzke, H.: Becherglasmodelle im Chemieunterricht – eine empirische Erhebung zu Modellvorstellungen zum Lösen und Ausfällen von Ionenverbindungen. Bachelorarbeit. Münster 2016
28. Sauermann D, Barke H-D (1998) Moleküle und Molekülstrukturen. Chemie für Quereinsteiger, Bd. 3 (Siehe auch im Internet: wikichemie. Zugegriffen: Mai 2018)
29. Barke H-D (2001) Raumvorstellung zur Struktur von Teilchenverbänden. In: Chemiedidaktik Heute. Lernprozesse in Theorie und Praxis. Springer, Heidelberg
30. Barke H-D (1980) Die Unverzichtbarkeit der Strukturmodelle für das Verständnis der chemischen Reaktion. PdN-Ch 29:372
31. Barke H-D, Wirbs H (2000) Chemische Symbole für kleinste Struktureinheiten. PdN-Ch 49(2):2
32. Heimann R (2002) Struktur-Eigenschafts-Beziehung organischer Stoffe im Chemieunterricht der Sekundarstufe I. PdN-CidS 51:36
33. Watson JD (1969) Die Doppel-Helix. Rowohlt, Hamburg
34. Firma Richter, August-Bebel-Weg 11, 09514 Lengenfeld (Erzgebirge), Tel. 037367 2449, E-Mail: RichterOtto@aol.com
35. Barke H-D (1996) Elementarzelle – Stiefkind neben dem Molekül? Chemkon 3(1):19

Scientific Language and Symbols

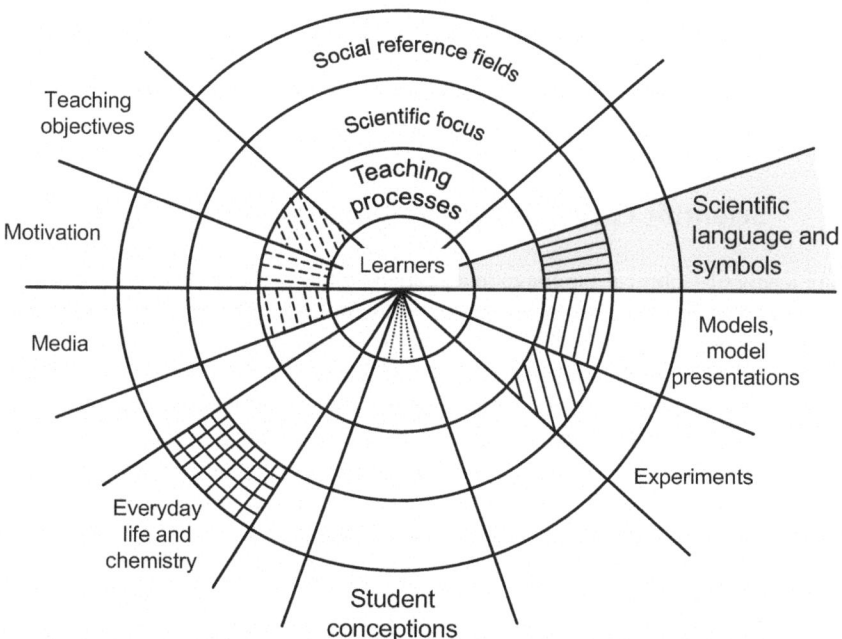

Formulas are the scholarly Latin of chemistry. Without them, communication across national borders would be unthinkable, and the representation of chemical processes would be as cumbersome as if we had to write our CV in cuneiform [1].

Peter von Zahn [1] emphasizes the great importance of chemical symbols, characterizing them as a unique means of communication for chemists—regardless of whether they work in Europe or America, in China or Japan. In this sense, it is also a goal of chemistry education to introduce young people to this symbolic language and to enable them to read and understand the achievements of natural sciences and technology in magazines. On the other hand, it is rightly demanded that scientists make their methods and findings understandable at the level of the educated layman:

> Researchers and inventors cannot live isolated in an ivory tower. They need a sounding board. They need a broad foundation of approval from those who ultimately benefit from technology [1].

Many technical terms are known in everyday language, but also terms of chemistry are part of everyday language, such as the terms substance, metal, acid, alkali, gas, combustion, etc. The meaning of such terms in scientific language is often different or more extensive, always more precise than in everyday language. In order to be able to speak appropriately with the learners about chemical facts, to formulate observations and to discuss explanations that are understood in the same way by all participants, the introduction to the use of scientific language and symbolism is essential, starting from everyday language. Which didactic problems need to be considered and which solutions are available will be reflected in the following.

8.1 Scientific Focus: Terms, Symbols, Quantities, Units

Many terms of technical language and nomenclature , of quantities and units have grown historically and have been redefined again and again by standard committees such as the *International Union of Pure and Applied Chemistry* (IUPAC) to achieve further precision and standardization. These problems of chemical nomenclature and historical changes in the meaning of terms, symbols, quantities and units are discussed for better understanding. In particular, it should be emphasized that chemical formulas are not abbreviations of substance names, but fundamentally inform about the quantitative composition and chemical structure of pure substances.

8.1.1 Système Internationale and derived units

The historical change of quantities and units is to be illustrated by three examples.

Length
The meter as a unit of length was defined in the eighteenth century as the 10,000,000th part of the earth quadrant (1/4 of the earth's circumference, current average value: 10,000.09 km). In 1889, the meter was more precisely defined by two marking lines on a platinum-iridium rod deposited in Sèvres near Paris, which then served as the "original meter" for the calibration of measuring devices in many European countries. In order to have an even more precise definition

available using new methods, the Physikalisch-Technische Bundesanstalt in Braunschweig (PTB) determined in 1960 as a meter the 1,650,763.73 times the wavelength of the orange-red radiation, which under specific conditions emanates from the isotope krypton-86.

Time

The mean solar second corresponded before 1964 to the well-known part of a day or a complete rotation of the earth around its axis. Due to the beginning of space travel, a greater accuracy was needed and the PTB defined the second based on the absorption of vibrations in intra-atomic processes of the isotope Cs-133: The time for 9,192,631,770 vibrations of this cesium isotope was set as the base unit for 1 s.

Thermal Energy

The original unit for thermal energy was the calorie. 1 cal describes the amount of energy that 1 g of water absorbs to heat up by 1 °C, more precisely: the amount of energy that must be supplied to heat from 14.5 °C to 15.5 °C. In order to be able to compare this amount of energy with other energies and to convert it in size calculation, physicists introduced the unit joule (J): $1 \text{ J} = 1 \text{ Nm} = 1 \text{ kg m}^2/\text{s}^2$. 1 J thus corresponds to the energy required to move a body of mass m = 1 kg with acceleration b = 1 m/s² one meter on a horizontal path. After this determination, the conversion for thermal energy is: 1 cal = 4.18 J.

With this redefinition based on the SI units, energies are to be converted, for example potential energy into heat energy: A stone of mass 1 kg, which falls 1 m downwards in the Earth's gravitational field, loses a certain amount of potential energy:

$$E(\text{pot}) = m \, g \, h = 1 \text{ kg} \cdot 9.81 \text{ m/s}^2 \cdot 1 \text{ m} = 9.81 \text{ kg m}^2/\text{s}^2 = 9.81 \text{ J}$$

This amount of energy is equivalent to the heat quantity E = 9.81: 4.18 = 2.35 cal; thus, with this energy, 2.35 g of water can be heated from 14.5 °C to 15.5 °C.

SI Units

The last example once again highlights the peculiarity of the new units: The unit J has been defined by a term of the basic units kg, m and s. Since an international conference in Paris in 1968, these units have been subject to a new definition: the *Système Internationale*. Therefore, the basic units are called **SI** units (Table 8.1). There are many other quantities that can be derived from the SI units or have been redefined on this basis: area, volume, density, pressure, speed, acceleration, concentration, force, energy, enthalpy, entropy, potential, voltage, current, electric charge, etc.

8.1.2 School-Relevant Sizes and Units

Many SI units are familiar to students from everyday life, such as the meter, the kilogram, or the second. However, they struggle with derived units for density or

Tab. 8.1 Basic quantities and SI units, their symbols and conversion factors

Quantity	Symbol	SI unit	Symbol	Definition
Length	l	Meter	m	Wavelength Krypton-86
Mass	m	Kilogram	kg	1 dm^3 Water (4 °C)
Time	t	Second	s	Frequency Caesium-133
Temperature	T	Kelvin	K	0 K = −273.15 °C
Current	I	Ampere	A	0.19 cm^3 Bang gas per 1 s
Amount of substance	n	Mole	mol	Avogadro constant N_A
Luminous intensity	I	Candela	cd	Spec. Amyl acetate flame

The following prefixes are commonly used to indicate multiples or fractions of SI units:

10^9	10^6	10^3	10^2	10^{-1}	10^{-2}	10^{-3}	10^{-6}	10^{-9}	10^{-12}
G	M	k	h	d	c	m	μ	n	p
Giga	Mega	Kilo	Hecto	Deci	Centi	Milli	Micro	Nano	Pico

mass and volume contents. The atomic masses and the resulting unit of the amount of substance, the mole, are completely unknown to them and must be carefully worked out in class and applied in a variety of ways.

Mass and Density

If we consistently base on SI units, all masses should be given in kg and densities in kg/m^3: The density of gold would then be 19,300 kg/m^3 (instead of the previously usual 19.3 g/cm^3). On the one hand, it makes sense to use g and mg for the order of magnitude of substance portions, on the other hand, the usual densities in g/cm^3 or g/mL for solids or liquids and in g/L for gases should be retained. In the literature, the lowercase letters l or ml are often used as symbols for liter or milliliter. However, because the capital letter L is more readable, the letters L or mL are preferred for this book.

A special density unit is the *Öchsle*—it is used by winemakers to measure the sugar content of the must of their grapes and to estimate the future alcohol content. Mr. Ferdinand Öchsle has defined in a simple way: Degrees Oechsle indicates how many grams a liter of must weighs more than a liter of pure water. So if the density of a must is 1080 g/mL, then the "must weight" is 80 °Oe. For berry selections, Oechsle grades up to 300 °Oe can be measured. The density can be determined by immersing a calibrated araeometer, or by measuring the refractive index of the must with a specific refractometer.

Mass and Volume Contents

Initial concentration data is often given in percentages. This includes mass percent (%) and volume percent (Vol.-%). Therefore, a 20% sugar solution contains 20 g of sugar in 100 g of sugar solution, and a 40 Vol.-% liquor contains 40mL of alcohol per 100 mL of liquor. If the densities of the solutions are known, mass contents can be converted into volume contents and vice versa.

Solubility is defined in a different way: it refers to the mass of a substance that just dissolves in 100 g of *water* to form a saturated solution. For example, 36 g of table salt saturates 100 g of water, thus producing 136 g of *solution* (masses can always be added, but not volumes!). If normalized to 100 g of saturated salt solution, the maximum mass content of 26.5% is obtained. These data roughly correspond to the salt solution of the Dead Sea in Israel, which has a density of 1.3 g/L: You can read a newspaper lying on your back in the Dead Sea without any swimming movements.

Atomic Mass

A completely different unit of mass than the kilogram is also common in chemistry: the unit u for atomic mass. The former "relative atomic weights" were dimensionless and were initially determined by comparing with the mass of the H-atom (Dalton), then with the mass of the O-atom (Berzelius) in the first atomic mass tables. In order to be able to include atomic masses in size calculations, the unit u has been defined: $1u = 1/12$ of the mass of the isotope C-12. It is linked with the SI unit mole or with the Avogadro constant ($N_A = 6 \cdot 10^{23}$) and with the unit g: $1g = N_A \cdot 1u$. Accordingly, $6 \cdot 10^{23}$ H-atoms or 1 mol H-atoms have the mass of 1g, an H-atom then weighs $1u = 1g / (6 \cdot 10^{23}) = 0.167 \cdot 10^{-23}$ g—an unimaginably small mass!

The molecular masses are derived in a known way by adding the masses of the involved atoms: Law of conservation of mass. The *molar mass M* is calculated in the same way. Examples:

$$m_{1\ C\text{-atom}} = 12\,u, \quad m_{1\ mol\ C\text{-atoms}} = 12\,g, \quad M(C\text{-atoms}) = 12\,g/mol$$

$$m_{1\ C_2H_5OH\text{-molecule}} = 46\,u, \quad m_{1\,mol\,C_2H_5OH\text{-molecules}} = 46\,g,$$
$$M(C_2H_5OH\text{-molecules}) = 46\,g/mol$$

Mole

The unit for the amount of substance n is the <u>M</u>ole with the unit symbol <u>mol</u>. This unit has undergone a particularly drastic change in meaning. In earlier decades, people spoke of the "mole as the molecular weight in grams" and used the numerical value of the relative molecular mass with the unit g: "1 mol of water weighs 18 g". Today, the following IUPAC definition applies: A mole is the amount of substance of a portion of substance that contains as many particles as 12 g of carbon, consisting exclusively of C-12 isotopes. The particles of the considered portion of substance need to be characterized more closely, they can be atoms, molecules, ions, electrons and other particles. So the particles to be counted must be named: "1 mol H_2O molecules weigh 18 g" [2].

The imprecise statement "1 mol of oxygen" does not reveal whether 16 g or 32 g of oxygen need to be weighed. The clear advantage of the IUPAC definition is that by stating "1 mol O atoms" or "1 mol O_2 molecules" the portion of oxygen is clearly determined, in the first case 16 g of oxygen is weighed, in the second case 32 g:

$$m_{1\ mol\ O\text{-atoms}} = 16\,g, \quad m_{1\ mol\ O_2\text{-molecules}} = 32\,g$$

The substance portion of 32 g of sulfur contains either 1 mol S atoms or 1/8 mol S_8 molecules:

$$m_{1\ mol\ S\text{-atoms}} = 32\,g, \quad m_{1\ mol\ S_8\text{-molecules}} = 256\,g$$

The only remaining problem is how to capture the smallest particles of salts. Derived from the molar mass $M(NaCl) = 58.5$ g/mol, this portion of substance contains 2 mol ions, namely 1 mol Na^+ ions and 1 mol Cl^- ions. In order to be able to assign only 1 mol particles to the molar masses as usual, Kremer [3] suggests defining the Na^+Cl^- ion group as the smallest countable unit according to IUPAC—and thus solves the problem. It also becomes clear that salts contain neither atoms nor molecules, but ions. The word ion *pair* is not chosen, but deliberately the word ion *group*, to also allow compounds that are made up of several

8.1 Scientific Focus: Terms, Symbols, Quantities, Units

types of ions: alum salt from $K^+Al^{3+}(SO_4^{2-})_2$ ion groups and about black iron oxide from $(Fe^{3+})_2(Fe^{2+})_1(O^{2-})_4$ ion groups.

Derived from the mole, the molar mass M (g/mol) is defined, the molar volume V_m (L/mol), the molar particle number N_A, the molar charge F (C/mol, 1 Coulomb = 1 C = 1 As), the substance concentration Molarity (mol/L) or Molality (mol/kg). The usual concentration in mol/L allows precise information according to the substance definition—namely the indication of the particles that are actually present in solutions. The usual "1-molar sulfuric acid" or the indication "$c_{H_2SO_4}$ = 1 mol/L" wrongly suggests the presence of H_2SO_4 molecules in diluted acid solution, it should be replaced by the indication:

$$c_{H^+(aq)} = 2\,\text{mol/L} \quad \text{bzw.} \quad c_{SO_4^{2-}} = 1\,\text{mol/L}$$

One can also imagine labels on storage bottles of acid solutions with the inscription: "$c_{H^+(aq)}$ = 2 mol/L (sulfuric acid)" or "$c_{H^+(aq)}$ = 2 mol/L (hydrochloric acid)". This signals that the solutions contain certain concentrations of hydronium ions and these are to be offset against the hydroxide ions of the alkalis during titration. If the accompanying ions such as the sulfate ions for the precipitation of barium sulfate are important, this information can be taken from the indication "sulfuric acid". The same can be done for alkaline solutions, indicating the concentrations of hydroxide ions on the labels: "$c(OH^-)$ = 0.1 mol/L (sodium hydroxide solution)".

Nurma Y. Indriyanti [4] has analyzed the difficulties students have with the concept of the mole in more detail and developed the proposal to prepare the concept of the mole in action-oriented teaching ("Experiential Learning"). For this purpose, each group of students received a bag of candies with the task of determining how many bags of the same candies and how many candies in total are in a large sealed container: The principle of "counting by weighing" was recognized by the learners and successfully transferred to the abstract concept of the mole. Such teaching has been carried out and evaluated both in Germany and in Indonesia [4].

Hörnig and Habelitz-Tkotz [5] propose the introduction of stoichiometry using examples of molecular reactions under the title "Chemical Calculation—Unpopular but Indispensable". They aim to remove the "stumbling block" that metal-nonmetal reactions are often chosen for initial calculations, thus introducing the difficulty of "ion groups" or "formula units" in metal oxides or metal sulfides. For example, the reaction of propane and oxygen to carbon dioxide and water can easily be described with the amounts of substances, and the significance of coefficients and indices can be visualized using a model drawing (Fig. 8.1).

Temperature

The unit °C, commonly used in everyday life, was derived by the Swedish astronomer Anders Celsius in 1742 from the two known fixed points: the freezing temperature of pure water (0 °C) and the boiling temperature at standard pressure (100 °C). After Lord Kelvin postulated the absolute zero point of temperatures in

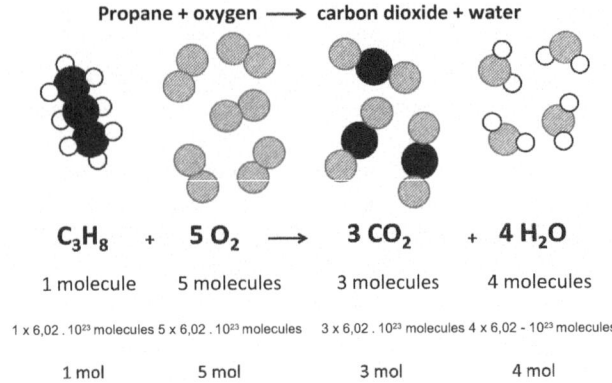

Fig. 8.1 Visualization of amounts of substances in molecular reactions [5]

1848, the scale of absolute temperatures was defined with the unit K (not °K!): 0 K = −273 °C or 0 °C = 273 K or 20 °C = 293 K (Fig. 8.2). Today, the Kelvin scale with the unit K is used for scientific calculations as the SI unit.

It should always be used in terms of the terms of Gay-Lussac's gas law, namely in the context of temperature and volume at constant pressure: The volume of a gas portion always increases with increasing temperature, formulated as a law: $V_1 : T_1 = V_2 : T_2$. This relationship is understood when the mathematical theorems of intersecting lines and the intersection of the volume line with the temperature line are used (Fig. 8.2). The figure also shows that historically, the intersection with the abscissa was obtained only by theoretically extending the empirically measured volume line and a certain minimum of temperatures had to be postulated

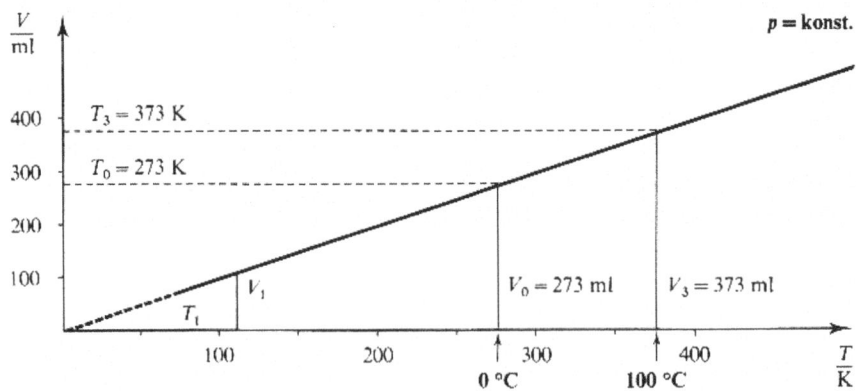

Fig. 8.2 The dependence of gas volume on temperature and Gay-Lussac's gas law [6]

completely hypothetically. Today, by using liquid hydrogen and helium, we can get very close to the absolute zero point and confirm it experimentally. The explanation based on the increasingly accepted Brownian molecular motion also gradually prevailed: The permanent particle motion in the gas portion decreases with cooling until it comes to a standstill at the absolute zero point. Therefore, a temperature lower than 0 K cannot be reached.

In the USA, temperatures are still given in degrees Fahrenheit in everyday life. To avoid negative values, in 1714, the German researcher Daniel Gabriel Fahrenheit in Gdansk set the zero point (0 °F) with the lowest achievable temperature of a salt-ice mixture at the time (−18 °C), and he chose the constant body temperature of humans (37 °C) as the second fixed point (100 °F). So when Californians talk about it being "100 degrees" hot outside again, they mean a temperature of 37 °C.

Pressure

Historically, Torricelli conducted his groundbreaking experiment on air pressure in 1648 and demonstrated that a 1m long tube filled with mercury shows a drop in the mercury column to 760 mm when the one-sided closed tube is opened under mercury (Sec. 2.1.4). As a result of these mercury experiments, Boyle enclosed a certain volume of air in a U-shaped glass tube in 1654 and changed the pressure on the gas by continuously adding more mercury (Fig. 8.3). In this way, he found that the volume becomes smaller the higher the pressure increases: $p_1 \cdot V_1 = p_2 \cdot V_2$ or $p \cdot V =$ constant. At that time, the unit Torr was adopted in honor of the Italian experimenter for normal air pressure and the unit atmosphere was defined as follows: 1 atm = 760 mm mercury column = 760 torr.

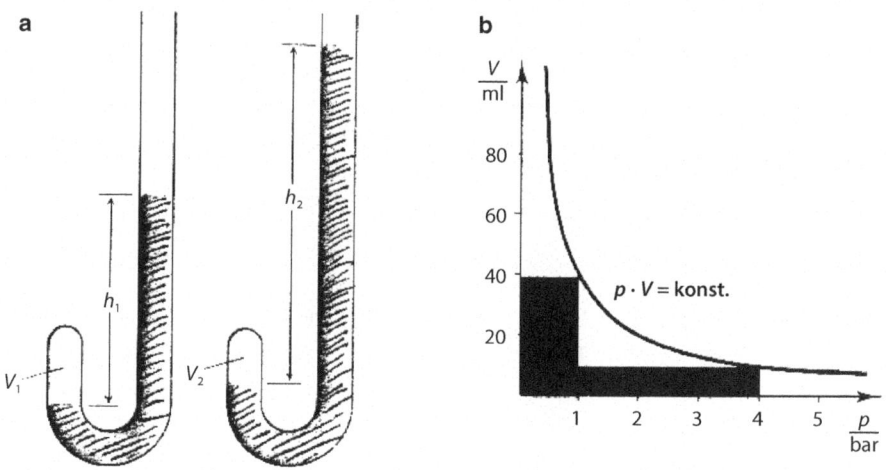

Fig. 8.3 a The dependence of gas volume on pressure and b the gas law of Boyle-Mariotte [6]

The current SI unit for pressure is the Pascal with the unit symbol Pa: 1 Pa = 1 N/m². Due to the SI unit m², the Pascal is the unit for a very small pressure, namely that of a 100-g chocolate bar in the gravitational field of the Earth on an area of 1 m². Therefore, the normal air pressure corresponds to the large number of 100,000 Pa or 1000 hPa (Hecto-Pascal). The conversions of the various pressure units are shown in Table 8.2.

Voltage and Current
The *current* has been defined with the SI unit Ampere (A): 1 A is the current that generates 0.19 mL of detonating gas per 1 s in electrolysis of water. In this process, a certain amount of charge Q flows, it is 1 Coulomb (C): 1 C = 1 A · 1 s. The number of electrons flowed is calculated to be $6.25 \cdot 10^{18}$ electrons (1 mol of electrons flow at Q = 96.485 C = 1 F).

When charge flows, energy is transported along metal wires by electromagnetic fields. It is transferred to a consumer, such as a light bulb, which converts the electrical energy into light and heat energy. Similarly, the potential energy of water flowing downhill is converted into kinetic energy at the water wheel of a historic mill. If the electrical charge does not flow from pole to pole, one can speak of "energy on demand" [7], of an electrical *voltage* between both poles. Therefore, such a voltage U depends on the energy W (W: work) transferred during the current flow and the amount of charge Q: U = W / Q. The unit of voltage is the volt (V): 1 V = 1 J/C.

In everyday life, electrical energy is measured in kilowatt-hours (kWh). An example makes this clear [7]: A heater, which is connected to the standard voltage of 220 V, is traversed by a current of I = 4.5 A for the time t = 1 h and releases the following energy: E = U · I · t = 220 V · 4.5 A · 3600 s = 220 J/C · 4.5 C/s · 3600 s = 3,600,000 J. This value represents the energy of one kilowatt-hour (1 kWh). The watt W is defined as the power that converts 1 J of energy per 1 s: 1 W = 1 J/s. Since one hour corresponds to 3600 seconds and the kilowatt is set at 1000 watts, the result is: 1 kWh = 3,600,000 J. These and similar calculations are not our topic—but they are important when the teacher teaches chemistry and physics at the same time and wants to make these units clear.

Table 8.2 Conversion factors for pressure units

Given/required	Pa	bar	hPa = mbar	atm	Torr
1 Pa	1	10^{-5}	10^{-2}	$9.8692 \cdot 10^{-6}$	$750.06 \cdot 10^{-5}$
1 bar	10^5	1	10^3	0.98692	750.06
1 mbar	10^2	10^{-3}	1	$0.98 \cdot 10^{-3}$	0.75
1 atm	101.325	1.013	1013.25	1	760
1 torr	133.322	0.00133	1.333	$1.316 \cdot 10^{-3}$	1

8.1.3 School-Relevant Terms

There are important terms and concepts that play a major role in introductory teaching and throughout chemistry teaching require careful reflection: formula, equation, particle, valency, energy, redox reaction, acid-base reaction, isomerism etc. Of course, there are further problem areas and even "home-made misconceptions" [8] in advanced teaching, the discussion of which would go beyond the scope of this book.

8.1.3.1 Formulas and Equations

It has become customary to refer to molecule symbols like CH_3COOH as formulas. If we assume that in the terminology of mathematics and physics, formulas are understood to be terms into which numerical values can be inserted and new numerical values result, then we should consistently stick to this meaning in all school subjects. In chemistry, the term "formula" can easily be replaced by "symbol": There are atom symbols, ion symbols, molecule symbols, structure symbols, reaction symbols, equilibrium symbols, etc.

Equations also have a different meaning than in mathematics and physics—chemical reaction *equations* can only mean that the sum of the atoms and consequently also the sum of the masses on the left and right of the reaction arrow are identical. Since substances or properties or energy contents on both sides of the arrow are definitely different, the term equation is misleading for beginners and should therefore be replaced by the *reaction symbol*. As long as a reaction symbol is used in words, it is also referred to as a *reaction scheme*:

$$\text{carbon (s)} + \text{oxygen (g)} \rightarrow \text{carbon dioxide (g)}$$

However, one cannot go so far as to completely ignore the expressions formula and equation commonly used among chemists—this would be a violation of the principle of connectable communication among each other. It is therefore suggested to discuss the usual terminology with older students and to use it synonymously. There is also nothing to prevent this problem from being addressed right at the introduction of the symbols and agreeing on a common language regulation.

8.1.3.2 Chemical Symbols

When we talk about the CH_3COOH symbol or the molecule symbol, we not only avoid the inaccurate term formula, but also use a basic concept of modern philosophy: "Symbols are not representations of their objects, but vehicles for the conception of objects. In the denotation of the ordinary type of symbol function, there must be four members: *subject, symbol, conception* and *object*" [9]. The chemist (subject) takes note of the written symbol CH_3COOH (symbol) and imagines the spatial structure of the acetic acid molecule made up of C-, H- and O-atoms (conception). He knows that the structure of this molecule (object) can be confirmed at any time by methods of instrumental analysis. These relationships can be read in more detail elsewhere [9].

The variety of symbols has grown with the advances in analytical chemistry, especially in the sense that the distinction between *infinite* and *finite* particle aggregates has required ever more precise symbols. Figure 8.4 shows, using the examples of calcium fluoride and ethanediol, with which information-poor symbols one historically had to work and which further information-rich structure symbols have developed.

Even today, we still work with information-poor symbols like NaCl, CaF_2 or Al_2O_3 in chemistry lessons and convey—in contrast to the predominantly used structure symbols for molecules in organic chemistry—no visualization of the structure of these compounds. Therefore, the symbols of the salts should at least show the structure from the involved ions, for example as follows: Na^+Cl^-, $Ca^{2+}(F^-)_2$ or $(Al^{3+})_2(O^{2-})_3$.

With the agreed symbol concept, it also becomes clear that symbols like C_2H_5OH do not belong to the *continuum area* and therefore should not be chosen as abbreviations for substance names (EtOH would be an arbitrary abbreviation for ethanol, often used by lab technicians for convenience). Symbols like C, H, O or C_2H_5OH refer to the *discontinuum area* and initially mean 1 C-atom, 1 H-atom, 1 O-atom or 1 C_2H_5OH molecule. In reaction symbols, such symbols can also stand for a large number of particles, in the special case for 1 mol of the corresponding particles.

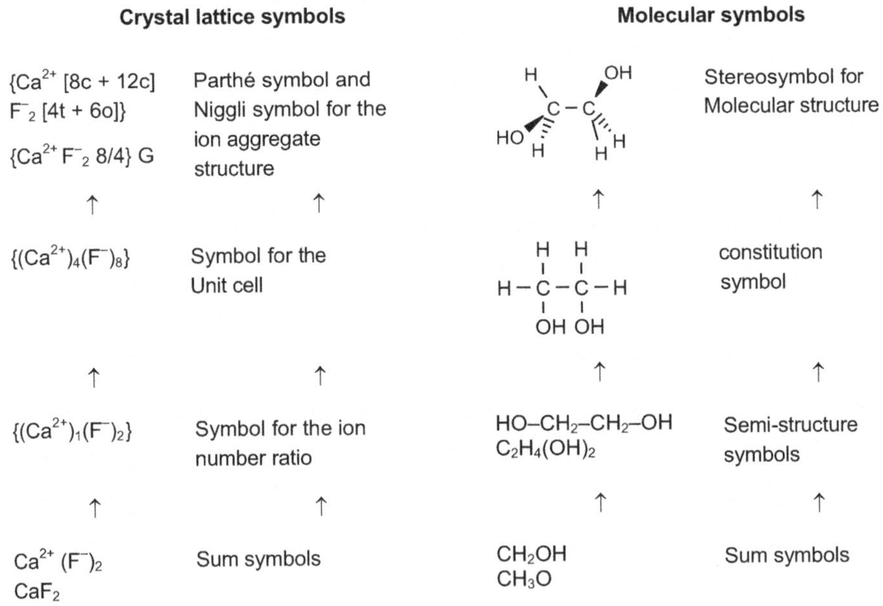

Crystal lattice symbols		Molecular symbols	
$\{Ca^{2+} [8c + 12c]$ $F^-_2 [4t + 6o]\}$ $\{Ca^{2+} F^-_2 \; 8/4\} \; G$	Parthé symbol and Niggli symbol for the ion aggregate structure	H, OH \\ C–C // HO, H / H H	Stereosymbol for Molecular structure
↑	↑	↑	↑
$\{(Ca^{2+})_4(F^-)_8\}$	Symbol for the Unit cell	H H | | H–C–C–H | | OH OH	constitution symbol
↑	↑	↑	↑
$\{(Ca^{2+})_1(F^-)_2\}$	Symbol for the ion number ratio	$HO-CH_2-CH_2-OH$ $C_2H_4(OH)_2$	Semi-structure symbols
↑	↑	↑	↑
$Ca^{2+}(F^-)_2$ CaF_2	Sum symbols	CH_2OH CH_3O	Sum symbols

Fig. 8.4 Types of chemical symbols (examples calcium fluoride and ethanediol)

To avoid as much as possible ambiguous designations for the *submicroscopic* area, atom, ion or molecule symbols should be linked with the known signs for masses, volumes or concentrations as follows:

$m_{1 \text{ mol H-atoms}} = 1\,\text{g}$, $V_{1 \text{ mol H}_2\text{-molecules}} = 22{,}4\,\text{L (standard cond.)}$, $c_{H^+(aq)} = 0{,}1\,\text{mol/L}$

It is then consistent to write out the names for observable substances in the *macroscopic* area—also on labels of storage bottles in chemical collections or for measurable properties of substances:

$\text{density}_{\text{ethanol}} = 0{,}79\,\text{g/mL}$, boiling $\text{temperature}_{\text{ethanol}} = 78\,°\text{C}$ (standard cond.)

Accordingly, teachers should from the beginning distinguish the scientific language in the macroscopic area from that in the submicroscopic area: from substances like water or ethanol on the one hand and from H_2O molecules and C_2H_5OH molecules on the other hand. Abbreviations for substance names like "H_2O for water" and "C_2H_5OH for ethanol" must be avoided.

8.1.3.3 Reaction Symbols

In reaction symbols ambiguity of those symbols cannot be avoided. As an example, the combustion reaction of ethanol with oxygen is noted:

$$C_2H_5OH\,(l) + 3O_2\,(g) \rightarrow 2CO_2\,(g) + 3H_2O\,(g); \quad \Delta H < 0$$

$$1\,\text{molecule} + 3\,\text{molecule} \rightarrow 2\,\text{molecule} + 3\,\text{molecule}$$

$$Z\,\text{molecules} + 3\,Z\,\text{molecules} \rightarrow 2\,Z\,\text{molecules} + 3\,Z\,\text{molecules}$$

$$1\,\text{mol molecules} + 3\,\text{mol molecules} \rightarrow 2\,\text{mol molecules} + 3\,\text{mol molecules}$$

If one wanted to use symbols unambiguously, then one would have to list the individual molecules (see first line under the reaction symbol). However, for stoichiometric considerations, large numbers of molecules are needed, which could possibly be marked by the letter Z (see second line) or directly with quantities in mol (see third line). Advanced students are expected to infer from the context whether a molecule symbol stands for exactly 1 molecule or for 1 mol of molecules.

As long as the discontinuum area is not yet underlying in the initial instruction and no considerations on stoichiometry or the structure of the substances play a role, in any case only *reaction symbols in words* should be used:

ethanol (l) + oxygen (g) $\rightarrow$ carbon dioxide (g) + water (g); exotherm

If one wants to consistently leave the symbol concept to the discontinuum area, one can introduce on the continuum area terms like *reaction scheme* or *word scheme*—in no case the incorrect term "word equation": The differences between the starting materials and reaction products should be in the foreground in the initial instruction! The meaning of the plus sign as "enumerating and" and the meaning of the reaction arrow must also be clarified: It can be read as "react to form".

8.1.3.4 Particles

The particle concept is often introduced in class as the first preliminary model idea, for example when the schematic structure of a sugar crystal from sugar particles or an ethanol solution from ethanol and water particles is to be illustrated. Of course, with the first introduction of a model, it is fundamentally necessary to reflect on the scientific model concept (Chap. 7). However, there are didactic problems in two respects.

As long as the particle concept is applied to examples such as metals or gases, there are no difficulties: A ball or a circle can be seen as a model for a metal particle or a gas particle respectively. If the particle concept is applied to salts or salt solutions, one encounters the limits of this model idea. According to the agreement to assign a ball as a model to a particle of a pure substance, the ball in the case of sodium chloride would have to stand for the ion group Na^+Cl^- or even for the unit cell $\{(Na^+)_4(Cl^-)_4\}$. Factually correct, however, salts or salt solutions are composed of at least two types of ions and a model should start from two types of balls. Because of this problem, it is recommended not to apply the particle concept to solid salt crystals or to salt solutions.

The second difficulty lies in the observation of very small particles in powder mixtures, suspensions or aerosols. Learners, for example, like to describe a mixture of iron and sulfur powder with "iron particles and sulfur particles", thus using the particle concept for visible crystals—knowing full well that each small crystal consists of billions of particles. If one insists on consistently assigning the concept of the smallest particle to the submicroscopic area of atoms, ions and molecules, then one should choose the more appropriate designations "crystals, grains, granules, droplets" or similar for tiny portions of substance.

8.1.3.5 Valency

A general concept of valency is still used, which was historically helpful for hypotheses on the composition of substances before the discovery of ions. As long as it was assumed in the 19th century that all compounds are made up of molecules, the statement "aluminium is trivalent" was interpreted to mean that Al atoms have three "valences" in the sense of Kekulé and accordingly form "molecular compounds", which were then logically marked with abbreviated symbols like AlF_3 or Al_2O_3.

This general concept of valency has become meaningless today according to Herdt [10] and is therefore dispensable. On the one hand, the concept of ions is available, and for the *ionic* compounds, the *charge numbers* of the ions can be used: for example, starting from Al^{3+}-, F^-- or O^{2-}-ions, formulating corresponding compound symbols like $\{(Al^{3+})_1(F^-)_3\}$ or $\{(Al^{3+})_2(O^{2-})_3\}$ and possibly shortening to AlF_3 or Al_2O_3. On the other hand, for many *molecular* substances, the *valencies* of the non-metal atoms that make them up are argued, as they are marked in the well-known molecule construction kits with push buttons or plastic rods: C atoms are imagined as tetravalent, H atoms as monovalent and corresponding molecule symbols like CH_4 or C_2H_6 are assumed. In both cases, charge number and valency are interpretable in terms of structural chemistry and lead to factually correct structural ideas—the historical concept of valency does not do this.

8.1.3.6 Energy

We distinguish between the terms energy, enthalpy and "free enthalpy". In terms of chemical reactions, we speak of *enthalpies* when constant pressure conditions are assumed, of *energies* when there are constant volumes. Since usual measurements in the school laboratory mostly take place at constant pressure, we can stick to the term enthalpy. However, this term is also not often used, we imprecisely speak of reaction heat or heat energy. The *free enthalpy* is discussed in connection with the "driving force" or "voluntariness" of chemical reactions, which depends on both the *enthalpy H* and the *entropy S*—the Gibbs-Helmholtz equation establishes quantitative relationships in this regard: $\Delta G = \Delta H - T\Delta S$.

School chemistry often tries to distinguish between *chemical and physical processes*. However, since enthalpies are also measured during melting, evaporating or dissolving, which in no way differ from the reaction enthalpies of usual reactions, it makes no sense to want to construct this demarcation: All processes that, in addition to the changes in substances, for example from solid to melt, also show energy changes of the system, should accordingly be referred to as chemical reactions.

For the *qualitative indication of energy turnover*, various reaction symbols are common, for exothermic reactions the symbol $\Delta H < 0$:

hydrogen (g) + oxygen (g) → water (l) + thermal energy

hydrogen (g) + oxygen (g) → water (l); exotherm (pr $\Delta H < 0$)

If—as in the first reaction symbol—the heat energy is placed with a plus sign in the row of symbols for substances, learners may come to the idea that the heat energy is a "heat substance" involved in the reaction, as it has also been historically discussed. It is therefore recommended to clearly separate the specification of substances from the specification of energy turnover by a semicolon, as shown by the second reaction symbol.

When making *quantitative statements of enthalpies*, it should be noted that numerical values in kJ apply to the substance conversion of the *formulated reaction symbol*, while the numerical values for ΔH_f (f: formation) refer to the indication of the formation enthalpies and thus always to 1 mol of formed particles—and are therefore marked with the unit kJ/mol:

$$2H_2 \text{ (g)} + O_2 \text{ (g)} \rightarrow 2H_2O \text{ (l)}; \quad \Delta H = -570\,\text{kJ};$$
$$H_2 \text{ (g)} + 1/2 O_2 \text{ (g)} \rightarrow H_2O \text{ (l)}; \quad \Delta H = \Delta H_f = -285\,\text{kJ/mol}$$

Learners often find it difficult to understand the given ΔH values and the concept of *chemical energy*. It is therefore necessary to provide a diagram that illustrates the different energies in a coordinate system (Fig. 8.5): The system "hydrogen/

Fig. 8.5 Chemical energy in the exothermic reaction of hydrogen and oxygen

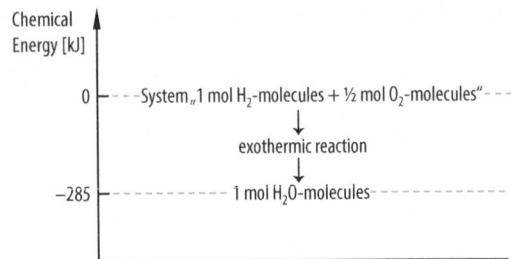

oxygen" is a relatively energy-rich system, while water is a comparatively energy-poor substance. Since the elements have been defined with the chemical energy of 0 J, a substance of lower chemical energy is present after the exothermic reaction of both gases to water, the measurement value for 1 mol of formed H_2O molecules is therefore defined as negative: $\Delta H_f = -285$ kJ/mol. Such enthalpies are measured in the laboratory using a calorimeter, a particularly simple possibility is shown by experiment V2.6 in Chap. 2.

8.1.3.7 Redox Reactions

When you ask advanced students about how the well-known iron nail-copper sulfate reaction leads to the formation of the red deposit on the iron nail, you might be surprised: "Copper oxidizes to copper oxide and deposits on the nail; iron is oxidized and copper sulfate is reduced; iron is oxidized, taking O from the $CuSO_4$; electrons are released and oxygen is absorbed" [11]. Although the respondents know the extended redox concept and electron transfer, they argue with the substances iron and copper sulfate, instead of appropriately naming the iron atoms, which are oxidized to Fe^{2+} ions by electron donation, and the Cu^{2+}(aq) ions, which are reduced to Cu atoms by electron uptake. On the other hand, they revert to the historical definition of oxygen transfer and look for "the oxygen that must be involved in red '**ox**' reactions: iron takes O from the $CuSO_4$".

It must be clear that the historical oxygen definition is so attractive and understandable that learners do not see why the electron transfer should be more explanatory when offered the extended definition. Therefore, it should be considered whether to omit the redox concept in introductory teaching and to explain the well-known metal oxide-metal reactions, such as the reaction "copper oxide + iron → copper + iron oxide", simply by stating that copper oxide is reduced and iron is simultaneously oxidized. Even more radically, Habelitz-Tkotz and Werner [12] demand the abandonment of the Lavoisier's redox definition: "To avoid the problems associated with the new definition when transitioning from the Lavoisier concept to the electron-theoretical redox concept, the introduction of the oxygen-related redox concept is omitted in introductory teaching. In the context of combustion reactions, the term oxidation is avoided and only—if at all—spoken of oxide formation or metal oxides" [12]. If necessary, students recount the history of the discovery of oxygen and the first definition of oxidation by Lavoisier in 1784.

8.1.3.8 Acid-Base Reactions

According to historical development, there are at least three different levels of the definition of the acid concept (and analogously also of the base concept):

1. The oldest view was *substance-related*: "An acid is a substance that specifically colors a plant indicator dye—such as the litmus dye—and dissolves limestone crystals with gas development" (Boyle 1663).
2. Then came *structure-related* definitions, as the compositions of different acids were gradually recognized: "Acids are hydrogen compounds in which the hydrogen can be replaced by metals" (Liebig 1838). This means that in the acetic acid molecule CH_3COOH, the H-atom of the acid group can be replaced by a metal atom, for example to CH_3COONa. The further structure-related postulate was: "Acids dissociate in water, there are H^+ ions and acid residue ions in aqueous solution" (Arrhenius 1884).
3. Finally, a *function-related* definition emerged, which no longer refers to substances, but exclusively to smallest particles such as molecules or ions: "*Particles* that can give off protons are called acids" (Brönsted 1923). Particles—such as H_2O molecules—are in this sense either acid or base depending on the reaction partner, or they fulfill both functions. It would be nonsensical to say "water can be acid or base": it always refers to the H_2O molecule, which can react as an acid-*particle* or as a base-*particle*.

The *Arrhenius definition* represented a revolutionary advancement over earlier definitions: The electrical conductivity and the specific changes in boiling and freezing temperatures were explained, the H^+(aq) ions could be held responsible for the acidic properties, the OH^-(aq) ions for the alkaline properties, and neutralization was described as the reaction of the H^+(aq) ions and OH^-(aq) ions to H_2O molecules. Because of these advantages, the Arrhenius definition is still used in many curricula today.

However, there were also open questions. It was completely unclear for what reasons acid molecules split into ions or by what forces they are split, or *dissociate*, and why some acids only do so to a certain degree and had to be described with the so-called *degree of dissociation*. Secondly, the substance ammonia and its alkaline solution in water could not be reconciled with the definition: OH^- ions do not formally split off from NH_3 molecules. *Ammonium hydroxide* NH_4OH was invented and the aqueous solution was described with the corresponding ions. However, this substance does not exist as a solid and does not provide the ions through *dissociation* in water like the other solid hydroxides. Even today, old bottles of this fictitious substance labeled "NH_4OH" can still be found in chemical collections.

Finally, the Arrhenius definition was limited to the *solvent water*. On the one hand, the neutralization reaction of gaseous ammonia and gaseous hydrochloric acid directly to the solid salt ammonium chloride was known—without the need for aqueous solutions and corresponding "dissociations in water":

$$NH_3 (g) + HCl (g) \rightarrow NH_4Cl (s).$$

On the other hand, there were substances that, for example, indicate acids and bases with common indicator dyes, but do not exist in aqueous solutions. For example, in liquid ammonia at the boiling temperature of -33 °C, the neutralization of ammonium salts and metal amides can be shown using universal indicator and formulated analogously to the neutralization in aqueous solution: NH_4^+ ions react with NH_2^- ions and form NH_3 molecules:

$$NH_4^+ + NH_2^- \rightarrow 2NH_3$$

With the *Brönsted definition*, these contradictions were eliminated: An HCl molecule each gives a proton (H^+ ion) to an NH_3 molecule, forming the ionic lattice from NH_4^+ ions and Cl^- ions, NH_4^+ ions each give a proton to NH_2^- ions and form NH_3 molecules.

These reactions are also excellent examples of the meaningful expansion of model ideas on the one hand, and the constructive coexistence of two theories on the other. Even today, the Arrhenius definition is valid in the area of aqueous solutions and cannot be called "wrong"—a problem that can also be discussed with students in class. With regard to the definition of Brönsted, it should only be emphasized that the term acid does not refer to substances, but rather to *acid particles*. This is shown by the examples in Table 8.3.

In the area of alkaline solutions, a distinction between substance and particle level is possible through terms, we speak of caustic soda or sodium hydroxide as substances, with *bases* we denote particles that are capable of accepting protons: OH^- or NH_2^- ions or NH_3 molecules. In the area of *acids*, it must be inferred from the context whether an acidic substance or an acid particle is meant.

Particles, however, cannot generally be classified into the categories "acids and bases"—depending on the reaction partner, certain particles react both as an acid and as a base: H_2O or NH_3 molecules, OH^- or HSO_4^- ions. They are also called *ampholyte particles* or *ampholytes*. It is useful to provide corresponding symbols for *protonations* and *conjugated acid-base pairs*. Under no circumstances should the statement—as it often comes up in examination discussions—be accepted that, for example, in the reaction of hydrochloric acid with sodium hydroxide, "the acid HCl donates a proton to the base NaOH": The strong acid "HCl molecule"

Table 8.3 Acid particles according to the Brönsted definition

Substances	Acid particles	Additional particles
Hydrochloric acid(aq)	H_3O^+(aq) ions	H_2O molecules, Cl^-(aq) ions
Nitric acid(aq)	H_3O^+(aq) ions	H_2O molecules, NO_3^-(aq) ions
Sulfuric acid(aq)	H_3O^+(aq) ions	H_2O molecules SO_4^{2-}(aq) ions
Sulfuric acid(l)	H_2SO_4 molecules	$H_3SO_4^+$ ions, HSO_4^- ions
Sodium bisulfate(s)	HSO_4^- ions	Na^+ ions (in the ion lattice)

completely protonates in the presence of H_2O molecules, in the solution there are the H_3O^+(aq) ions as acid particles, each donating a proton to OH^-(aq) ions of the sodium hydroxide solution.

The mixing of the concepts of Arrhenius and Brönsted theory is also unfavorable. Thus, one teaches the proton transfer with respect to the Brönsted theory, but speaks of the "degree of dissociation" instead of the factually appropriate degree of protonation for weak acids. Or one writes the Brönsted reaction symbol $HCl + H_2O \rightarrow H_3O^+ + Cl^-$ for the Arrhenius theory and not historically appropriate $HCl \rightarrow H^+ + Cl^-$, as would be appropriate for the dissociation theory: The ion symbol H_3O^+ belongs to the Brönsted theory, while with respect to the Arrhenius theory, one should argue with the simple H^+ ions or with H^+(aq) ions.

In chemistry lessons, one rightly tries to formulate complete reaction symbols through "same types and numbers of atoms on the left and right of the reaction arrow" ("balanced equations"). For a real understanding of chemistry, one should additionally ask which particles are responsible for the reaction. One then recognizes, for example, certain types of ions responsible for the reaction and illustrates the reaction with corresponding ion symbols:

1. Sodium hydrogen sulfate reacts exothermically with water to form a strongly acidic solution—here the hydrogen sulfate ion reacts as an acid particle and transfers a proton to the H_2O molecule:

$$HSO_4^- + H_2O \rightarrow H_3O^+(aq) + SO_4^{2-}(aq).$$

If one considers that the sum of all charges to the left and right of the reaction arrow is the same, then one has a reaction symbol that accurately describes the acidic properties of the solution. Without ion symbols, the reaction is usually formulated as follows:

$$2NaHSO_4\ (s) \rightarrow aq \rightarrow Na_2SO_4(aq) + H_2SO_4(aq).$$

However, this symbol does not show the responsible hydronium ions in aqueous solution or the proton transfer of an acid-base reaction, but provokes misconceptions among learners such as the formation of "Na_2SO_4 molecules and H_2SO_4 molecules".

2. Sodium carbonate reacts with water to form a strongly alkaline solution—here the carbonate ion reacts as a base particle by taking up a proton from the water molecule:

$$CO_3^{2-} + H_2O \rightarrow HCO_3^-(aq) + OH^-(aq).$$

This reaction would have to be described without ion symbols as follows, provoking misconceptions of Na_2CO_3 molecules and NaOH molecules:

$$Na_2CO_3\ (s) + 2H_2O \rightarrow H_2CO_3(aq) + 2NaOH(aq)$$

3. The famous iron nail is covered with a copper layer in blue copper sulfate solution:

$Cu^{2+}(aq)\text{-Ion} + 2e^- \rightarrow Cu\text{-Atom}; \quad Fe\text{-Atom} \rightarrow Fe^{2+}(aq)\text{-Ion} + 2e^-;$

insgesamt: $\quad Cu^{2+}(aq) + Fe \rightarrow Fe^{2+}(aq) + Cu$

If one omits ion symbols and thus the crucial transition of electrons, then the corresponding reaction symbol leads to ideas of a "partner exchange":

$$CuSO_4(aq) + Fe \rightarrow FeSO_4(aq) + Cu$$

4. The precipitation reaction of barium chloride and sodium sulfate solutions to solid barium sulfate is easy to describe with ions, because when barium and sulfate ions meet, the insoluble white salt always forms—regardless of the accompanying ions:

$$Ba^{2+}(aq) + SO_4^{2-}(aq) \rightarrow Ba^{2+}SO_4^{2-}\ (s).$$

The complete, but not very meaningful reaction symbol

$$BaCl_2(aq) + Na_2SO_4(aq) \rightarrow BaSO_4\ (s) + 2NaCl(aq)$$

hardly shows the essence of the precipitation. It suggests a "double partner exchange" that associates the coming together of the sodium and chloride ions, which is not factually correct: These two types of ions do not react, they remain in solution ("spectator ions").

Real understanding of chemistry can therefore only be developed if acid-base reactions with proton transfers and redox reactions with the transfer of electrons are described. However, this requires the ion symbols and it is recommended to use them as early as possible and to familiarize learners with them and the corresponding reaction equations.

8.1.3.9 Isomerism
Isomers are molecules of the same empirical composition, but with different spatial arrangements of atoms. There are various *types of isomerism*: In terms of constitutional isomerism , we distinguish positional, functional isomerism, tautomerism and valence isomerism, regarding stereoisomers it can be cis-trans isomers, mirror image or conformational isomers (Fig. 8.6). The concept of isomerism is particularly important for understanding carbohydrates:

- D-Glucose (open-chain) and D-Fructose (open-chain) are constitutional isomers .
- D-Glucose (open-chain) and L-Glucose (open-chain) are mirror-image stereoisomers (enantiomers).
- α-D-Glucose (cyclic) and β-D-Glucose (cyclic) are non-mirror-image stereoisomers (diastereoisomers).
- Polymeric α-D-Glucose (starch) and polymeric β-D-Glucose (cellulose) are diastereoisomers.

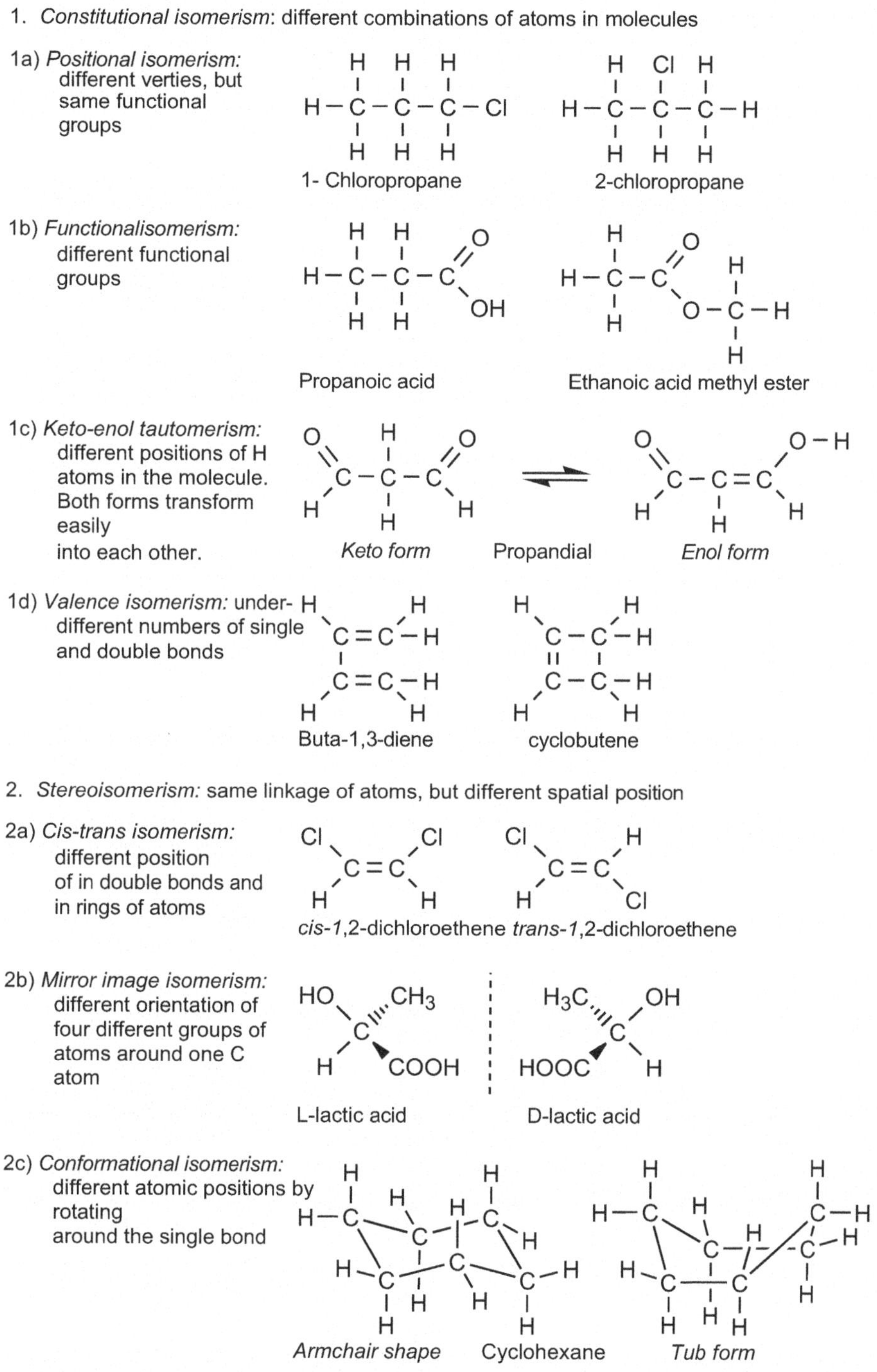

8.1 Scientific Focus: Terms, Symbols, Quantities, Units

1. *Constitutional isomerism*: different combinations of atoms in molecules

1a) *Positional isomerism:* different verties, but same functional groups

1- Chloropropane   2-chloropropane

1b) *Functionalisomerism:* different functional groups

Propanoic acid   Ethanoic acid methyl ester

1c) *Keto-enol tautomerism:* different positions of H atoms in the molecule. Both forms transform easily into each other.

Keto form   Propandial   Enol form

1d) *Valence isomerism:* under different numbers of single and double bonds

Buta-1,3-diene   cyclobutene

2. *Stereoisomerism:* same linkage of atoms, but different spatial position

2a) *Cis-trans isomerism:* different position of in double bonds and in rings of atoms

cis-1,2-dichloroethene trans-1,2-dichloroethene

2b) *Mirror image isomerism:* different orientation of four different groups of atoms around one C atom

L-lactic acid   D-lactic acid

2c) *Conformational isomerism:* different atomic positions by rotating around the single bond

Armchair shape   Cyclohexane   Tub form

Fig. 8.6 Different isomerisms and corresponding technical terms [13]

The differences between the pairs to be compared are to be illustrated with structure symbols and spatial molecular models, the resulting properties are to be discussed. It should be made clear to the students that molecules that look similar can have very different effects in our organism:

- D-Glucose is edible and essential, L-Glucose is toxic.
- We can nourish ourselves with starch, but not with cellulose from wood.
- Ruminants digest cellulose because they have the enzyme cellulase.

Regarding mirror-image isomerism , the thalidomide tragedy can be pointed out: Between 1958 and 1962, the sleeping pill thalidomide was on the market, which consisted of a mixture of molecules of both mirror-image forms. Only one of the two enantiomers acted as a sleeping pill, the mirror-image molecules caused fruit damage and caused severe malformations in newborns' arms and legs. Thanks to the Münster human geneticist Widukind Lenz, that the cause of the increased malformations was discovered by him in 1961—too late for about 10,000 children!

8.2 Teaching Processes: Everyday Language → Scientific Language → Symbolic Language

All teaching must start with the children's experience (Dewey).

All new experiences that learners make in class are organized with the help of existing ideas (Ausubel).

These well-known statements are empirically confirmed today by learning psychologists who represent constructivism, in that each individual constructs his or her knowledge structure on the basis of prior experiences and the existing cognitive structure. These learning psychological facts apply particularly to the extension of everyday language to scientific language and finally to chemical symbolic language.

8.2.1 Linking Everyday Language and Scientific Language

As soon as a new topic is started in chemistry class, initial explanations of experiments or natural phenomena should be paraphrased in everyday language and initial reaction symbols should be used in words before sum symbols or structure symbols follow (Table 8.4). Once a certain level of sceintific language has been successfully learned, corresponding terms should be used in new descriptions to link existing knowledge with new aspects of knowledge and thus expand the existing cognitive structure of the learners to the new knowledge structure. If terms such as acid solution and lye, acidic and alkaline, or H^+ and OH^- ion are present in the sense of the Arrhenius concept, they can be picked up again when

8.2 Teaching Processes: Everyday Language → ...

Table 8.4 Descriptions of chemical processes at the level of everyday language, scientific language and chemical symbol language

Everyday language	Scientific language	Symbol language
Lime dissolves in water until something remains undissolved at the bottom	Saturated solution of calcium hydroxide is in equilibrium with the ground body, solubility at 20 °C: 0.96 g/L water	$Ca^{2+}(OH^-)_2 \rightleftarrows Ca^{2+}(aq) + 2OH^-(aq)$ $L(Ca(OH)_2) = 8 \cdot 10^{-6}$ $c(Ca^+) = 0.013$ mol/L; $c(OH^-) = 0.026$ mol/L (Standard conditions)
Lime water tastes soapy	Calcium hydroxide solution turns universal indicator paper blue, the pH value is greater than 7	For 0.005-molar calcium hydroxide solution: $c(OH^-) = 0.01$ mol/L = 10^{-2} mol/L, pOH = 2, pH = 12
Limestone Quicklime	Calcium carbonate Calcium oxide	$\{Ca^{2+}CO_3^{2-}\}$, $CaCO_3$ $\{Ca^{2+}O^{2-}\}$, CaO
Reaction of quicklime with water forms slaked lime	Calcium oxide (s) + water → Calcium hydroxide (s); exothermic	$CaO + H_2O \rightarrow Ca(OH)_2$; $\Delta H < 0$ $O^{2-} + H_2O \rightarrow 2\, OH^-$; $\Delta H < 0$ (Protolysis: O^{2-} ion is base, H_2O molecule is acid)

introducing the Brönsted concept, linked in this new sense and compared with old meanings. If terms like donor and acceptor are known from descriptions of acid-base reactions, they can also be used in redox reactions and both types of reactions can be conceptually combined under "donor-acceptor reactions".

8.2.1.1 Concept Networks

Sumfleth [14] has described the networking of basic concepts of chemistry teaching, proposed corresponding concept networks, and empirically investigated them (Fig. 8.7)—"systematization aids" are shown regarding the topics "acid-base reactions" and "types of reactions". However, schemas of this kind are only real learning aids if the learners recognize and understand the different meanings of the arrows between the concepts.

For learning objective control, students can be asked to create such concept networks or also *Concept Maps*. For this, learners are asked to sensibly arrange given terms, to provide related terms with a relationship arrow and to label them appropriately. Further elaborations on this evaluation method are outlined by Behrendt [15].

8.2.1.2 Spiral Curriculum

Expansion of meanings of certain terms, models or symbols and associated progressive abstraction are also often referred to as *spiral curricular* approach or learning in the form of a *curriculum spiral*. The beginning of such a spiral should always include the use of everyday language, the ascent from learning level to learning level should consistently ensure the linking of existing terms with new terms. Examples from the area of "solubility" and "acids" are shown in Fig. 8.8. Beyond these examples, Schmidkunz and Büttner present the entire "Chemistry

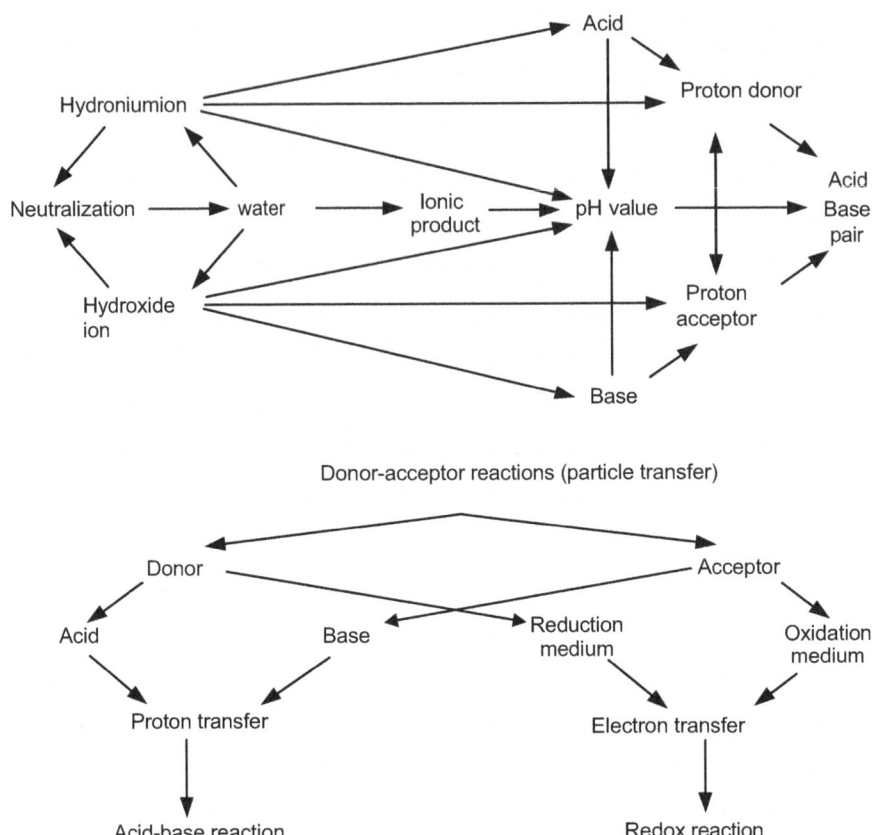

Fig. 8.7 Concept networks for "acid-base reactions" and "types of reactions" [14]

Teaching in the Spiral Curriculum" [16] for better coordination of teaching topics (Fig. 4.8 in Chap. 4).

8.2.1.3 Concept Levels

In Chap. 7, the work with mental models in human consciousness and with concrete visual models such as sphere packing and lattice models was reflected, a schema for thinking in models was depicted (Figs. 7.1 and 7.2 in Chap. 7). The three levels discussed there also correspond to three concept levels: terms of the substance level, the level of mental models, and the level of concrete models. An example of the dissolution of sugar in water and an interpretation based on the particle model is shown in Fig. 8.9. In classroom discussions, it should be achieved as far as possible to sensibly link terms of one level: a) sugar dissolves in water to form a sugar solution that tastes sweet, b) we imagine as a model idea

8.2 Teaching Processes: Everyday Language → …

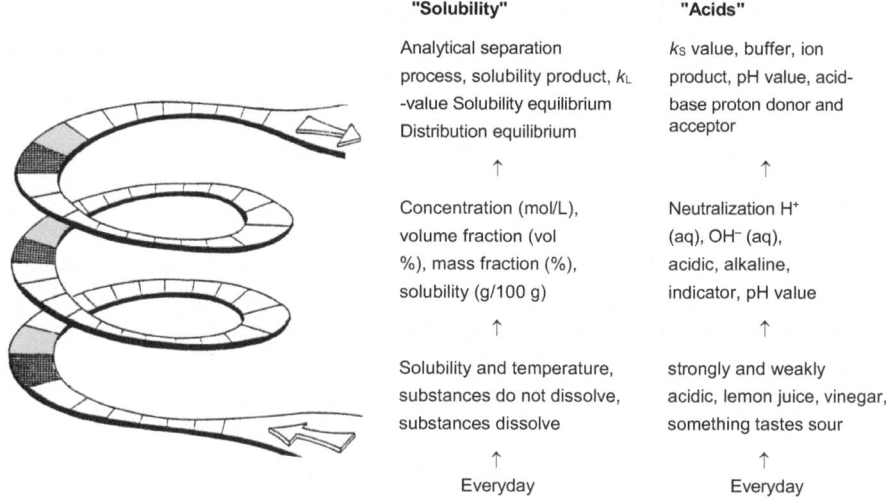

Fig. 8.8 The terms "solution" and "acid" in the curriculum spiral

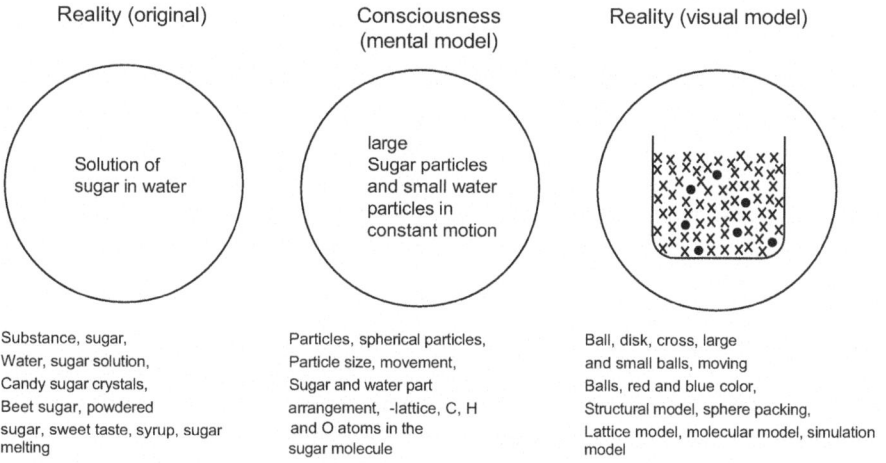

Fig. 8.9 Terms at the level of substances, mental models, factual models

strongly moving sugar and water particles, c) for visualization, red balls and blue crosses can be chosen for a model drawing. If terms from the levels are mixed at random, the result is meaningless connections and incorrect model ideas such as "sugar solution consists of red balls and blue crosses" or "sugar particles are red and taste sweet".

8.2.2 The Chemical Symbol Language

> When we ask adults what they remember from their school chemistry lessons, we often encounter a difficulty that also frequently plagues our students. The answer is usually: "Oh, there were formulas", and one is proud if one still knows what H_2SO_4 is—not what it means—because usually not much chemical knowledge has remained. A high-ranking official of a secondary school authority, who was even a scientist, once told me: "Keep away from me with the educational value of chemistry, it's just formula stuff" [17].

Elsewhere, Scheible [17] even states: "The chemical formula has brought us into disrepute". These quotes are intended to show that it seems to be difficult in chemistry lessons to foster an understanding of chemical symbols.

8.2.2.1 Teaching Without Formulas

Initially, the problems with symbolic language can be circumvented in introductory lessons by writing out everyday names or names for substances, formulating reaction symbols in words, using symbols (s), (l), (g) or (aq) for this:

hydrogen (s) + oxygen (g) → carbon dioxide (g);

exotherm

hydrogen chloride (g) + water (l) → hydrogen chloride(aq)(= hydrochloric acid);

exotherm

hydrochloric acid + magnesium (s) → magnesium chloride (aq) + hydrogen (g);

exotherm

The differences between the symbols (l) and (aq) must initially be explained by referring, for example, to liquid hydrogen chloride(l) with a boiling point of -85 °C and distinguishing it from the aqueous solution hydrogen chloride(aq), which is called hydrochloric acid.

8.2.2.2 Symbols at the Level of the Particle Model

Once the particle model has been introduced, appropriate models can be built or model drawings can be offered to understand chemical processes based on this model idea, in addition to reaction symbols in words (Fig. 7.11 and 8.9). The names of the particles must be written out in the legend of such a drawing: sugar particles, water particles. Often students want to show that they know the H_2O symbol and formulate inappropriately "H_2O particles": The H_2O symbol clearly belongs to the level of Dalton's atomic model and has no place in the model idea of smallest particles!

To demonstrate the spatial extent of crystals with visual models, dense sphere packings can also be built at the level of the particle concept, as has already been suggested (V7.13 in Chap. 7). The connection between, for example, shapes of sugar crystals ("rock candy") can be pointed out and compared with sphere packing models in which each sphere corresponds to a sugar particle: identical shapes, straight edges or smooth surfaces of the original are reflected in suitable sphere packing models.

For the particle model for air, Harsch and Heimann [18] have developed simple experiments that can be vividly interpreted with corresponding particle images. The authors recommend working with corresponding models or model drawings as early as possible, as empirical surveys on the particle model of air have revealed serious misconceptions [19–21]. Students also have such misconceptions: If you describe the distribution of perfume in the air and ask for the model idea of the involved particles, you often find circles or spheres for perfume particles—but not for the also present air particles and their mixture with perfume particles!

8.2.2.3 Symbols at the Level of Dalton's Atomic Model

The philosophy published by Dalton in 1808 included the linking of the concept of elements with the concept of atoms: There are as many different types of atoms as there are elements, different types of atoms have different masses. With the novel *atomic masses*, first *atomic mass tables* were developed, first *atomic symbols* were used. Dalton's idea brought the millennia-old search of Greek natural philosophers for the basic building blocks of matter to a first result and he was able to propose new, composite atomic associations by *combining atoms*. Unfortunately, he had assigned the atomic association HO to the water molecule; in the first tables, this resulted in the relative atomic masses of 1 and 8 for the H atom and the O atom instead of 1 and 16. Only Avogadro's hypothesis, in conjunction with the experimental findings of water syntheses, led to the factually appropriate symbol H_2O for a water molecule. With the knowledge of these molecule symbols, ideas could emerge as they are depicted today in almost every textbook (Fig. 8.10).

Combinations of metal atoms of one type result in the simplest case in densest packings of metal atoms in a *metal lattice*. In models for metal lattices, spheres can be stacked into sphere packings and coordination numbers such as 12 and 8 can be determined (Chap. 7). Furthermore, it is possible to substitute metal atoms in the metal lattice with atoms of another metal, thus conceptually obtaining *alloy*

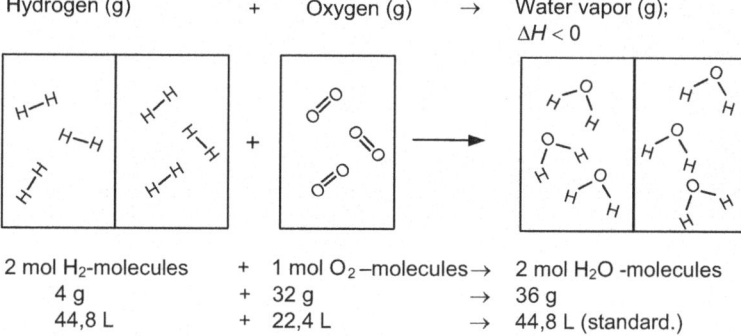

Fig. 8.10 Model ideas for the rearrangement of atoms to water molecules

crystals, which can be illustrated in models by packings with equally sized spheres of two colors. Hints on structure and model building of exemplary metal and alloy crystals can be found in Sauermann and Barke [22].

Combinations of non-metal atoms usually lead to *molecules*: H_2, O_2, Cl_2, P_4 or S_8 in elemental substances, H_2O, HCl, H_2SO_4, P_4O_{10} or H_3PO_4 in compounds. The valences of the non-metal atoms are predetermined in spheres of usual molecule model kits by snap buttons or rods. They are suitable for constructing molecule models in the form of ball-and-stick or space-filling models and for illustrating rearrangements of atoms in chemical reactions (Fig. 8.10). Similar to this example, many reactions of organic school chemistry can be illustrated with molecule models, visualized with drawings of the indicated type, symbolized with structure symbols (Fig. 8.1).

If one wants to complete the "model kit of basic building blocks of matter" according to today's knowledge, the *ions* are added to the atoms: *Extended Dalton's atomic model*. The learners are offered a model representation of the most important atoms and ions in the form of the periodic table, as shown in Fig. 8.11. Some ion symbols have been omitted for clarity (for example Cu^{2+} or Pb^{2+})—they can be added if necessary. On the internet, differently colored periodic tables of the "atoms and ions as basic building blocks of matter" can be downloaded: www.educhem.eu [23].

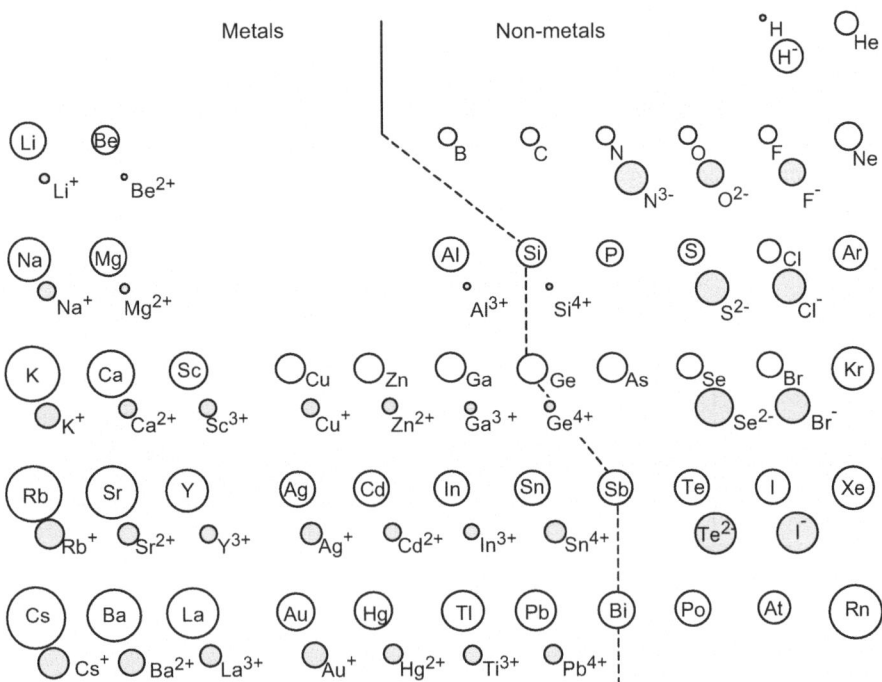

Fig. 8.11 Atoms and ions as basic building blocks of matter [23]

8.2 Teaching Processes: Everyday Language → ...

By mental combinations of ions "left and right in the PSE" (Periodic System of Elements), salt structures can be generated as mental models and chemical symbols of salts can be derived from the charge numbers of the ions: Thus, the learners mentally combine Al^{3+} ions and O^{2-} ions into an electrically balanced ion lattice with the ion number ratio 2 : 3 and find corresponding symbols $\{(Al^{3+})_2(O^{2-})_3\}$ or Al_2O_3.

Wirbs [24] was able to show in an empirical survey with 9th grade students of a secondary school that almost all participants were able to solve a given task scheme (see Fig. 8.17) without great difficulty and with symbols like $Ca^{2+}(Cl^-)_2$ or $Al^{3+}(F^-)_3$ proved that they imagine ions as the smallest particles for salt crystals [24]—and not mistakenly molecules, as other studies show [8].

It is generally possible to systematically combine atoms and ions (Table 8.5), if certain agreements are made [22]:

- Metal atoms "left and left in the PSE" can be combined with *unoriented bonding abilities* to mentally create large atomic aggregates: crystals of metals and alloys. They are illustrated by sphere packing.
- Non-metal atoms "right and right in the PSE" bind with *directed* bonding, resulting in molecules or atomic lattices (for example, graphite and diamond lattices). They can be made clear with molecule models or lattice models.
- Ions "left and right in the PSE" bind *unoriented* to ion aggregates, which are also called *ion lattices*. They can be visualized with sphere packing of at least two different sizes of spheres of different colors.

Based on the Extended Dalton model, *chemical reactions* can be adequately described for cases where the *type of particle does not change*. Thus, metal-metal reactions can be vividly illustrated as alloys, as can non-metal-non-metal reactions to volatile substances (Fig. 8.10), or solution and precipitation reactions between salts can be appropriately described, such as the dissolution of sodium hydroxide in water (Fig. 8.12). The precipitation reaction of a sparingly soluble salt from corresponding salt solutions can also be visualized, such as the precipitation of white barium sulfate (Fig. 7.16 in Chap. 7): Ions of the salt solutions are merely rearranged and fixed in the ionic lattice of barium sulfate.

In these cases, the type of particle does not change during reactions. In metal-non-metal reactions or electrolyses, atoms change to ions or vice versa ions to atoms. These changes in the type of particle can only be described with the model

Table 8.5 Linking rules for metal atoms, non-metal atoms and ions (Fig. 8.11)

Linking according to location in the PSE	Particle type	Type of bond	Structure
"left and left"	Metal atoms	Spatially unoriented	Metal lattice
"left and right"	Ions	Spatially unoriented	Ion lattice
"right and right"	Non-metal atoms	Spatially directed	Molecules, atomic lattice

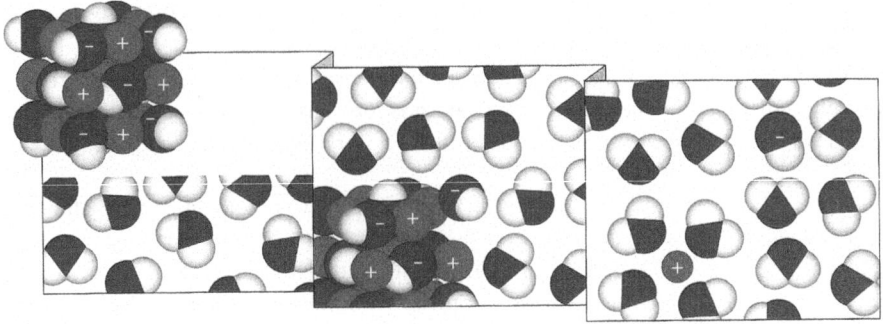

Fig. 8.12 Model concept for the dissolution of sodium hydroxide in water [13]

concept of the structure of atoms from atomic nucleus and shell and are then to be interpreted as redox reactions.

8.2.2.4 Symbolism at the Level of the Core-Shell Model

As long as *structures* of atom and ion associations are in the foreground, they can be sufficiently illustrated at the level of Dalton's model concept—a differentiated atom model is not necessary. If, beyond the distinction between spatially undirected and directed bonding ability, the *chemical bond* is to be taught, the core-shell model including the shell or energy level model of the atomic shell cannot be avoided. On the one hand, terms such as atomic nucleus, proton, neutron, and isotope must be introduced, on the other hand, the electrons and electron shells with K-, L- and M-shell, electron pair and charge cloud in the sense of the *electron pair repulsion model*. The discussion of the chemical bond leads to the distinction between ionic bond and electron pair bond, which can be delimited with the help of the concept of *electronegativity*. Polarized bonds and dipole molecules can also illustrate hydrogen bonding and van der Waals forces.

Based on the core-shell model, it is finally possible to describe the structure and bonding of simply structured substances from elements of the first three periods of the periodic table and to interpret chemical reactions with the *change of particle type*:

Acid-base reactions	Transfer of protons
Redox reactions	Transfer of electrons
Complex reactions	Exchange of ligands in complexes
Substitution reactions	Exchange of atoms or atom groups in molecules

8.2.2.5 Concepts at the Level of the Orbital Model

Simple core-shell models are sufficient as long as one discusses the structure of substances that are composed of elements from the first three periods: Up to the element calcium, the successive incorporation of electrons from shell to shell can be explained. As soon as one wants to sufficiently explain the construction of atoms of the subgroups or transition metals, one needs the differentiation of the main shells into s-, p- d- and f-*subshells* and the distribution of electrons there.

If one also aims to convey the wave-particle duality of electrons and wants to understand them as matter waves, the theory of *wave mechanics* needs to be reflected. In its sense, electron systems are described as standing waves, each standing wave is assigned a specific energy state, and quantum numbers characterize the electron system: principal quantum number n, azimuthal quantum number l, magnetic quantum number m, spin quantum number s.

Energy states can be calculated from the quantum numbers using *wave functions*, these calculations lead to specific probability densities or areas of residence of electrons: s-, p-, d- and f-electron clouds or orbitals. One distinguishes between atomic and molecular orbitals, bonding and antibonding orbitals: It should be noted that these orbitals are obtained purely formally and mathematically by combinations of wave functions, that the densities of individual electron clouds are not measurable, but only the total electron densities of an atom or molecule. Conversely, the structure and bonding of electron systems can be predicted by mathematical combinations of wave functions: *Molecular Modelling* or molecule design have become successfully possible.

For mediation processes in chemistry teaching, it should thus become clear: It is difficult to make the orbital model tangible because it is based on formal mathematical assumptions and on formal combinations of wave functions—in this sense, an electron is to be understood as a mathematical term. The electron can also be grasped neither as a wave nor as a particle: Depending on the experimental arrangement one chooses, one can confirm one or the other description. The wave-particle duality accordingly expresses that the electron behaves neither exclusively as a wave nor exclusively as a particle. Because of this fundamentally lacking tangibility, teachers must carefully consider to what extent they should better choose phenomena that can be explained with simpler model ideas. Only when experimental findings cannot be explained otherwise, the wave mechanical atom conception may play a role.

The wave-particle duality also plays an important role in light. On the one hand, light behaves as an electromagnetic wave (for example, in interference phenomena at the optical grating), on the other hand, light is described with corpuscles (photons in the experiment for the photoelectric effect). This duality is dealt with in physics teaching at the high school level—the chemistry teacher can possibly refer to it when he wants to work out the electron dualism.

8.2.3 Derivation of First Chemical Symbols in Teaching

Because of the outstanding and international importance of chemical symbols as a unique means of communication among chemists, the introduction and use of symbols in chemistry teaching is of particular importance. Therefore, the manner of their introduction has been extensively argued in all chemistry education journals at all times: The historical-empirical derivation of formulas from the comparison of mass ratios and the foundation of structure models are currently the most discussed ways.

8.2.3.1 Historical-Empirical Derivation of Chemical Symbols

At the level of the Dalton model, it is historically possible to derive the "empirical formula", the sum symbol, from mass ratios of the reacting substances and by comparing with corresponding atomic masses. As with classical elemental analysis, the masses of the substances used and the masses of the products formed after complete reaction are weighed. After converting the masses into quantities of substance, the numerical ratio of atoms in the starting substance and thus the composition can be given as an empirical formula. If the substance is made up of molecules, a molecular mass determination follows. To get to the molecular structure, specific investigations using spectroscopic methods must be carried out and evaluated in the sense of structure elucidation.

The classical analysis path is abstract because it calculates empirically determined mass ratios with theoretical atomic masses and leads to the atomic number ratio via a mass comparison. Most learners at the secondary level I are overwhelmed by this. Kaminski and Jansen [25] therefore suggest using the atomic numbers present per 1 mg of substance and calculating directly. This is a sensible simplification, but the problem remains that these are naturally numbers on the order of 10^{17} and students are not practiced in dealing with powers of ten.

The use of analytical balances for the derivation of chemical symbols is and remains historical—corresponding analysis methods were developed by Liebig and Berzelius in the middle of the nineteenth century and were necessary for a long time. From 1960 onwards, the methods of instrumental analysis became available in chemical institutes: spectral analysis, X-ray structure analysis, atomic absorption spectrometry, UV, IR, NMR spectroscopy and mass spectrometry. Today, analyses of substances are carried out exclusively with these and other investigation methods. These also include all kinds of separation methods, such as the methods of fractional distillation, chromatography and electrophoresis.

8.2.3.2 Derivation of Symbols from Structural Models

With the help of X-ray structure analysis, for example, the structure of crystalline substances can be determined. From computer printouts or spatial representations of the structure, bond angles, bond lengths, lattice constants or all forms of symbols can be determined: both structure and sum symbols. If you show Laue diagrams in chemistry lessons using some examples and illustrate the underlying diffraction phenomena with laser beam model experiments, the *principle* of X-ray structure analysis can also be grasped by learners [26].

If you base your desired substance on the finished structure model of a molecule or crystal lattice determined by experts, the students can imagine the structure and derive the chemical symbol from the model by counting the numerical ratio of atoms or ions [27]. If you choose the well-known cubic section from the sodium chloride

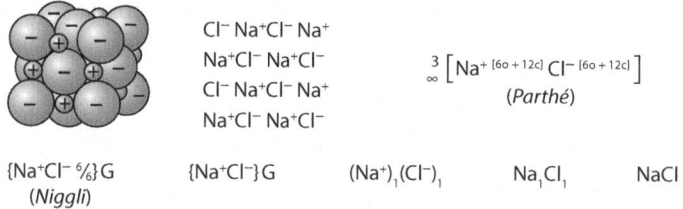

$Cl^-\,Na^+Cl^-\,Na^+$
$Na^+Cl^-\,Na^+Cl^-$
$Cl^-\,Na^+Cl^-\,Na^+$
$Na^+Cl^-\,Na^+Cl^-$

$^{3}_{\infty}\left[Na^{+\,[6o+12c]}\,Cl^{-\,[6o+12c]}\right]$
(*Parthé*)

$\{Na^+Cl^-\,{}^6\!/_6\}G$ (*Niggli*) $\{Na^+Cl^-\}G$ $(Na^+)_1(Cl^-)_1$ Na_1Cl_1 NaCl

Fig. 8.13 Derivation of the NaCl symbol from the sodium chloride structure

structure (Fig. 8.13), the expert can clarify the Parthé and Niggli symbols, the learner can derive the symbols like Na^+Cl^- or NaCl by further shortening (Fig. 8.13).

This teaching approach also corresponds to the didactic demand to insert the level of structural ideas for understanding chemical symbols, as was already explained with Fig. 7.9 in Chap. 7 as the "Chemical Triangle" and is shown again slightly changed in Fig. 8.14: first structural ideas, then chemical symbols!

This teaching approach corresponds to some considerations on modern structural analysis: Its results lead in all cases to the *smallest structural units* of substances. In the case of molecular substances, the *molecule* inevitably proves to be such a unit and the chemical symbol is given for the molecule. For example, the symbols CH_3COOH or C_6H_6 result for acetic acid or benzene molecules, respectively, and no one would shorten these symbols to CH_2O or C_1H_1 or even CH.

This agreement can be transferred to the symbolism of salt crystals or ion lattices, the smallest structural unit should also be symbolized here: the *unit cell* . If we take the example of sodium chloride, its unit cell can be described with the lattice symbol $\{(Na^+)_4(Cl^-)_4\}$ or Na_4Cl_4 (Fig. 7.3 in Chap. 7). In the process of mediation, symbols such as Na_4Cl_4, Li_8O_4 or Zn_4S_4 for structural units could be formulated accordingly, starting from demonstration and counting of unit cell models [27]—just like the symbols CH_3COOH or C_6H_6 for molecules as smallest units.

In the second step, the reduction to the usual—albeit least informative—symbols such as NaCl, Li_2O or ZnS is allowed, because they are internationally

Appearances	Substances	→	Chemical reactions
Continuum			
Discontinuum	↓		↓
Structural concepts	Models for the smallest structural units	→	Structural models for substances before and after the reaction
	↓		↓
Chemical symbols	Symbols for the smallest structural units	→	Regrouping symbols, reaction symbols

Fig. 8.14 Smallest structural units for understanding chemical symbols [27]

common. In this way, students learn the meaning of the different symbols by concrete counting on the model, imagine the unit cell or the endless crystal lattice in the case of crystals, and develop a modern understanding of chemistry. Last but not least, tests on *spatial ability* have shown that the majority of students in secondary level I are able to successfully count two-dimensional unit cells [28].

8.3 Learners: Student Conceptions of Chemical Structures and Symbols

Abbreviations by letters are very familiar to all students—for example, NATO or UNO for corresponding political societies. For this reason, students may misunderstand that chemical symbols are similar formal abbreviations of substance names: Examples like NaCl, CaO or MgO reinforce this view. If students have no structural conceptions, they cannot interpret the indices in symbols like H_2O or Al_2O_3 in terms of structural chemistry and learn these symbols by heart. Thus, they remain—as was also historically the case—with misunderstood conceptions and consider the chemical symbolism as a secret language of chemists.

8.3.1 Conceptions of Combustion

Already in Chap. 2 the conception of a student about magnesium combustion was presented and discussed as an example: This student had formulated the correct reaction symbol "$2Mg + O_2 \rightarrow 2MgO$", but as a conception he had written: "Magnesium consists of two types of particles, one evaporates during combustion, the other remains as magnesium oxide", and made a corresponding drawing (Fig. 2.3). A survey of about 300 young people showed that almost all gave a correct reaction symbol, but 70% of the subjects noted factually inappropriate conceptions or drew incorrect models [29]. It becomes clear that everyday conceptions, which young people have formed over years from observing combustions in their living environment, cannot be converted into scientific conceptions with the one-time formulation of reaction symbols. Only demonstration or construction of corresponding models for the structure of substances before and after combustion promote on the one hand the correct understanding of the combustion process, on the other hand the reaction symbol becomes comprehensible and thus understandable (see also Fig. 8.1).

8.3.2 Concepts of the Ion Idea

In a group of students who had already acquired the concepts of ion and redox reaction in the 10th grade of a high school and had described precipitation reactions with reaction symbols, the following test was conducted [9]:

8.3 Scientific Focus: Student Conceptions of Chemical Structures and Symbols

1. The reaction of nickel oxide with aluminum and the corresponding bright flash of light were shown, and the students were asked to draw their models of the structure of nickel oxide and aluminum crystals, and provide related reaction symbols in words, structures, and formulas.
2. Concentrated solutions of calcium chloride and sodium sulfate were mixed together and white crystals were observed as a precipitate. Model drawings of the two solutions and the reaction symbols in words, structures, and formulas were to be provided. In addition, the students had been clearly shown by examples that by "structures" the marking of ion symbols in the case of existing ions is meant, the marking of molecular structures in the case of existing molecules.

Only a few students solved the tasks completely. The statistics show [9] that almost all participants correctly formulate reaction symbols in words, but only a maximum of 20% structures or formulas. Up to 80% mark conceptions as they have been selected in Fig. 8.15. The central result of the study is that students very often describe the formation of ions from the corresponding atoms incorrectly: They do not accept "ready-made ions" in salt solutions, but postulate their respective formation from the atoms. They also do not note the ions of the metal oxides, but resort to molecules or offer ion and molecule descriptions in parallel (Fig. 8.15, dashed box). Some students actually asked with which *notation* they should record the answers—for them, the impression has arisen as if the question "ions or molecules" had to do with a "notation"!

These conceptions are "*homemade*" [8] and have occurred due to deficiencies in teaching—there are no original everyday conceptions as they were previously presented for combustion. It seems advantageous to provide learners with a list of atoms and ions as "basic building blocks of matter" as shown in Fig. 8.11. Further surveys on incorrect conceptions regarding other school topics can be found in Barke [8].

8.3.3 Concepts of Stoichiometry

Schmidt [30] was able to show through empirical surveys in the field of stoichiometric calculations that only a small part of the learners manage to acquire these skills. He identified the following "misconceptions": No distinction between equation coefficients and formula indices, for example between 2 O and O_2, no distinction between the molar ratio and the mass ratio, equality of the amounts of reactants and products, equality of the volumes of reactants and products in gas reactions, among others [30].

Schmidt found these "misconceptions" by constructing and evaluating specific multiple-choice tasks with suitable distractors [30]: They were designed in such a way that the students had to deal with numbers that fit both the correct answer and a wrong answer when solving the tasks.

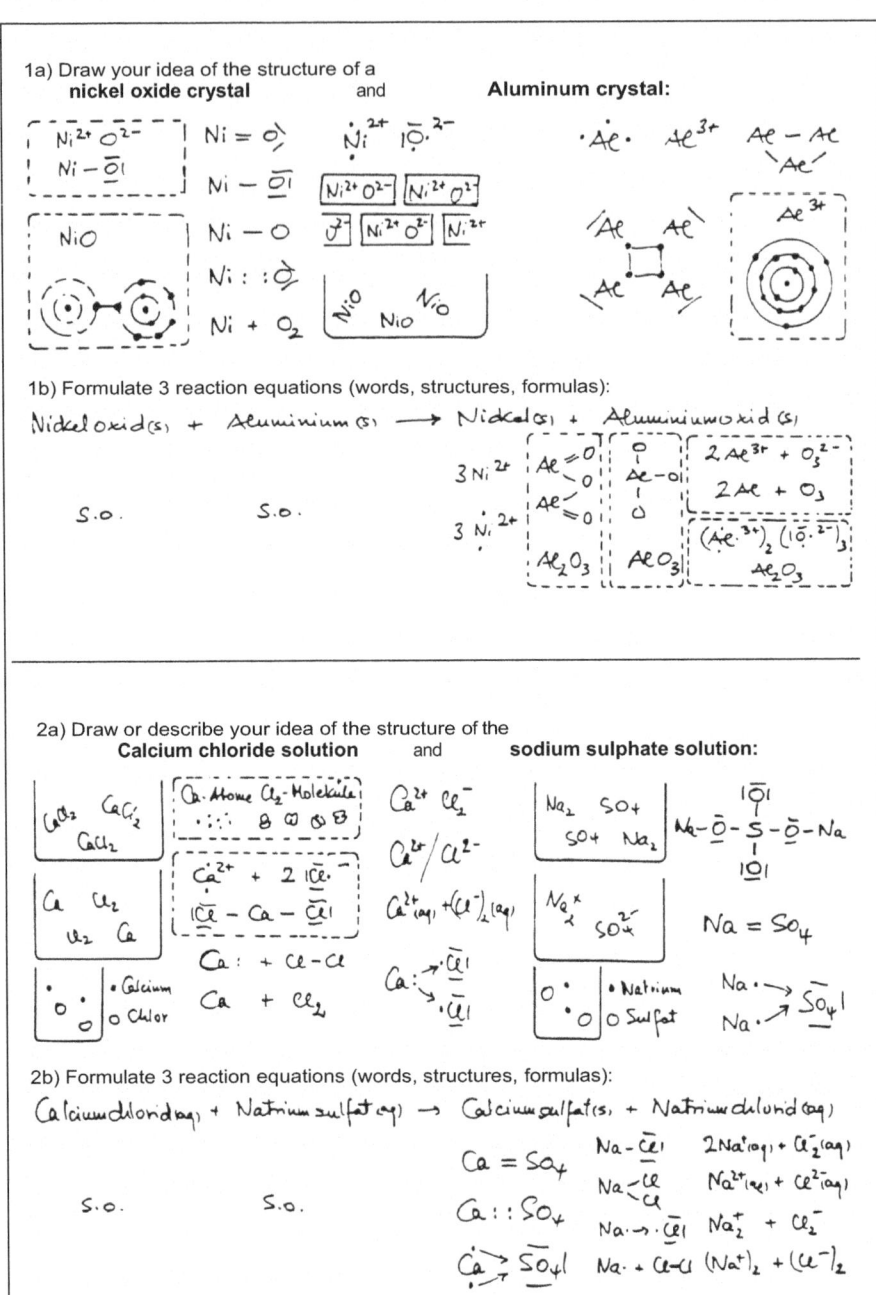

Fig. 8.15 Selected incorrect answers from learners of the 10th grade of a high school (answers framed in dashed lines come from the same student) [9]

As an example, "Task 91,000" is quoted [30]: "In 2 g of a compound, 1 g of copper is contained, the rest is sulfur. Which chemical formula fits these specifications? CuS (A), CuS_2 (B), Cu_2S (C) or Cu_2S_2 (D)"? The correct solution (B) should be found: since S atoms are about half as heavy as Cu atoms, the numerical ratio Cu atoms: S atoms = 1:2 must be.

Schmidt tried to find certain strategies according to which the students make typical mistakes and arrive at wrong solutions: "In chemistry lessons, you cannot avoid these misconceptions. They should not be suppressed, but made aware to the students and overcome by thinking together with them about the mistakes in their strategies". If it is not possible in class to solve stoichiometric tasks in a comprehensible way, then the students might turn away from chemistry or deselect chemistry for the upper level: "Perhaps stoichiometry is the crossroad at which it is decided whether a student finds his way into chemistry or not. It is therefore important to know where the difficulties come from" [30]. In order to eliminate these difficulties as early as possible in the initial chemistry lessons, the START concept was developed and successfully tested [31].

8.3.4 Laboratory Jargon and Misconceptions

Even the way lecturers and teachers speak can generate misconceptions. For example, learners hear that "two hydrogens and one oxygen form two waters", or as observed in Tanzania: "two hydrogen gas and one oxygen gas form two water". With these statements, they rightly ask themselves whether two grams or two milliliters of hydrogen are meant—or even two bubbles of the gas. The expert will correctly deduce that probably "2 H_2 molecules and 1 O_2 molecule are meant, which react to 2 H_2O molecules"—but young learners can never and never come to the correct description of the formation of water from the elements with the "laboratory jargon". This so-called laboratory jargon may be a welcome abbreviation of speech for experts—but it has no place in the teaching process for young learners. There are many more "jargon statements" that need to be reflected upon.

In the area of *acid-base reactions*, students often hear: "Hydrochloric acid gives off one proton, sulfuric acid two protons". In the last example, is pure sulfuric acid meant, which contains H_2SO_4 molecules and can indeed give off two protons each to H_2O molecules? In diluted sulfuric acid, there are no more H_2SO_4 molecules, there the H_3O^+ ions are proton donors—as it also applies to hydrochloric acid. Learners all too gladly take the substances and argue with them, although they have learned in lectures and seminars that the Brönsted theory presupposes the molecules or ions that can each give off or take up one or two protons. In Table 8.6, a series of further nebulous statements of laboratory jargon are compiled, which should be corrected to the scientific concept [32].

The last example in particular clearly shows that when using the respective reacting molecules or ions, it becomes apparent that the hydronium ions or hydroxide ions added to the buffer solution convert into water molecules, thereby

Table 8.6 Examples of laboratory jargon for acids and bases, appropriate terminology based on the Brönsted theory (Proton = H^+ ion, HAc = $HOOCCH_3$ molecule) [32]

Laboratory Jargon	Appropriate Formulation (Brönsted)
1. Acid-Base Definitions (also historical)	
Acids contain hydrogen, which can be replaced by a metal during neutralization: CH_3COOH becomes CH_3COONa (Liebig, 1834)	Molecules or ions are acid particles or proton donors. They contain H atoms like HAc molecules, which can be released as H^+ ions, for example to H_2O molecules: $HAc + H_2O \rightarrow H_3O^+(aq) + Ac^-(aq)$ In neutralizations, HAc molecules and H_3O^+ ions react with OH^- ions of the sodium hydroxide solution, $Na^+(aq)$ ions do not participate in the reaction
Hydrochloric acid dissociates into ions: $HCl \rightarrow H^+ + Cl^-$	In hydrochloric acid, $H_3O^+(aq)$ ions and $Cl^-(aq)$ ions are present, the $H_3O^+(aq)$ ions are the acid particles or proton donors.
Sulfuric acid dissociates into ions: $H_2SO_4 \rightarrow 2H^+ + SO_4^{2-}$ (Arrhenius, 1884)	In pure sulfuric acid, H_2SO_4 molecules are the proton donors: during the reaction with water, protolysis occurs, in diluted sulfuric acid, the $H_3O^+(aq)$ ions are the acid particles (also $HSO_4^-(aq)$ ions).
Sodium hydroxide dissociates: $NaOH \rightarrow Na^+ + OH^-$ (Arrhenius, 1884)	In solid sodium hydroxide, Na^+ and OH^- ions are already present in the ionic lattice, they are separated by H_2O molecules, forming hydrated ions: $Na^+OH^- \rightarrow aq \rightarrow Na^+(aq) + OH^-(aq)$
The self-dissociation of water produces H^+ and OH^- ions in equilibrium: $H_2O \rightleftarrows H^+ + OH^-$	The autoprotolysis of H_2O molecules produces the ions in the protolysis equilibrium: $H_2O + H_2O \rightleftarrows H_3O^+ + OH^-$
Water is an ampholyte—it can be an acid or a base	H_2O molecules are ampholytes: Depending on the reaction partner, they can take up or release an H^+
Sodium bicarbonate is an ampholyte	$HCO_3^-(aq)$ ions are ampholytes: they can take up or release an H^+ ion
The concentration of water is: $c = 55.5$ mol/L	The concentration of H_2O molecules in water is: $c = 55.5$ mol H_2O molecules/Liter
Strong acids have a low pH value, weak acids a relatively high pH value	Strong acids are fully protolyzed, weak acids like HAc molecules only to a very small extent, there is an equilibrium: $HAc + H_2O \rightleftarrows H_3O^+(aq) + Ac^-(aq)$ Solutions of strong acids also take on high pH values like 5 or 6 when diluted
Citric acid is a weak acid	The HCit molecule is a weak acid, HCit molecules are in equilibrium with ions: $HCit + H_2O \rightleftarrows H_3O^+(aq) + Cit^-(aq)$
Ammonia is a weak base	The NH_3 molecule is a weak base, NH_3 molecules are in equilibrium with ions: $NH_3 + H_2O \rightleftarrows NH_4^+(aq) + OH^-(aq)$
2. Example Carbonate-Acid Reaction	
Sodium carbonate reacts with hydrochloric acid, releasing carbon dioxide gas: $Na_2CO_3(s) + 2HCl(aq) \rightarrow 2NaCl(aq) + H_2O + CO_2(aq, g)$	Base particles and acid particles react: $CO_3^{2-} + 2H_3O^+(aq) \rightarrow H_2CO_3 + 2H_2O$ $Na^+(aq)$ and $Cl^-(aq)$ are spectator ions, they do not react, H_2CO_3 molecules react further: $H_2CO_3(aq) \rightleftarrows H_2O + CO_2(aq, g)$

(continued)

Table 8.6 (continued)

Laboratory Jargon	Appropriate Formulation (Brönsted)
Calcium carbonate reacts with citric acid: $CaCO_3(s) + 2HCit(aq) \rightarrow CaCit_2(aq) + H_2O + CO_2(aq, g)$	$CO_3^{2-} + 2HCit(aq) \rightarrow H_2CO_3 + 2Cit^-(aq)$ $Ca^{2+}(aq)$ are spectator ions, they do not react. H_2CO_3 molecules react further (see above)
3. Neutralization of Acid Solutions	
Example Hydrochloric Acid-and Sodium Hydroxide solution: HCl is the acid, NaOH the base, the salt sodium chloride and water are formed: $HCl + NaOH \rightarrow NaCl + H_2O$. Neutralization means salt formation.	$H_3O^+(aq)$ ions of hydrochloric acid are acid particles and $OH^-(aq)$ ions of sodium hydroxide solution are base particles. Both react to form H_2O molecules: $H_3O^+(aq) + OH^-(aq) \rightarrow 2H_2O$ Accompanying ions $Na^+(aq)$ and $Cl^-(aq)$ do not react.
Example Acetic Acid and -Sodium Hydroxide soliution: Acid HAc and base NaOH react, the salt sodium acetate and water are formed: $HAc + NaOH \rightarrow NaAc + H_2O$	In acetic acid solutions, there are two types of acid particles that react with OH^- ions: 1. HAc molecules + $OH^-(aq)$ ions $\rightarrow H_2O + Ac^-(aq)$ 2. $H_3O^+(aq) + OH^-(aq) \rightarrow 2H_2O$ Accompanying ions $Na^+(aq)$ and $Ac^-(aq)$ do not react
4. Acetic Acid-Acetate Buffer	
The aceiticc acid buffer buffer is a mixture of acetic acid and sodium acetate solution. When acid is added, the acetate solution reacts: $NaAc + HCl \rightleftarrows NaCl + HAc$ When base is added, the acetic acid reacts: $HAc + NaOH \rightleftarrows NaAc + H_2O$ pH values do not change	In the buffer solution (pH 4.8), HAc molecules and $Ac^-(aq)$ ions are present in approximately equal concentration. Added $H_3O^+(aq)$ ions react to form H_2O molecules, the pH value remains: $H_3O^+(aq) + Ac^-(aq) \rightleftarrows H_2O + HAc$ Added $OH^-(aq)$ ions react to form H_2O molecules, the pH value does not change: $OH^-(aq) + HAc \rightleftarrows H_2O + Ac^-(aq)$

neutralizing their effect as an acid or base—the pH value thus does not change. In general, when using the Brönsted definition, it is always necessary to decide which molecules or ions react as acid particles and which as base particles. Learners thus train their interpretation at the submicro level and understand chemistry. In particular, beaker models (Fig. 7.18 in Chap. 7) make the argument even more vivid.

Lab jargon is also common when it comes to *redox reactions*. For example, the reaction of iron with copper sulfate solution is interpreted as "iron is oxidized, it gives off two electrons, copper sulfate is reduced and takes up two electrons" [32]. The smallest portions of iron already give off billions of electrons—the iron atom is of course meant, which gives off two electrons, and at the same time a Cu^{2+} ion of the salt solution can take up two electrons. Similar examples are listed in Tab. 8.7 [32].

The symbols H^+ and e^- suggest independently existing protons and electrons, which switch from one particle to another. These model ideas are

Tab. 8.7 Examples of lab jargon for redox reactions and appropriate terminology [32]

Lab jargon	Appropriate formulation
1. Redox definitions (historical)	
In the combustion of magnesium, magnesium is oxidized and takes up oxygen: $2Mg(s) + O_2(g) \rightarrow 2MgO(s)$	Mg atoms are oxidized, O atoms are reduced: 2Mg atoms $\rightarrow$ 2Mg^{2+} ions + 4e$^-$ O_2 molecule + 4e$^-$ $\rightarrow$ 2O^{2-} ions Mg^{2+} and O^{2-} ions form Mg^{2+}O^{2-} ion lattice.
In the reaction of copper oxide with magnesium, copper oxide is reduced and gives off oxygen, magnesium is oxidized, takes up oxygen: $CuO(s) + Mg(s) \rightarrow Cu(s) + MgO(s)$	Cu^{2+} ions are reduced, Mg atoms are oxidized: Mg atom $\rightarrow$ Mg^{2+} ion + 2e$^-$ Cu^{2+} ion + 2e$^-$ $\rightarrow$ Cu atom O^{2-} ions merely change the ion lattice
2. Metal-acid reactions	
Magnesium reacts with acid solutions, gaseous hydrogen escapes: $Mg(s) + 2HCl(aq) \rightarrow MgCl_2(aq) + H_2(g)$ $Mg(s) + H_2SO_4(aq) \rightarrow MgSO_4(aq) + H_2(g)$	Mg atoms are oxidized, H$^+$ ions are reduced: Mg atom + 2H$^+$(aq) $\rightarrow$ Mg^{2+} ion + H$_2$ molecule Mg^{2+}(aq) ions go into solution and form a magnesium chloride solution with the Cl$^-$(aq) ions of the hydrochloric acid, with the SO$_4^{2-}$(aq) ions a magnesium sulfate solution is formed.
Magnesium reacts with pure sulfuric acid, gaseous hydrogen sulfide is released: $4Mg(s) + 5H_2SO_4(l) \rightarrow H_2S(g) + 4MgSO_4(s) + 4H_2O$	Mg atoms are oxidized, S atoms of the H$_2$SO$_4$ molecules are reduced to S atoms in H$_2$S molecules: 4Mg atoms $\rightarrow$ 4Mg^{2+} ions + 8e$^-$ H$_2$SO$_4$ molecule + 8H$^+$ + 8e$^-$ $\rightarrow$ H$_2$S + 4H$_2$O (S atom with oxidation number +VI goes to S atom with -II)
3. Metal-Salt Solution Reactions	
Iron reacts with copper chloride solution, iron is oxidized, copper chloride is reduced: $Fe(s) + CuCl_2(aq) \rightarrow Cu(s) + FeCl_2(aq)$	Fe atoms are oxidized, Cu^{2+} ions are reduced: Fe atom + Cu^{2+} ion $\rightarrow$ Cu atom + Fe^{2+} ion Cl$^-$(aq) ions of the salt solution are accompanying ions.
Copper reacts with silver nitrate solution, copper is oxidized, silver nitrate is reduced: $Cu(s) + 2AgNO_3(aq) \rightarrow 2Ag(s) + Cu(NO_3)_2(aq)$	Cu atoms are oxidized, Ag$^+$ ions are reduced: Cu atom + 2Ag$^+$ ions $\rightarrow$ 2Ag atoms + Cu^{2+} ion Nitrate ions of the salt solution are accompanying ions
4. Potassium Permanganate-Hydrochloric Acid Reaction	
In the reaction, gaseous chlorine is formed, potassium permanganate is reduced: $KMnO_4(s) + 4HCl(aq) \rightarrow 1.5\,Cl_2(g) + MnO_2(s) + KCl(aq) + 2H_2O$	Mn atoms of the MnO$_4^-$ ion are reduced (+VII $\rightarrow$ +IV), Cl$^-$ ions of the hydrochloric acid are oxidized to Cl atoms (-1 --> 0): 3Cl$^-$ ions $\rightarrow$ 3Cl atoms + 3e$^-$ MnO$_4^-$-Ion + 4H$^+$(aq) + 3e$^-$ $\rightarrow$ MnO$_2$ + 2H$_2$O
5. Oxygen Corrosion	
Iron corrodes in moist air to form iron hydroxide, iron is oxidized in the process: $2Fe(s) + 2H_2O + O_2(aq) \rightarrow 2Fe(OH)_2(s)$	Fe atoms are oxidized, O atoms are reduced: 2Fe atoms $\rightarrow$ 2Fe^{2+} ions + 4e$^-$ 2H$_2$O + O$_2$ + 4e$^-$ $\rightarrow$ 4OH$^-$ ions. These ions form the ionic lattice Fe^{2+}(OH$^-$)$_2$

preliminary—they are to be expanded in later lessons when atoms, ions, and molecules are described with electron clouds. Using the reaction of HCl- and H_2O molecules as an example, a model is proposed that starts from the transition of protons from electron cloud to electron cloud (Fig. 8.16). Christen and Baars [33] write about this: "There are no free, independently existing H^+ ions (as postulated by Arrhenius in 1883); the proton initially bound to the HCl molecule by a pair of electrons detaches and 'slips' into one free cloud of the H2O molecule" [33].

Electrons can indeed be regarded as particles due to the wave-particle duality, but they rather represent elementary charges. These charges are emitted from electron clouds of atoms, ions, or molecules and absorbed by others—but they always form charge clouds with measurable electron density, never point-like structures. Therefore, the Bohr atomic model with the nucleus and "orbiting electrons" around it is always very preliminary, as are all images that are supposed to indicate a particle with a point or the symbol e^-.

In many cases, electrical charge is not transferred, but only shifted: "The redox process is an electron shift" [33]. For example, hydrogen and oxygen react to form water, the reaction is referred to as a redox reaction: $2H_2 + O_2 \rightarrow 2H_2O$. Although oxidation numbers can simulate an electron transfer, in reality, the electron pairs or electron clouds of the O-atoms are merely shifted: From nonpolar covalent bonds in the O_2 molecule, strongly polar electron pairs shifted towards the O-atom in the formed H_2O molecule—an electron shift is the result. For young learners, however, the agreement remains: They may regard protons and electrons as the smallest particles and work out proton or electron transfers as described.

Pilot study on laboratory jargon.
We wanted to investigate to what extent students of chemistry teaching use laboratory jargon uncritically or are able to reflect on corresponding statements. A questionnaire was developed [34], which is intended to show to what extent students relativize quoted statements on laboratory jargon and mark one of the four offered multiple-choice answers that is correct. An example:

Laboratory jargon: "Hydrochloric acid releases a proton"

a. Hydrochloric acid can be deprotonated.
b. Hydrochloric acid can also accept protons.
c. In hydrochloric acid, H_3O^+(aq) ions are present, they can release protons.
d. HCl molecules are present in hydrochloric acid, they release protons [34].

The desired answer is of course (c): "H_3O^+(aq) ions are present, they can release protons". We also offered the popular answer (d) to see to what extent the idea of HCl molecules in diluted hydrochloric acid is present. Because it sounds so scientific, (a) is an attractive distractor, while (b) is just fake.

The correct answer (c) was satisfactorily chosen by 40% of the participants, the drastic misconception (d) was only marked by 5%. However, the answer (a) reached the majority of 55%: Many students think of the scientifically good sound

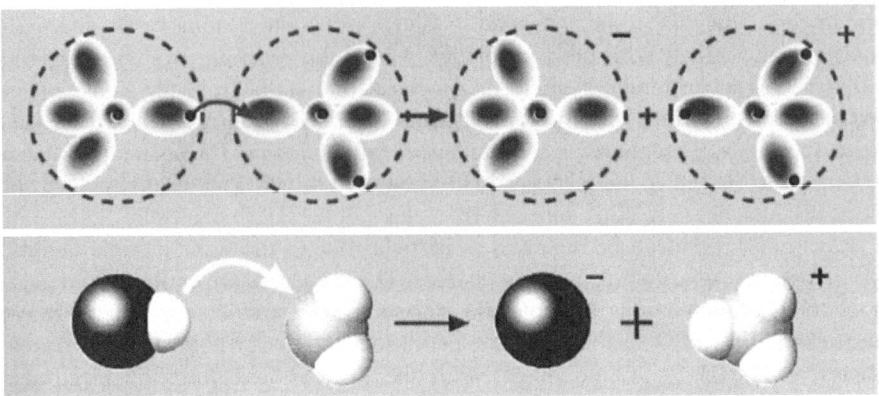

Fig. 8.16 Model representation of proton transfer between electron clouds [33]

Name of the salt	Ions involved	Numerical ratio of the ions	Sum symbol
Calcium fluoride	Ca^{2+}, F^-	$\{(Ca^{2+})_1(F^-)_2\}$	CaF_2
Calcium nitride			
Barium chloride			
Aluminum fluoride			
Lithium oxide			
Sodium hydroxide			
Calcium hydroxide			
Magnesium nitrate			
Sodium carbonate			
Calcium sulphate			
Aluminum sulfate			
Potassium aluminum Sulfate			

Fig. 8.17 Worksheet for formulating symbols for ion lattices

of "deprotonation"—but do not reflect on how the "substance hydrochloric acid solution might be deprotonated". Overall, in the ten tasks of the questionnaire, half of all students answered correctly—at least the other half is subject to laboratory jargon [34]. The entire questionnaire is task A8.5 in section 8.5.

8.4 Social Reference Fields: Laypeople and Chemical Jargon

Scientific language and symbols serve within a science as a means of simple and rational communication, as a communication tool with high information content, which is understood equally worldwide, regardless of cultural circles, national languages, scripts, and social systems. At the same time, however, every technical language creates a demarcation against all those who are not familiar with it. The high level of development of the various languages even results in serious problems for the communication between the sciences: Only those who master the scientific language of a specific science can communicate with its representatives about factual issues.

In addition, scientific language complicates communication with the general public and understanding of scientific problems [35]. It often generates mistrust towards experts in public discussion when they are unable to translate technical terms into everyday language. This situation has led to the development of *popular science publications* that aim to make science understandable to laypeople.

Teachers should see themselves in their role not only as competitors of these popularizations in magazines, radio, television, and the internet, but above all as constructive mediators who respond to the questions of the students. In particular, teachers are called upon to reflect on the current information about the advertising of many products on the internet, on television and radio with the learners. When commercials for medications and ointments increasingly announce that the products are "chemical-free", but the study of the list of ingredients then specifies specific chemicals, teachers can have these contradictions discussed by the students and help them become informed consumers. They may also integrate everyday dialogues [36] as a cross-curricular stylistic device into the subject teaching to give learners aids to understand and critically view the scientific facts presented in the media: an ambitious goal of teaching chemical language !

8.5 Exercise Tasks

A8.1
In chemistry, a variety of symbols ("formula types") have been agreed upon. Specify different types for a) three crystalline solids, b) three volatile substances. What information content is hidden behind the different symbols?

A8.2
In the development of the chemical symbol language, it makes sense to first teach learners in everyday language, then switch to scientific language, and finally introduce chemical symbols. Explain this path using two factual situations.

A8.3
Chemical symbols can look very different at various stages of the curriculum spiral: They can be words, sum symbols, or structure symbols. Choose two different reactions and formulate reaction symbols at these three mentioned levels.

A8.4
Combine mentally "ions left and right in the PSE" (Fig. 8.11) and give three examples of salt crystals. Draw a cube in perspective and into it corresponding space lattices. Determine chemical symbols for ion lattices based on the charge numbers of ions using examples from the worksheet in Fig. 8.17.

A8.5
Work on the following questionnaire about problems with lab jargon in teaching and chemistry lessons. Find the correct answer and discuss in particular the alternative answers in terms of the argumentation pro substances and contra smallest particles like atoms, ions, and molecules.

Multiple-Choice Questionnaire on Lab Jargon
Which answer would be the correct expression? Please tick the box.

1. *Lab jargon: "Carbon dioxide consists of carbon and oxygen"*
   a) CO_2 consists of one C and two O.
   b) Carbon dioxide consists of carbon and oxygen.
   c) CO_2 consists of one part carbon and two parts oxygen.
   d) CO_2 molecules consist of one C atom and two O atoms.

2. *Lab jargon: "Hydrochloric acid gives off a proton"*
   a) Hydrochloric acid can be deprotonated.
   b) Hydrochloric acid can also accept protons.
   c) In hydrochloric acid, $H_3O^+_{(aq)}$ ions are present. They give off protons (H^+ particles).
   d) HCl molecules are present in hydrochloric acid, they give off protons.

3. *Lab jargon: "The self-dissociation of water yields H^+ and OH^- ions in equilibrium"*
   a) The equilibrium of water yields protons and hydroxide ions.
   b) Water can dissociate both H^+ ions and OH^- ions.
   c) The autoprotolysis of H_2O molecules yields H_3O^+ ions and OH^- ions.
   d) H_2O yields protons and hydroxide ions in autoprotolysis.

8.5 Exercise Tasks

4. *Lab jargon: "Ammonia is a weak base"*
   a) The NH_3 molecule is a weak base, NH_3 molecules are in equilibrium with corresponding ions.
   b) Ammonia solution is weakly concentrated.
   c) NH_3 molecules react completely to NH_4^+ molecules.
   d) Ammonia forms ammonium chloride with an HCl molecule.

5. *Lab jargon: "The concentration of water is $c = 55.5$ mol/L"*
   a) The concentration of H_2O is 55.5 mol/L.
   b) The concentration of H_2O molecules in water is: $c = 55.5$ mol H_2O molecules/L.
   c) Water consists of 2 mol hydrogen and 1 mol oxygen.
   d) Water is 100% composed of hydrogen and oxygen.

6. *Lab jargon: "Sodium hydroxide dissociates in water into Na^+ ions and OH^- ions"*
   a. NaOH molecules dissociate in water into Na^+ ions and OH^- ions.
   b. Solid NaOH consists of Na^+ and OH^- ions, in water $Na^+_{(aq)}$ and $OH^-_{(aq)}$ ions are formed.
   c. Na^+OH^- ion pairs of solid sodium hydroxide separate into individual ions.
   d. In water, Na atoms transfer electrons to OH groups, forming ions.

7. *Lab jargon: "Hydrochloric acid neutralizes with sodium hydroxide to water and salt"*
   a. Neutralization means salt formation.
   b. After neutralization, equal concentrations of acid and base are present.
   c. $H^+Cl^-_{(aq)} + Na^+OH^-_{(aq)} \rightarrow H_2O_{(l)} + Na^+Cl^-_{(aq)}$
   d. $H^+_{(aq)} + Cl^-_{(aq)} + Na^+_{(aq)} + OH^-_{(aq)} \rightarrow H_2O + Na^+_{(aq)} + Cl^-_{(aq)}$

8. *Lab jargon: "Strong acids have a low pH value, weak acids a higher pH value"*
   a. Strong/weak acids are highly/weakly concentrated.
   b. The pH value indicates the concentration of the acid.
   c. The pH value indicates the concentration of H^+ ions.
   d. Weak acids have a pH value between 3 and 6.

9. *Lab jargon: "Indicator papers show the strength of an acid"*
   a. Indicator papers show whether an acid is strong.
   b. Indicator papers show strong or weak acids.
   c. Indicator papers show how concentrated an acid is.
   d. Indicator papers can show whether acid or base is present.

10. *Lab jargon: "Water is an ampholyte, it can be acid and base"*
    a. The H_2O molecule is an ampholyte particle, it can accept a proton (H^+) and it can donate a proton (H^+).
    b. Water can be both acid and base.
    c. H_2O is acid and base at the same time, the molecules dissociate into H^+ and OH^- ions.
    d. H_2O is acidic, basic, or neutral.

References

1. v. Zahn P (1981) Freund und Helfer oder heimlicher Feind ? Chemie im Kreuzfeuer der öffentlichen Meinung. CU 12:1
2. Dörrenbächer A (1995) IUPAC-Regeln und DIN-Normen im Chemieunterricht. Aulis, Köln
3. Kremer M (2012) Grundbildung in den naturwissenschaftlichen Fächern. MNU-Themenreihe Bildungsstandards. Seeberger, Neuss
4. Indriyanti NY (2017) Das Molkonzept durch Experiential Learning. Eine empirische Studie zu Chemieunterricht in Deutschland und Indonesien. Chemkon 24:64
5. Hörnig J, Habelitz-Tkotz W (2015) Chemisches Rechnen – unbeliebt, aber unverzichtbar. PdN-CiS 2(64):23
6. Barke H-D (1998) Strukturorientierte Einführung in die Allgemeine und Anorganische Chemie Vol. 1. Schüling, Münster
7. Dorn F, Bader F (1989) Physik in einem Band. Schroedel, Hannover (Neubearbeitung)
8. Barke H-D (2006) Chemiedidaktik – Diagnose und Korrektur von Schülervorstellungen. Springer, Heidelberg
9. Barke H-D (1988) Chemiedidaktik zwischen Philosophie und Geschichte der Chemie. Peter Lang, Frankfurt
10. Herdt Chr (2015) Bindigkeit und Ionenladung. Eine Alternative zur stöchiometrischen Wertigkeit. PdN-CidS 2(64):14
11. Barke H-D (2012) Der einfache und erweiterte Redoxbegriff. Schülervorstellungen und deren Prävention im Chemieunterricht. PdN-CiS 4(61):11
12. Habelitz-Tkotz W, Werner E (2015) Die Redoxreaktion – ein bekanntes Problemfeld im Chemieunterricht mit hausgemachten Stolpersteinen. PdN-CidS 2(64):5
13. Asselborn W, Jäckel M, Risch KT (1998) Chemie heute Sekundarbereich II. Schroedel, Hannover
14. Sumfleth E et al (1989) Stoffe: Eigenschaften und Reaktionen, Modelle: Teilchenanordnungen und -umordnungen. Eine mit Lernhilfen gestützte Einführung in die Chemie. MNU 42:411
15. Behrendt H (1997) Concept mapping. Schülerinnen und Schüler konstruieren eigene Begriffsnetze. NiU-Physik 8:18
16. Schmidkunz H, Büttner D (1985) Chemieunterricht im Spiralcurriculum. NiU-P/C 33:19
17. Scheible A (1969) Ist unser Chemieunterricht noch zeitgemäß? MNU 22:449
18. Harsch G, Heimann R (2006) Von der Luft zu den „Lüften" – Experimente und Teilchenbilder zur Entwicklung eines tragfähigen Gasbegriffs im Chemieanfangsunterricht. MNU 59:406, 478
19. Heimann R, Harsch G, Katzorke J (2006) Die Vorstellungen von Zehntklässlern im Zusammenhang mit Gasen. Chim Did 32:32
20. Heimann R, Merge V, Harsch G (2009) Teilchenvorstellung – zwei Studien zum Umgang mit dem Teilchenbegriff in der Sekundarstufe I. PdN-CidS 58:34
21. Bellmann M et al (2011) Schülervorstellungen zum Teilchenmodell der Luft. Eine empirische Untersuchung an Gymnasien in den Jahrgangsstufen 5–10. Schüling, Münster

22. Sauermann D, Barke H-D (1998) Chemie für Quereinsteiger. Schüling, Münster
23. Rölleke, R., Hilbing, C.: Das Periodensystem der Atome und Ionen. www.chemischdenken. de. Zugegriffen: Mai 2018
24. Barke H-D, Wirbs H (2016) Ionenbegriff erarbeiten, üben und auf Alltagsbezüge anwenden. PdN-CidS 4(65):15
25. Kaminski M, Jansen W (1994) Die Ermittlung der chemischen Formel im Anfangsunterricht. NiU-Chemie 25:12
26. Barke H-D, Rölleke R (1999) Max von Laue: ein einziger Gedanke – zwei große Theorien. PdN-Ch 48:16
27. Barke H-D, Wirbs H (2000) Chemische Symbole für kleinste Struktureinheiten. PdN-Ch 49(2):2
28. Barke H-D, Sopandi W (2006) Raumvorstellung und Chemieverständnis – sie korrelieren! PdN-Chemie 55:41
29. Barke H-D (1982) Probleme bei der Verwendung von Symbolen im Chemieunterricht. NiU-P/C 30:131
30. Schmidt HJ (1990) Stolpersteine im Chemieunterricht. Diesterweg, Frankfurt
31. Harsch G, Heimann R, Benmokhtar S, Wagner A (2014) Das START-Konzept – Teilchenmodelle und Formelsprache im Chemieanfangsunterricht. Aulis, Hallbergmoos
32. Barke H-D (2016): Donor-Acceptor reactions: Good bye to the Laboratory Jargon. AJCE 6
33. Christen HR, Baars G (1997) Diesterweg-Sauerländer. Chemie, Frankfurt
34. Barke H-D (2018) Laboratory jargon of lecturers and misconceptions of students. AJCE 8
35. Becker H-J (1988) Verbraucherfragen im RIAS-Telefonstudio: Gegenstand fachdidaktischer Forschung? Chim Did 14:69
36. Becker H-J (1995) Ein Alltagsdialog über „Joghurt" – Chance für fächeraufweitenden Chemieunterricht. PdN-Ch 44:17

Everyday Life and Chemistry

9

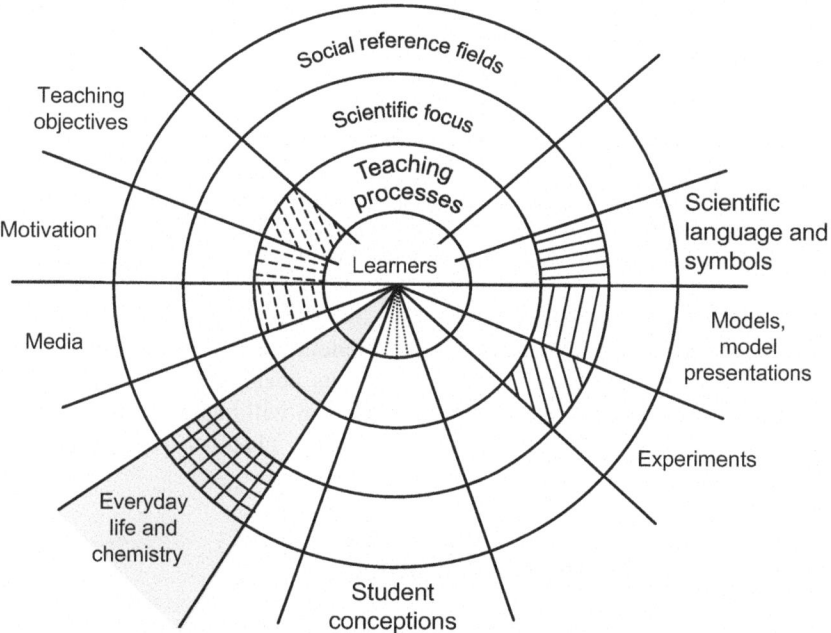

Sarah hears a lot about sulfur dioxide and its catalytic oxidation to sulfur trioxide in her chemistry class, about the contact process for its production, and the importance of sulfuric acid for chemical engineering. One evening, her father reads something about the increase in acid rain in the newspaper and asks Sarah: "You've been studying chemistry for a while now. Tell me—what is acid rain? How does your teacher explain this phenomenon?" Sarah's answer: "I don't know, the teacher didn't say anything about it."

© The Author(s), under exclusive license to Springer-Verlag GmbH, DE, part of Springer Nature 2025
H.-D. Barke et al., *Chemistry Didactics Compact*,
https://doi.org/10.1007/978-3-662-70080-8_9

David Waddington [1] thus caricatured the usual chemistry teaching in Great Britain from his point of view. German students often miss the connection to everyday life in their lessons. In a survey of high school students in grades 9-11, they commented after a chemistry lesson with concrete everyday references:

> Chemistry lessons don't seem so pointless when you can apply the material in everyday life.
> If there were practical references, then chemistry lessons wouldn't be an abstract complex of formulas.
> Such everyday references are good for general education and also benefit those who do not choose a career in chemistry later [2].

The German Chemical Society (GDCh) states:

> The task of chemistry teaching is to make the central importance of chemical knowledge for today's world understandable and tangible. The connection between chemistry and the learners' life area must be used or established to prepare for responsible handling of the environment. Learners must be enabled to meaningfully include knowledge from chemistry and technology [3].

For the subject of chemistry, the core curriculum for secondary level I of the grammar school in North Rhine-Westphalia has stated since 2011:

> All content areas with their focuses are mandatory, as is working in professional, related contexts. If contexts other than those suggested are chosen, they must be equivalent and the subject conference must decide on this in a uniformly binding manner [4].

The great importance of using or establishing everyday references has always been clear to most teachers, textbook authors, or guideline experts—the extent of everyday chemistry in the overall curriculum and its position in relation to subject systematics are rather controversial. This discussion will begin in Sect. 9.2—first, some aspects of learners' experiences with everyday chemistry will be presented from their perspective.

9.1 Learners: Curiosity and Interest

Young people have a natural interest in learning more about themselves as well as the objects and processes of their immediate world. Chemistry lessons can connect to this curiosity with meaningful questions from everyday life and work on them in an age-appropriate manner—related questions are raised and reflected upon.

1. **From which areas do learners' everyday experiences come?** Certainly, these are initially experiences from the parental home, from the kitchen, bathroom, garage or garden. In an overview, Pfeifer, Häusler and Lutz [5] summarize many areas from which possible everyday experiences of young people can originate (Fig. 9.1).

Fig. 9.1 Areas of the everyday life of young people from which experiences originate [5]

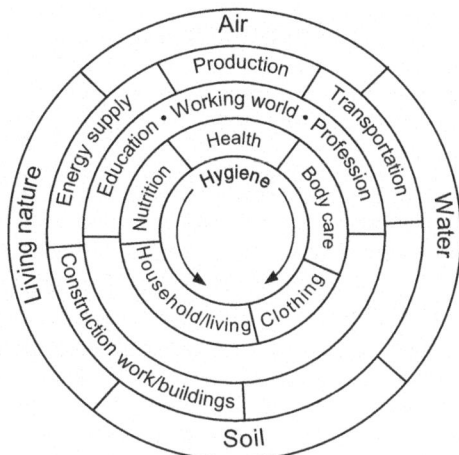

2. **How does the specific environment of the learners influence their ideas?** If there is a large industrial complex in the place of residence and many fathers or mothers work there, certain ideas about this industrial plant have developed. If young people live in the countryside, they will bring a different attitude to questions of agriculture, fertilizers or pesticides than young people from the city.
3. **What material phenomena do students experience daily?** They often experience the concept of destruction when they talk about spot "removers", ink "killers" or electricity "consumption" (Chap. 2). They also hear something daily about environmental problems regarding food, water, air and soil and develop corresponding attitudes against "chemistry", which must be problematized in class. In particular, the discussion about climate change must be included, which affects the children and young people living today and which they will probably experience particularly as adults.
4. **How are everyday substances brought closer to students by the adult world?** The internet, television, radio, magazines and newspapers convey a one-sided image of substances and materials through their advertising, which shape the students in a certain way. Such imprints should be found out and discussed in class. If a baker advertises in the bakery window with the statement "We bake without chemistry", one should ask whether the baking powder used does not contain sodium bicarbonate and citric acid.
5. **How can factual criticism and changes in behavior be promoted?** A chemistry lesson that adequately considers everyday and environmental topics would—in addition to an education in this regard by the parents—contribute to the ability to criticize and change behavior. Prepared excursions to facilities such as sewage treatment plants, recycling stations or waste incineration plants are also suitable for revising preconceived opinions and developing new ideas.

9.1.1 Student Interests

Another question arises from the previous set of questions: What interests do students have in certain topics of everyday and environmental chemistry? The answer to this question can be interesting for teachers because a planned project, which is often the only project in the school year, should address the interests of the students—otherwise the teacher might teach against the interests of all students, or against those of the boys or against those of the girls.

To answer this question, a questionnaire [6] was developed (Fig. 9.2) and used in 1986 at high schools around Hannover and in 1995 at high schools in Jena with students in grades 9-11 and evaluated by gender. The result of the evaluation is

---

1. Specifically, I would like the following topics to be covered in lessons on "*Everyday life and the environment*" (please assign all numbers 1-4 to exactly four favorite topics, 1 means "most desired", etc.):

☐ Food	☐ Wc cleaners	☐ Gasoline, fuels
☐ Food Preservatives	☐ Detergents	☐ Cement, building
☐ Alcohol, -drinks	☐ Cosmetics	☐ Fertilizers

2. Specifically, I would like the following topics to be covered in lessons on "*environmental protection*" (assign numbers 1-4 for exactly four topics):

☐ Pollutants in the water	☐ Acid rain and forest dieback	☐ Treatment of household waste
☐ Pollutants in the air	☐ Exhaust gas and catalytic converters	☐ Recycling paper and glass
☐ Pollutants in the soil	☐ Overfertilization and eutrophication	☐ Treatment of waste oil

3. Specifically, I would like the following topics to be covered in the "*chemical engineering*" lessons (assign numbers 1-4):

☐ Produce photos	☐ Adhesives	☐ Batteries and accumulators
☐ Galvanizing	☐ Explosives	☐ Fuel cell
☐ Coloring	☐ Metal alloys	☐ Rocket propulsion

4. Specifically, I would like the following topics to be covered in the "*chemical industry*" lessons (assign numbers 1-4): The factory production of:

☐ Steel and metals	☐ Sugar from beets	☐ Plastics
☐ Gasoline and heating oil	☐ Salt in the mine	☐ Dyes and pigments
☐ Sulphuric acid	☐ Paper made from wood	☐ Pharmaceutica

Fig. 9.2 Excerpt from a questionnaire to explore the interest situation [2]

shown in Table 9.1. It becomes clear that there are areas that are equally interesting for boys and girls: topics such as food, alcohol or explosives, the production of photos or paper are among them [6].

Other topics are only of interest to boys or only to girls, many topics show very low interest in both boys and girls (Table 9.1). Since the interests depend very much on the region of the school, each teacher should ideally conduct his or her own specific survey on site and can then assess which project topic is suitable for the lesson or which excursion is desired by the students. They must expect that a project topic about the sugar factory at the school location or the salt mine at the school location is not interesting enough for the teenagers, as the survey at the Gymnasium in Lehrte near Hannover showed, where both large companies existed at the time.

Table 9.1 Results of a survey to explore interests in 1995 in Jena, comparison with results from 1986 in Hannover, grades 9-11 [6]

1.	**Very interesting for boys and girls:**	
	Food	
	Alcohol, beverages	
	Producing photos	(also in 1986)
	Explosives	(also in 1986)
	Paper from wood	(also in 1986)
2.	**Very interesting only for girls:**	
	Cosmetics	(also in 1986)
	Treatment of household waste	(also in 1986)
	Dyeing	(also in 1986)
	Medicines	(also in 1986)
3.	**Very interesting only for boys:**	
	Gasoline, fuels	(also in 1986)
	Waste gases, catalysts	(also in 1986)
	Rocket propulsion	(also in 1986)
	Gasoline and heating oil	
4.	**Moderate interesting for boys and girls:** Food Preservatives, water pollution, air pollution, acid rain and forest dieback, recycling of paper and glass, battery and accumulator, fuel cell, steel and metals, plastics	
5.	**Not or nearly not interesting for boys and girls:** Toilet cleaner (also in 1986), detergent, cement and building materials (also in 1986), fertilizers (also in 1986), soil pollution (also in 1986), over-fertilization and eutrophication (also in 1986), treatment of used oil (also in 1986), galvanizing (also in 1986), metal alloys (also in 1986), sulfuric acid (also in 1986), sugar from beets (also in 1986), salt from the mine (also in 1986)	

9.1.2 Household Chemicals and Interest

Another study aimed to determine the extent of teenagers' interest in chemistry and whether this interest could be influenced if substances from the household were used in class instead of laboratory chemicals. Wanjek [7] planned the unit "Acids and Bases" for several school classes of grade 9 at a comprehensive school in Münster: In student experiments, household chemicals were tested with universal indicator, the use of acidic and alkaline cleaners was examined, and the neutralization of these solutions was worked out. The entire lesson was taught by various teachers of the school and lasted only six hours. The teenagers were asked before and after about their interest—also in student experiments.

The survey results of about 300 subjects show that there is indeed interest in chemistry, however, it drops significantly among girls at some schools in North Rhine-Westphalia from grade 8 to grade 10 (Fig. 9.3 (1)). In chemistry courses of grade 11, it is naturally more present again, because students with little interest have dropped chemistry. At a school in Thuringia, it was also found that boys are more interested in chemistry lessons and girls are moderately interested (Fig. 9.3 (2)).

Further surveys relate to the interest before and after the teaching unit "Acids and Bases". Here too, before the lesson, the girls expressed a lack of interest, but after the lesson, it increased noticeably (Fig. 9.3 (3) and (4)). The influence of student experiments was significant: both girls and boys expressed a high interest in student experiments (Fig. 9.3 (5)). When asked about the combination of household chemicals and student experiments, even a higher interest of the girls than that of the boys emerged: For the girls, the household chemicals played the decisive role in their interest (Fig. 9.3 (6)).

9.1.3 Attitudes towards Chemistry and Chemistry Education

In a study with a large number of participants, Müller-Harbich, Wenck and Bader [8] found that students hardly differentiate between attitudes towards chemistry and chemistry education:

> Students who have a positive attitude towards chemistry education also show an open-minded attitude towards chemistry. On the other hand, particularly female students with a positive affective attitude towards environmental problems have a negative attitude towards chemistry and vice versa. The attitude observed in the girls corresponds to the common opinion: Those who are environmentally committed reject chemistry [8].

To determine attitudes towards chemistry, Heilbronner and Wyss [9] had young people from Switzerland draw pictures about chemistry. The disasters in chemical companies at the end of the 70s were reflected in the results: Two thirds of the pictures show negative motives related to environmental destruction, threats to humans and animal testing. The authors' conclusion:

9.1 Learners: Curiosity and Interest

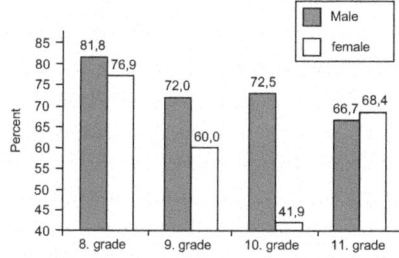

(1) Interest in chemistry as a school subject among young people in North Rhine-Westphalia

(2) Interest of young people in grade 9 in Thuringia in chemistry as a school subject

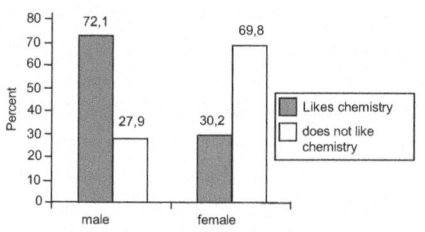 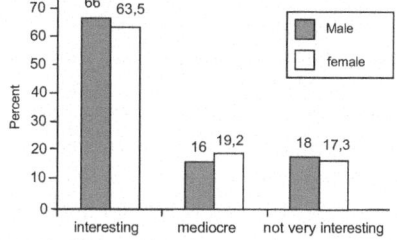

(3) Interest of young people in Münster in the subject of chemistry *before* lessons on acids and alkalis using everyday chemicals

(4) Interest in chemistry among young people in Münster *after* lessons on acids and alkalis using everyday chemicals

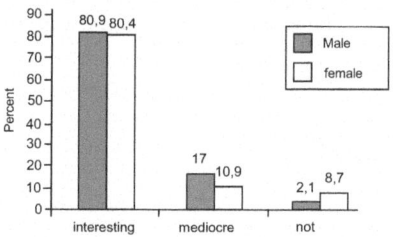 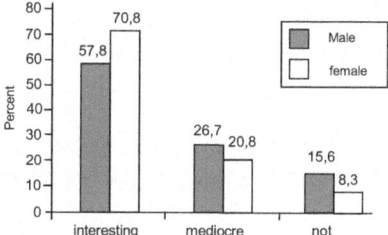

(5) Interest of young people in Münster in school experiments on the subject of in Chemistry

(6) Interest of young people in Münster in the combination of student experiments and everyday chemicals on the subject of acids and alkalis

Fig. 9.3 Results of an interest survey before and after a teaching unit [7]

> The chemistry teacher is probably the only one who has to face a class at the beginning of his lessons that has already formed an opinion about the value or rather disvalue of the subject that is now coming their way [9].

To check at the end of the 1990s to what extent this thesis can still stand, young people in the Münster area were asked to "draw their picture" of chemistry [10].

Some examples of about 160 artworks are shown in Fig. 9.4. Many pictures show both positive and negative motives at the same time, so the question of to what extent positive and negative attitudes are reflected is not based on the number of submitted pictures, but on the number of motives contained. Hilbing [10] found that only 35% of the boys and 17 % of the girls had painted exclusively motives that predominantly reflect negative attitudes. Compared to the study from Switzerland, the percentage of about 66% negative motives has almost halved.

In the accompanying questionnaire for grades 6-9 regarding the attitude towards *Chemistry* [10], it was found that 65% of the boys express themselves positively, but only 33% of the girls. In response to questions about the attitude towards *Chemistry Education*, only half of the positive statements from the boys as before, namely 31%, and only 18% from the girls. So the attitudes towards chemistry are unexpectedly more positive than those towards chemistry education. The result correlates with the motives of the pictures that were painted by the same young people: Here too, a generally positive basic attitude towards *Chemistry* became clear: Successful *Chemistry Education* should be able to build on this positive basic attitude. The negative attitude towards chemistry education must therefore result from the way in which the school subject chemistry is taught: Chemistry teachers and chemistry educators have to find ways and means to improve this perceived negative state.

In a 2013 study on the task "Draw your picture of chemistry" with 320 participants from grades 5–10, Pietsch [11] found that 87% of the respondents painted at least one positive motive and 38% only positive motives. On the other hand, at least one negative motive was painted by 45%; however, only negative motives were painted by only 5% of the respondents. Therefore, there is currently a generally positive basic attitude of students towards chemistry—the prerequisites for successful chemistry education seem to be good.

9.2 Scientific Focus: Subject Systematics versus Everyday Chemistry

Questions have always been discussed in the planning of chemistry lessons regarding what proportions of everyday chemistry should be implemented to what extent: Whether the introduction to facts should take place with an everyday phenomenon or whether the conveyed facts should be applied to everyday references at the end, or even whether the entire chemistry lesson should be planned and implemented in relation to everyday life. In particular, the position of pure subject systematics is controversial among many chemistry teachers and chemistry educators. Friedrich [12] summarizes this discussion as follows:

> If students come to chemistry lessons in grades 5/6 with a high level of motivation and great interest, a permanent loss of interest in the subject of chemistry can be observed in the following years. The reasons for this are varied, but well known: The contents of chemistry are difficult, complex, unattractive and cannot be anchored with things from the everyday and life world of the students. The curricular content to be conveyed is too much based on scientific findings, but not on the interests of the students and their life world.

9.2 Subject Focus: Subject Systematics versus Everyday Chemistry 323

Fig. 9.4 Selection of results for the task: "Draw your picture of chemistry" [10]

So how can the problem be solved, or at least minimized? Everyday relevant, student-oriented subject areas must be built up curricularly—at least in secondary level I—in order to enable as many students as possible to access chemistry and the minimum scientific standards. But: Student orientation or everyday relevance alone must not be the determinant for the selection of teaching content; basic subject systematics must be worked out and must not be neglected, if this is done in an everyday relevant way—very nice. However, if a life-worldly location is primarily not present, this must not be the argument for not dealing with basic subject-specific principles [12].

This argumentation for and against everyday chemistry runs through the entire chemistry education literature. In order to get to know the scientific side of the topic better, some aspects will be presented below.

9.2.1 Everyday Phenomena and Chemistry

Chemistry lessons can help students to work on and understand chemical reactions from everyday life and the world of life in a professional way. In particular, it is possible to "undress" everyday phenomena of their "packaging" and to "translate" them into chemical processes. Some examples:

Making coffee:	Extraction
Removing stain:	Solubility
Washing:	Emulsifying and Dispersing
Ink eraser:	Redox reaction
Black and white photography:	Redox reaction, Complex reaction
Polishing silver:	Redox reaction
Effervescent tablets:	Acid-base reaction
Baking powder:	Acid-base reaction
Descaler:	Acid-base reaction
Mason's mortar:	Acid-base reaction
Oven spray:	alkaline hydrolysis

In chemistry lessons, the reciprocal relationships between chemical knowledge, technological progress and individual living habits can also be reflected and traced in their historical development. Example topics are: Soaps, detergents and cosmetics, preservation of food, fertilization and pest control, medicines and pharmaceuticals etc.

9.2.2 Technical Interpretations, Experiments

If one bases on the overview of Pfeifer, Häusler and Lutz [5] on the experience areas of young people (Fig. 9.1), according to the overview a multitude of facts or everyday chemicals are to be reflected: They are presented below and sketched with reaction symbols. It is indicated whether redox reactions (RR), acid-base reactions (SBR) or complex reactions (CR) are underlying, which "packaging" can accordingly "undress a phenomenon". Experiments regarding presented everyday

9.2.2.1 Hygiene: Example "Bathroom Chemicals"

Drain Cleaner "NaOH/Al-Type" (Sect. 9.6: V9.1)
This cleaner is intended to decompose organic substances through a strong alkaline reaction and remove blockages in washbasins or toilets. The addition of aluminum shavings is intended to react with water to produce hydrogen or ammonia—a vortex effect occurs and increases the effect:

$$Al\,(s) + 3H_2O + OH^-(aq) \rightarrow [Al(OH)_4]^-(aq) + \frac{3}{2}H_2 \qquad RR, KR$$

Hydrogen is initially detectable as a gas, but "nascent" H-atoms react with nitrate ions of the sodium nitrate, which is added for safety reasons:

$$8H + NO_3^-(aq) \rightarrow NH_3 + OH^-(aq) + 2H_2O \qquad RR$$

Toilet Cleaner "HSO$_4^-$-Type" (Sect. 9.6: V9.2)
This cleaner contains solid sodium hydrogen sulfate (NaHSO$_4$), which reacts strongly acidic with water and converts lime residues (CaCO$_3$) from tap water splashes:

$$HSO_4^- + H_2O \rightarrow H_3O^+(aq) + SO_4^{2-}(aq) \qquad SBR$$
$$CO_3^{2-} + 2H_3O^+(aq) \rightarrow 3H_2O + CO_2 \qquad SBR$$

Sanitary Cleaner "HOCl/Cl$^-$-Type" (Sect. 9.6: V9.3)
This solution, also known as bleach lye, forms free O-atoms ("nascent O-atoms"), which attack organic material and therefore act as a bleach or germicide:

$$HOCl(aq) + H_2O \rightarrow H_3O^+(aq) + Cl^-(aq) + O \qquad SBR, RR$$

When the acid concentration is increased, dissolved and gaseous chlorine is formed; due to the toxicity of free chlorine, the label warns against combining this cleaner with an acidic cleaner—such as sodium hydrogen sulfate:

$$HOCl(aq) + Cl^-(aq) + H_3O^+(aq) \rightarrow Cl_2(aq, g) + 2H_2O \qquad SBR, RR$$

Oxi-Cleaner "Sodium Percarbonate-Type" (Sect. 9.6: V9.4)
This cleaner releases O-atoms and therefore also acts as a bleach or germicide. It contains a sodium carbonate-hydrogen peroxide compound, which reacts with water as follows [13]:

$$2Na_2CO_3 \cdot 3H_2O_2 \rightarrow aq \rightarrow 4Na^+(aq) + 2CO_3^{2-}(aq) + 3H_2O_2$$
$$H_2O_2 \rightarrow H_2O + O \quad \text{or} \quad 2H_2O_2 \rightarrow 2H_2O + O_2 \qquad RR$$

9.2.2.2 Body Care: Example "Deodorants"
Deodorant "Al^{3+}-Type" (Sect. 9.6: V9.5)
Some deodorant substances—such as the "Anti-Perspirant Hydrofugal"—react on the basis of aluminum chloride hexahydrate. Both acidic reactions and the presence of aluminum ions have a germicidal effect:

$$\{[Al(H_2O)_6](Cl)_3\} + H_2O \rightarrow H_3O^+(aq) + [Al(H_2O)_5OH]^{2+}(aq) + 3Cl^-(aq) \quad \text{SBR, KR}$$

Since the involvement of aluminum compounds is controversial, new deodorants based on aluminum-free organic substances have been developed.

9.2.2.3 Health: Example "Mineral Tablets"
Mineral Tablets "Ca and Mg Type" (Sect. 9.6: V9.6)
Calcium and magnesium supplements are offered both as chewable tablets and as effervescent tablets. The latter contain the carbonates mixed with citric acid crystals (abbreviated: HCit). During the reaction, gaseous carbon dioxide bubbles out, effective Ca^{2+}(aq) or Mg^{2+}(aq) ions are released and can be absorbed by the body:

$$MgCO_3\,(s) + 2HCit\,(s) \rightarrow Mg^{2+}(aq) + 2Cit^-(aq) + H_2O + CO_2 \quad \text{SBR}$$

9.2.2.4 Nutrition: Example "Table Salts"
Table Salt "Iodine Type" (Sect. 9.6: V9.7)
In today's table salts, in addition to the actual sodium chloride, mineral salts are often contained in small concentrations, such as calcium carbonate, sodium phosphate, sodium fluoride or sodium iodate ("iodized salt"). They serve not only as supplementary nutrients (promotion and maintenance of teeth), but also as a technical means to improve flowability. If you acidify an iodized salt solution and add a solution of potassium iodide, a brown-colored iodine solution appears. If the brown coloration is not visible, starch solution can indicate even the smallest iodine concentrations by the well-known blue coloration:

$$IO_3^-(aq) + 5I^-(aq) + 6H^+(aq) \rightarrow 3I_2\,(aq, brown) + 3H_2O \quad \text{RR}$$

9.2.2.5 Household: Example "Baking Agents"
Baking Powder "Soda Type" (Sect. 9.6: V9.8)
Baking powder is usually used for baking bread and cakes. Its task is to develop a gas in the heat, which provides the dough with cavities and creates the loose bread structure. In most cases, sodium bicarbonate ("soda", NaHCO$_3$) and solid acids such as citric acid (abbreviated HCit) or sodium dihydrogen phosphate (NaH$_2$PO$_4$) are used. The gas carbon dioxide forms in both cases:

$$HCO_3^- + HCit \rightarrow Cit^-(aq) + H_2O + CO_2\,(g) \quad \text{SBR}$$
$$HCO_3^- + H_2PO_4^- \rightarrow HPO_4^{2-}(aq) + H_2O + CO_2\,(g) \quad \text{SBR}$$

Baking Powder "Hartshorn Salt Type" (Sect. 9.6: V9.9)
If an ammonium salt such as ammonium carbonate ("hartshorn salt", $(NH_4)_2CO_3$) is used, then—in addition to carbon dioxide and steam—gaseous ammonia is also produced. In this case, only flat pastries should be made so that this gas can escape:

$$2NH_4^+ + CO_3^{2-} \rightarrow 2NH_3 + H_2O + CO_2 \qquad \text{SBR}$$

9.2.2.6 Housing: Example "Heating Fuels"
Many apartments are heated by fossil fuels: Either lignite or hard coal is burned in the fireplace or stove, heating is done with natural gas from the gas line or with propane or heating oil from storage tanks in the house. In all cases, the chimney sweep checks at certain intervals both the levels of soot, carbon monoxide and carbon dioxide as well as exhaust gas temperatures and exhaust gas losses, in order to ensure optimal and thus environmentally friendly combustion as far as possible:

$$C\,(s) + O_2\,(g) \rightarrow CO_2\,(g); \Delta H = -393\,\text{kJ} \qquad \text{RR}$$
$$CH_4\,(g) + 2O_2\,(g) \rightarrow CO_2\,(g) + 2H_2O\,(g); \Delta H = -890\,\text{kJ} \qquad \text{RR}$$

9.2.2.7 Clothing: Example "Textile Decolorizers"
Textile Decolorizer "Dithionite Type" (Sect. 9.6: V9.10)
Sodium dithionite is often used as a "reduction bleach" for removing stains or decolorizing textiles. The alkaline solution forms "nascent H atoms", which destroy organic oxygen compounds (for example, dyes or inks):

$$Na_2S_2O_4 + 2OH^-\,(aq) + H_2O \rightarrow 2Na^+\,(aq) + SO_4^{2-}\,(aq) + SO_3^{2-}\,(aq) + 4H \quad \text{RR}$$

9.2.2.8 Leisure: Example "Black and White Photography"
Developer "Hydroquinone-Type" (Sect. 9.6: V9.11)
Color photography is very complex to describe, in contrast, black and white photography is relatively simple: It involves reactions of the silver bromide applied to the photo paper with light and the removal of the unexposed silver bromide. The exposure leads to invisible silver nuclei, the development with alkaline hydroquinone solution produces visible amounts of finely dispersed silver and thus black colored areas on the photo paper:

$$Ag^+Br^- \rightarrow Ag\,(s, \text{silver germ}) + Br\,(\text{desolved in silver bromide}) \quad \text{RR}$$
$$2AgBr + (C_6H_4)(OH)_2 + 2OH^-\,(aq) \rightarrow$$
$$2Ag + (C_6H_4)O_2 + 2H_2O + 2Br^-\,(aq) \qquad \text{RR}$$

Fixer "Thiosulfate-Type" (Sect. 9.6: V9.12)
Fixing is necessary because after development, unexposed silver bromide would stick to the photo paper and later darken in the light. It is removed by complex formation using sodium thiosulfate solution ($Na_2S_2O_3$):

$$AgBr\,(s, \text{unexposed}) + 2S_2O_3^{2-}\,(aq) \rightarrow [Ag(S_2O_3)_2]^{3-}\,(aq) + Br^-\,(aq) \qquad \text{KR}$$

9.2.2.9 Work Environment: Example "Metal Processing"
Etching Chemical "Fe^{3+}-Type" (Sect. 9.6: V9.13)
For the production of circuit boards for electronic components, copper-coated plastic plates are used. To create specific conductor tracks for the flow of electricity, corresponding lines on the plate are protected by wax and the remaining copper layer is dissolved. One method is the use of iron(III) chloride solution:

$$Cu + 2Fe^{3+}(aq) \rightarrow Cu^{2+}(aq) + 2Fe^{2+}(aq) \qquad RR$$

9.2.2.10 Energy Supply: Example "Accumulators"
Accumulator "Type Pb/PbO_2" (Sect. 9.6: V9.14)
Accumulators are capable of supplying power and being recharged. The most well-known is the "lead-acid battery" in the car. It provides the electrical energy to start the starter, which in turn starts the engine. The electrodes in the charged state consist of metallic lead or reddish-brown lead dioxide—they are surrounded by a 20% sulfuric acid solution:

$$\text{negative pole:} \quad Pb \rightarrow Pb^{2+}(PbSO_4) + 2e^- \qquad RR$$

$$\text{positive pole:} \quad PbO_2 + 4H^+(aq) + 2e^- \rightarrow Pb^{2+}(PbSO_4) + 2H_2O \qquad RR$$

Accumulator "Type Cd/Ni"
Nickel-cadmium cells, which can be recharged with a charger, are commonly used for household power supply, thus reducing the waste of conventional batteries. The electrodes in the charged state consist of finely dispersed cadmium (negative pole) and solid nickel(III) oxide hydroxide (positive pole):

$$\text{negative pole:} \quad Cd + 2OH^-(aq) \rightarrow Cd(OH)_2 + 2e^- \qquad RR$$

$$\text{positive pole:} \quad 2NiOOH + 2H_2O + 2e^- \rightarrow 2Ni(OH)_2 + 2OH^-(aq) \qquad RR$$

Accumulator "Type Lithium-Ion" (Sect. 9.6: V9.15)
In this accumulator, lithium ions are stored or released at both poles (Fig. 9.5 [14]). For student experiments, Oetken and Hasselmann [15] developed a model accumulator that starts from two graphite electrodes for both poles and is based on the "intercalation" of both lithium ions and perchlorate ions in graphite. The *charging process* in a suitable electrolyte with lithium perchlorate as the conducting salt is formulated as follows [15]:

$$\text{negative pole:} \quad C_n + x\,Li^+ + x\,e^- \rightarrow Li_x^+ C_n^{x-} \qquad RR$$

$$\text{positive pole:} \quad 2C_n + x\,ClO_4^- \rightarrow C_n^{x+}(ClO_4^-)_x + x\,e^- \; [15] \qquad RR$$

Both authors developed an experimental kit [16], with which essential observations on the lithium-ion battery with two graphite electrodes can be reproduced.

9.2 Subject Focus: Subject Systematics versus Everyday Chemistry

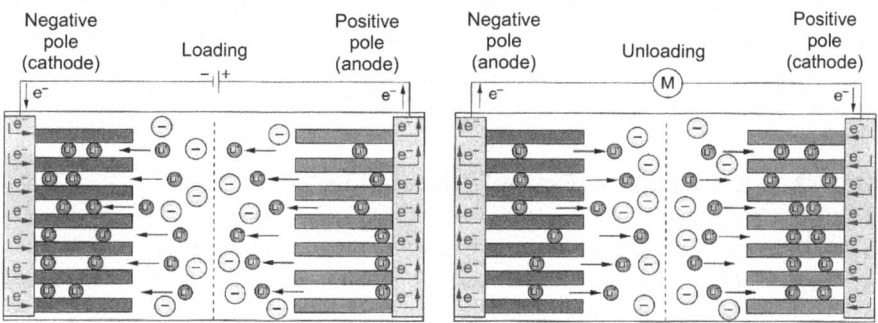

Fig. 9.5 Model concept for the lithium-ion accumulator. (With kind permission from Schroedel publishing house [14])

9.2.2.11 Construction Measures: Example "Setting of Mortar"
Rapid Cement "Ca(OH)₂ Type" (Sect. 9.6: V9.16)

The mason's lime is chemically called calcium hydroxide (Ca(OH)₂), it can be produced by the so-called "slaking" of quicklime—by the reaction of calcium oxide (CaO) with water. When the mason's lime sets, the reaction of calcium hydroxide and carbonic acid to calcium carbonate (CaCO₃) takes place in the presence of water—depending on wall thicknesses and temperatures, it takes months or years until the setting is finished:

slaking of quicklime:

$$CaO + H_2O \rightarrow Ca(OH)_2 \text{ oder } O^{2-} + H_2O \rightarrow 2OH^- \qquad \text{SBR}$$

bonding of slaked lime:

$$2OH^- + H_2CO_3 \rightarrow CO_3^{2-} + 2H_2O \qquad \text{SBR}$$

Destruction of Buildings by Acid Rain

Calcium carbonate (for example in the natural stone of buildings) is attacked by "acid rain" and converted to gypsum containing crystal water. Since the volume increases in the process, the natural stone weathers very strongly on the surface:

$$CaCO_3 + 2H_3O^+(aq) + SO_4^{2-}(aq) \rightarrow CaSO_4 \cdot 2H_2O + H_2O + CO_2 \qquad \text{SBR}$$

9.2.2.12 Service: Example "Firefighting"

There are various types of fire extinguishers, most of which work on the basis of compressed carbon dioxide. If the fire brigade lays a foam carpet on the runway of an airfield as a precaution, this involves the reaction of solid aluminium and sodium bicarbonate (NaHCO₃) with the extinguishing water, which contains a suitable foaming agent:

$$Al^{3+} + 6H_2O \rightarrow [Al(H_2O)_5OH]^{2+}(aq) + H^+(aq) \qquad \text{SBR, KR}$$

$$H^+(aq) + HCO_3^- \rightarrow H_2CO_3 \rightarrow H_2O + CO_2 \qquad \text{SBR}$$

9.2.2.13 Transportation: Example "Alcohol Tests"
Alcotest "Chromat-Type"
To control the blood alcohol content of drivers in road traffic, the police today use test devices based on infrared spectroscopy. In addition, there are test tubes that serve to estimate the alcohol content in the breath. The test tube contains yellow potassium chromate crystals (K_2CrO_4) mixed with white sodium hydrogen sulfate ($NaHSO_4$). In the presence of moist alcohol vapor of the breath, a reduction to green-colored chromium(III) compounds, i.e., a color change from yellow to green, occurs:

$$2CrO_4^{2-}(aq) + 2HSO_4^-(aq) \rightarrow Cr_2O_7^{2-}(aq) + H_2O + 2SO_4^{2-}(aq) \qquad \text{SBR}$$

$$3CH_3CH_2OH + Cr_2O_7^{2-}(aq) + 8H^+(aq) \rightarrow 3CH_3CHO + 2Cr^{3+}(aq) + 7H_2O \qquad \text{RR}$$

9.2.2.14 Production: Example "Fertilizers"
In addition to natural fertilizers (manure, slurry), there are mineral fertilizers. They are obtained from the salt deposits underground: potassium, calcium and magnesium salts, nitrates, phosphates, etc. On the other hand, nitrates and ammonium salts are artificially produced by the Haber-Bosch synthesis and subsequent reactions:

$$N_2 + 3H_2 \rightarrow 2NH_3 \qquad \text{RR}$$

$$2NH_3 + 3\tfrac{1}{2}O_2 \rightarrow 2NO_2 + 3H_2O \qquad \text{RR}$$

$$4NO_2 + 2H_2O + O_2 \rightarrow 4HNO_3 \qquad \text{RR}$$

$$NH_3 + HNO_3 \rightarrow NH_4^+NO_3^- \qquad \text{SBR}$$

Insoluble calcium phosphate from above-ground deposits is converted with pure sulfuric acid to soluble dihydrogen phosphates, which are then suitable for fertilization purposes:

$$Ca_3(PO_4)_2 + 2H_2SO_4 \rightarrow 2CaSO_4 + Ca(H_2PO_4)_2 \qquad \text{SBR}$$

9.2.2.15 Air: Example "Smog"
In an inversion weather situation, a warm layer of air lies like a lid on the cold air on the ground: The layers of air do not mix sufficiently, gases such as sulfur dioxide, nitrogen oxides, carbon monoxide, soot and dust (smoke + fog: smog) cannot escape. They therefore significantly burden breathing. Nitrogen oxides are primarily produced by the reaction of air in the hot, high-speed car engine:

$$N_2 + O_2 \rightarrow 2NO; \; 2NO + O_2 \rightarrow 2NO_2 \qquad \text{RR}$$

If the car's exhaust catalyst works, finely dispersed platinum powder as a catalyst on the ceramic body reduces proportions of nitrogen oxides and carbon monoxide:

$$NO + CO \rightarrow \frac{1}{2}N_2 + CO_2 \qquad RR$$

Eleni Daoutsali [17] was able to demonstrate the effectiveness of today's car exhaust catalysts to many students in an experimental teaching sequence and convey through model drawings how the expert imagines the reaction of CO-molecules on the platinum surface (Fig. 9.6): CO- and O_2-molecules are separated under the influence of the Pt-atoms on the platinum surface and re-form into CO_2-molecules.

Today, the industry works with nano-platinum particles on the ceramic body, thereby saving a large part of precious metal. Using the example of nano-platinum particles, Daoutsali also reflected on the current branch of nanotechnology research, pointed out the available nanochemistry experiment kits and demonstrated some products of nanochemistry from them, such as the lotus effect, electrically conductive glass plates or impregnated wood surfaces that cannot be ignited with the burner [18].

9.2.2.16 Water: Example "Drinking Water"

To disinfect drinking water, either chlorine or ozone is used—in both cases, "nascent O-atoms" oxidize the contained organic impurities:

$$Cl_2 + H_2O \rightarrow 2H^+(aq) + 2Cl^-(aq) + \{O\} \qquad RR$$
$$O_3 \rightarrow O_2 + \{O\} \qquad RR$$

9.2.2.17 Soil: Example "Soil Acidification"

Acid rain is formed by industrial and car exhaust gases in the air: Droplets of hydrochloric acid, sulfuric acid or nitric acid solution are formed and cause acidification of the soils when it rains. This has the consequence that fine roots are

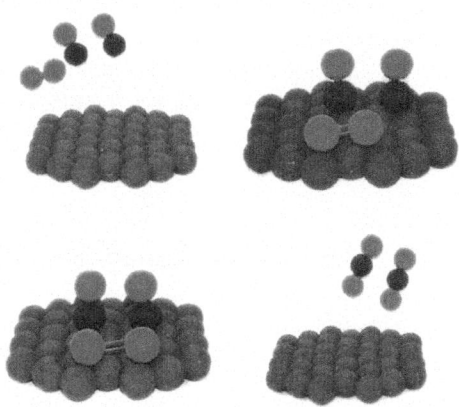

Fig. 9.6 Functioning of car exhaust catalysts [17]

damaged, carbonates of nutrient salts are dissolved and washed out. On the other hand, solid aluminum salts, in which Al^{3+}-ions are bound and thus harmless, can be dissolved and in this form attack trees and plants ("forest dieback"):

$$Al(OH)_3 + 3H^+(aq) \rightarrow Al^{3+}(aq) + 3H_2O \qquad \text{SBR}$$

9.2.2.18 Nature: Example "Climate Change"

The *natural* greenhouse effect, due to the presence of carbon dioxide and water in the atmosphere, ensures that the pleasant average temperature of 18 °C prevails on Earth. A possible climate change for the next decades is mainly caused by the *anthropogenic* greenhouse effect of too high concentrations of carbon dioxide and methane in the atmosphere: A part of the energy radiated from the Earth into space, which the sun provides, is absorbed by vibration-capable CO_2- and CH_4-molecules, temperatures on Earth rise, climate change can be the result.

Nina Harsch [19] was able to show through empirical surveys of student ideas that knowledge about the greenhouse effect, the ozone problem and acid rain is insufficient among young people in Germany and also in many other countries and should be promoted in chemistry lessons. She has made teaching suggestions for about 10 lessons, implemented and evaluated in classes 10 and 11 [20]. These suggestions also include numerous experiments and models that very well illustrate the addressed teaching content [20].

9.3 Teaching Processes: Subject Systematics plus Everyday Chemistry

> The term everyday chemistry includes all chemical processes and the substances and materials affected by them that play a role for us in everyday life. However, this would result in an unmanageable abundance of subject areas that would have to be differentiated according to individual interests. It is therefore clear that a chemistry class related to everyday life in an unreflected sense cannot be an alternative to a clearly structured, understandable subject class. On the other hand, the greatest efforts are necessary to bridge the repeatedly identified gap between chemistry lessons and everyday life, i.e., to develop strategies on how everyday life and relational learning can be brought together.

Lutz and Pfeifer [21] have formulated these demands and offered their proposed solutions. Which strategies for merging chemistry and everyday life are possible for chemistry lessons is up for discussion. Before that, a variety of methods will be presented with which mediation processes for everyday chemistry can be realized.

9.3.1 Methods for Mediation Processes

The mediation between everyday life and chemistry can happen in class in various ways and thus contribute to the variety of methods:

9.3 Teaching Processes: Subject Systematics plus Everyday Chemistry

- Learning through active action or experimenting in *action-oriented teaching*: Taking and analyzing water or soil samples (for example, using the Aquamerck boxes), making various mortar mixtures and testing the setting, etc.
- *Excursions* to extracurricular learning locations: Visiting the regional sewage treatment plant, the drinking water treatment plant, the waste processing and recycling station, preparing and conducting interviews, photo reports, posters, etc.
- Learning through lectures and discussions with *extracurricular experts*: Inviting a firefighter, food inspector, master painter or technician from the industry. Excursions to the corresponding businesses presented.
- Learning with *audiovisual media* or via *multimedia paths*: Designing learning materials by the learners themselves, critical review of the material and development of new learning environments, demonstration and reflection of self-designed media.
- Learning through *role-playing*: Topics that are difficult to deal with in experimental teaching are conveyed through role-playing, such as the question "Meat or grains"? [22]: The specified role texts are distributed, the roles of young people are played.
- Learning in *project teaching* or also in *project-oriented teaching* [23]: For a project "Water and Environment" [24], whose occasion was the visit of a sewage treatment plant, topics are distributed, worked out by project groups for posters, these are presented and exhibited.

9.3.2 Complete Curricula Based on Everyday Chemistry

For several years, there have been two teaching works from the Anglo-American language area that do not primarily orient themselves on the chemical subject systematics, but primarily base themselves on topics of everyday and environmental chemistry. Figs. 9.7, 9.8 and 9.9 show the tables of contents and allow conclusions about the intended teaching.

ChemCom, Chemistry in the Community [25] is a curriculum from the USA (Fig. 9.7), which provides professional information for each of the selected everyday topics as needed. For example, in the first topic "The Quality of Our Water" under "Measurement and the Metric System", the conversions from *inches* and *ounces* to common units of the metric system are discussed, at another point laboratory activities such as filtering, under "Molecular View of Water" models of the water molecule, the H_2O formula and other element, compound and reaction symbols. Under "Electrical Nature of Matter", the dipole moment of the water molecule is introduced, the concept of ions is laid out and tests for the detection of certain ions in bodies of water and in drinking water are carried out experimentally. These professional information are therefore completely devoid of the usual factual structure, especially since they are intended for introductory teaching.

Salters Advanced Chemistry—an English work—is divided into three volumes. *Chemical Storylines* [26] provides the everyday and environmental

Preface
To You, the Student
Safety In the Laboratory
Supplying Our Water Needs
  I The Quality of Our Water
  II A Look at Water and Its Contaminants
  III Investigating the Cause of the Fish Kill
  IV Water Purification and Treatment
  V Putting It All Together: Fish Kill in Riverwood-Who Pays?
Conserving Chemical Resources
  I Use of Resources
  II Why We Use What We Do
  III Can We Continue to Use Things Up?
  IV How Much Do We Have and for How Long?
  V The Promise of New Materials
Petroleum: To Build or to Burn?
  I Petroleum in Our Lives
  II Petroleum: What Is It? What Do We Do with It?
  III Petroleum as a Source of Energy
  IV Making Useful Materials from Petroleum
  V Alternatives to Petroleum
  VI Putting It All Together: Choosing Petroleum Futures

Understanding Foods
  I Foods: To Build or to Burn?
  II Food as Energy
  III Foods: The Builder Molecules
  IV Essential Materials Needed in Small Amounts
  V Food Additives
  VI Putting It All Together: Nutrition in Many Cultures

Preface
To You, the Student
Safety in the Laboratory
Nuclear Chemistry in Our World
  I Nuclear Phenomena in Everyday Life
  II Discovering Atomic Energy
  III Radioactive Decay
  IV Nuclear Energy: Power of the Universe
  V Living with Benefits and Risks
  VI Putting It All Together: Separating Fact from Fiction
Chemicals, Air, and Climate
  I Life in a Sea of Air
  II Investigating the Atmosphere
  III Atmosphere and Climate
  IV Human Alteration of the Atmosphere
  V Putting It All Together: Is Air a Free Resource?
Chemistry and Health
  I Maintenance of Health
  II Your Body as a Chemical Factory
  III The Chemistry of Exercise
  IV The Chemistry of Personal Health Care
  V Chemical Control: Drugs and Toxins in the Human Body
  VI Putting It All Together: Assessing Personal and Public Health Risks
The Chemical Industry: Promise and Challenge
  I A New Industry for Riverwood
  II An Overview of the Chemical Industry
  III Nitrogen Products and Their Chemistry
  IV Chemical Energy   Electrical Energy
  V Pharmaceuticals
  VI Putting It All Together: The Role of Chemical Industry Past, Present, and Future

Acknowledgments

Fig. 9.7 Table of contents "ChemCom. Chemistry in the Community" [25]

topics underlying the teaching (Fig. 9.8). In addition, there is the second volume *Chemical Ideas* [27], which offers chemical facts in a subject-systematic way and serves as a source of information for the first volume (Fig. 9.9). A third volume *Activities and Assessment* [28] provides instructions for laboratory experiments and examination tasks for all topics and is in the hands of the teacher.

In the lessons on the everyday topics of the first volume, cross-references to text sections of the second volume are now mentioned at suitable points in order to work out the appropriate professional-chemical information from these texts. On the other hand, there are references to the corresponding experiments and regulations of the third volume. With this approach, it is ensured that the students in the volume *Chemical Ideas* always recognize the information embedded in the subject structure. Since it is the "Advanced Level", i.e., advanced teaching after a two-year introductory course at the "Ordinary Level", the additional information for these students is also easier to recognize and process than is possible in the "ChemCom" curriculum, which was designed for introductory teaching in chemistry.

The English work "Salters Advanced Chemistry" has been offered in German translation since 2012: "Salters Chemie, Chemical Storylines, Kontexte" [29] and "Salters Chemie, Chemical Ideas, Theoretical Basics" [30].

Introduction for students

The Elements of Life
- EL1 What are we made of?
- EL2 Take two elements ...
- EL3 Looking for patterns in elements
- EL4 Where do the chemical elements come from?
- EL5 The molecules of life
- EL6 Summary

Developing Fuels
- DF1 Petrol is popular ...
- DF2 Getting energy from fuels
- DF3 Focus on petrol
- DF4 Making petrol – getting the right octane rating
- DF5 Trouble with emissions
- DF6 Methanol – the key to future fuels?
- DF7 What other solutions are there?
- DF8 Summary

From Minerals to Elements
- M1 Chemicals from the sea
- M2 Copper from deep in the ground
- M3 Mining Cornish tin
- M4 Summary

The Atmosphere
- A1 What's in the air?
- A2 Screening the Sun
- A3 Ozone: A vital sunscreen
- A4 The CFC story
- A5 How bad is the ozone crisis?
- A6 Trouble in the troposphere
- A7 Keeping the window open
- A8 Focus on carbon dioxide
- A9 Coping with carbon
- A10 Summary

Medicines by Design
- MD1 Alcohol in the body
- MD2 The drug action of ethanol
- MD3 Medicines that send messages to nerves
- MD4 Enzyme inhibitors as medicines
- MD5 Targetting bacteria
- MD6 Summary

Visiting the Chemical Industry
- VCI1 Introduction
- VCI2 The operation of a chemical manufacturing process
- VCI3 People
- VCI4 Raw materials and feedstock preparation
- VCI5 The best conditions for the process
- VCI6 Safety matters
- VCI7 Environmental issues
- VCI8 Costs
- VCI9 Location

The Polymer Revolution
- PR1 Designer polymers
- PR2 The polythene story
- PR3 Towards high density polymers
- PR4 Conducting polymers – breaking the rules
- PR5 The invention of nylon
- PR6 Kelvar
- PR7 Taking temperature into account
- PR8 Throwing it away... or not?
- PR9 Summary

What's in a Medicine?
- WM1 The development of modern ideas about medicines
- WM2 Medicines from nature
- WM3 Identifying the active chemical in willow bark
- WM4 Instrumental analysis
- WM5 The synthesis of salicylic acid and aspirin
- WM6 Delivering the product
- WM7 Development and safety testing of medicines
- WM8 Summary

Using Sunlight
- US1 The Sun: sustainer of life
- US2 Exciting light
- US3 Photosynthesis
- US4 Fuel for the future
- US5 Solar cells
- US6 The hydrogen economy
- US7 Summary

Engineering Proteins
- EP1 Christopher's story
- EP2 Protein building
- EP3 Genetic engineering
- EP4 Proteins in 3-D
- EP5 Giving evolution a push
- EP6 Enzymes
- EP7 Summary

The Steel Story
- SS1 What is steel?
- SS2 How is steel made?
- SS3 A closer look at the elements in steel
- SS4 Rusting
- SS5 What happens inside a 'tin' can?
- SS6 Recycled steel
- SS7 Summary

Aspects of Agriculture
- AA1 What do we want from agriculture?
- AA2 The world at your feet
- AA3 How does the soil remain fertile?
- AA4 Competition for food
- AA5 Summary

Colour by Design
- CD1 Ways of making colour
- CD2 The Monastral Blue story
- CD3 Chrome Yellow
- CD4 Chemistry in the art gallery
- CD5 At the start of the rainbow
- CD6 Chemists design colours
- CD7 Colour for cotton
- CD8 High-tech colour
- CD9 Summary

The Oceans
- O1 The edge of the land
- O2 Wider still and deeper
- O3 Oceans of energy
- O4 A safe place to grow
- O5 Summary

Index

Fig. 9.8 Table of contents "Salters Advanced Chemistry – Chemical Storylines" [26]

Introduction for Students

Chapter 1 Measuring amounts of substances

1.1 Amount of substance
1.2 Balanced equations
1.3 Using equations to work out reacting masses
1.4 Calculations involving gases
1.5 The Ideal Gas law
1.6 Concentrations of solutions

Chapter 2 Atomic structure

2.1 A simple model of the atom
2.2 Radioactive decay
2.3 Electronic structure: shells
2.4 Electronic structure: sub-shells and orbitals

Chapter 3 Bonding, shapes and sizes

3.1 Chemical bonding
3.2 The size of ions
3.3 The shapes of molecules
3.4 Structural isomerism
3.5 Geometric isomerism
3.6 Optical isomerism

Chapter 4 Energy changes and chemical reactions

4.1 Energy out, energy in
4.2 Where does the energy come from?
4.3 Entropy and the direction of change
4.4 Energy, entropy and equilibrium
4.5 Energy changes in solutions

Chapter 5 Structure and properties

5.1 Ions in solids and solutions
5.2 Forces between molecules: induced dipoles
5.3 Forces between molecules: permanent dipoles and hydrogen bonding
5.4 The structure and properties of polymers
5.5 Water: hydrogen bonding in action
5.6 Bonding, structure and properties: a summary

Chapter 6 Radiation and matter

6.1 Light and electrons
6.2 What happens when radiation interacts with matter?
6.3 Radiation and radicals
6.4 Where does colour come from?
6.5 Ultra-violet and visible spectroscopy
6.6 Chemistry of colour
6.7 Infra-red spectroscopy
6.8 Nuclear magnetic resonance spectroscopy
6.9 Mass spectrometry

Chapter 7 Equilibrium in chemistry

7.1 Chemical equilibrium
7.2 Equilibrium constants
7.3 Partition equilibrium
7.4 What decides the position of equilibrium?
7.5 Ion exchange
7.6 Chromatography
7.7 Solubility equilibria

Chapter 8 Acids and bases

8.1 Acid-base reactions
8.2 Weak acids and pH
8.3 Buffer solutions

Chapter 9 Redox

9.1 Oxidation and reduction
9.2 Redox reactions and electrode potentials
9.3 Predicting the direction of redox reactions
9.4 The effect of complexing on redox reactions

Chapter 10 Rates of reactions

10.1 Factors affecting reaction rates
10.2 The effect of concentration on rate
10.3 The effect of temperature on rate
10.4 Catalysis

Chapter 11 The Periodic Table

11.1 Periodicity
11.2 The s block: Groups 1 and 2
11.3 The p block: Group 4
11.4 The p block: Group 5
11.5 The p block: Group 7
11.6 The d block: characteristics of transition elements
11.7 The d block: complex formation

Chapter 12 Organic chemistry: frameworks

12.1 Alkanes
12.2 Alkenes
12.3 Arenes
12.4 Reactions of arenes

Chapter 13 Organic chemistry: modifiers

13.1 Halogenoalkanes
13.2 Alcohols and ethers
13.3 Carboxylic acids and their derivatives
13.4 The –OH group in alcohols, phenols and acids
13.5 Esters
13.6 Oils and fats
13.7 Amines, amides and amino acids
13.8 Azo compounds

Chapter 14 Organic synthesis

14.1 Planning a synthesis
14.2 A summary of organic reactions

Index

Fig. 9.9 Table of contents "Salters Advanced Chemistry – Chemical Ideas" [27]

9.3.3 Chemistry in Context

Following the successes of Anglo-American curricula such as "Chemistry in the Community" [25], "Salters Advanced Chemistry" [26] and "Chemistry in Context" [31], a working group consisting of Ilka Parchmann, Bernd Ralle and Reinhard Demuth [32] was formed, which has adopted these ideas and implemented them for chemistry teaching in secondary level II in the German-speaking area: "Our concept *Chemistry in Context* aims to

- reach a broad field of students;
- contribute to the development of a rational understanding in dealing with real-life problem situations;
- show the contribution of chemistry to general education;
- train independent learning in dealing with new methods and media;
- stimulate interest in dealing with chemical issues" [33].

The book lists the following *contexts* in the table of contents: Jack of all trades Alcohol; Cleaning and Care; Fuels in Discussion; Carbon Dioxide in Focus; Rust without Rest; Mobile Energy Sources; Stone Age—Iron Age—Plastic Age; Waste becomes valuable; Food for 8 billion; Chemistry in Humans; Miracles of Medicine; A Mouthful of Chemistry; The World is Colorful. These 13 contexts, with many colorful pictures, graphics and tables, take up about 200 pages. The chemically oriented *basic concepts* fill the remaining 350 pages: Similar to any textbook on chemistry, there is information on atomic structure and chemical bonding, on structure and properties, on energy, on equilibrium, on donor-acceptor reactions.

As soon as you study a context in the first part of the book, you come across references to the technical information and can look this up in the second part of the book. If you look at the production of wine in the first context "Jack of all trades Alcohol" and come across the term "alcohols", the reference "B2—Alcohols" appears. There you will find the homologous series of alkanes, alkenes and alkynes as well as alcohols, their molecular symbols and systematic names. So as soon as students want to study the chemical background beyond the interesting stories and pictures in the contexts of wine and beer production, they look up "B2" specifically in the second part of the book and find a lot of information. They learn—motivated by the meaningful context—the topic "alcohols".

A special *diversity of methods* is intended to "enable the creative activity of students and train self-responsibility for learning. The use of modern information and communication technologies is also taken into account more than before in this concept. The teaching units are usually divided into four phases: "encounter phase, curiosity phase, elaboration phase, deepening and networking phase" [33]. Concrete examples of some "learning cycles" can be found in a course structure that was tested for grade 11 in Lower Saxony (Table 9.2).

Table 9.2 Chemistry in Context: Example of a course structure in the 11th grade [33]

Course theme	Contextual content	Chemical content
Alcohol	Wine and beer production Use of alcohol Historical experiments with alcohol Difference Ethanol/Methanol Physiological effect of alcohol	Alcoholic fermentation Properties of Ethanol ⇒ Insights about the *substance* alcohol Elementary analysis Structure-property relationships ⇒ Insights about the *molecule* Ethanol Homologous series/Nomenclature (Dehydration Ethene/Alkenes) (Hydrogenation of Ethene, Ethane/Alkanes) Blood alcohol content
Soaps and detergents	The soap boiler—Washing in the past and today "Clean and pure"—Detergents in advertising Detergents—what's inside? Modern surfactants	Structure of a surfactant molecule Surface tension, wetting, dirt removal Micelle formation Surfactant production and properties Function of zeolites Detection, isolation and properties of selected accompanying substances Surfactants from renewable resources (Transition to the topic "Fats and Oils")
Fats and oils	History of margarine production Fats and oils as renewable resources Fat as an energy and food source	Structure and differences of fats and oils Carboxylic acids, esters Saponification of fats Soap—a surfactant Transesterification (Networking with fuels) (Calorific values of fats)
Fuels in development	Gasoline and crude oil Fuel properties Environmental pollution (e.g. ozone, greenhouse effect) Alternative fuels Biological and atmospheric carbon cycle	Alkanes, alkenes Isomerism Distillation, cracking (Gas chromatography) Carbohydrates (Halogenated hydrocarbons)
Optional addition		
Plastics	Structure of plastics (PE, polyester) Biodegradable plastics	Bonding and reaction types Reaction mechanisms

9.3.4 Chemistry for Life

The mentioned works such as "Chemistry in Context" are intended for secondary level II because there the knowledge about donor-acceptor reactions and mechanisms of organic chemistry is offered for the explanation of many chemical processes in everyday life. Two decades ago, Freienberg, Krüger, Lange, and Flint [34] discussed the question to what extent everyday chemistry can also be included in the teaching of secondary level I and is just as important for the

9.3 Teaching Processes: Subject Systematics plus Everyday Chemistry

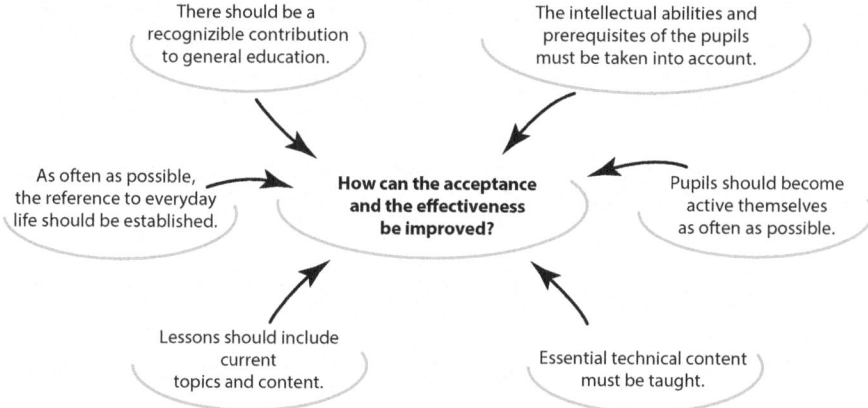

Fig. 9.10 Aspects for the inclusion of everyday references in the teaching of secondary level I [34]

motivation of students. They therefore ask quite directly: "Chemistry for life, even in secondary level I—is that possible?" [34]. As an answer they formulate: "If one initially limits the interpretation of reactions to the lowest, more phenomenologically oriented level, it becomes clear that one can indeed carry out and interpret experiments with everyday substances. However, limiting oneself to the phenomenological level does not mean that one does not later revisit selected reactions in the sense of a spiral curriculum and then consider them at higher levels of interpretation" [34]. The authors also reflect on the extent to which this contributes to general education or whether the teaching should include current topics and contents of the students' world of life (Fig. 9.10).

As examples for their approach regarding the "Chemistry for Life", the following teaching contents are described and illustrated with experiments:

- Treatment of acids and bases on a phenomenological level [34]
- Oxygen from oxy-cleaners [13]
- Rennie not only clears the stomach—antacids in chemistry teaching of secondary level I [35]
- Introduction to organic chemistry [36]
- Redox reactions without oxygen?—An extension of the redox concept [37].

9.3.5 NRW Curricula and New Textbooks

For several years now, special core curricula for the subject of chemistry [4] have been developed for the state of North Rhine-Westphalia. Thus, section 4 states: "Content areas and professional contexts for the subject of chemistry: The teaching is structured by *content areas* and *professional contexts* that are related thematically.

They enable a student-oriented development of scientific facts, the development and use of professional competencies, and the communication and reflection of scientific statements. Suitable contexts usually meet the following criteria:

- They offer students opportunities to develop competencies and to apply acquired competencies in different areas in a meaningful and successful way.
- They contribute to the development of basic concepts.
- They gain special significance through their relation to the learners' experiences.
- They offer diverse opportunities for action for an active learning process.
- They connect perspectives and procedures of the subjects chemistry, physics, and biology."

In this context, the following should be noted: "All content fields with their focal points are binding, as is working in professional, coherent contexts" [4]. In earlier curricula, the importance of everyday life and the world of life was pointed out, but it was left to the teachers whether they actually establish such references in the classroom or not. The new curricula leave no room for maneuver in this respect: The contexts are obligatory[4]. The first teaching unit "Substances and Substance Changes" is used as an example to show how the ministry envisages the implementation of the teaching (Table 9.3).

According to these new curricula, textbook authors have reformulated the current textbooks and introduced each topic with a reference to everyday life. Thus, the topics in textbooks of past decades were "Mixtures and Pure Substances" [38], and the melting and boiling temperatures or densities and solubilities of laboratory chemicals were immediately determined experimentally (Fig. 9.11a).

In the 2012 textbook "Chemistry today SI" [39], the same introductory chapter is called "We Investigate Food" and first compares colors, smell and taste of foods such as fruits and vegetables (Fig. 9.11b). Then "sweet foods" like jams, nougat and syrup are contrasted with "sour foods" like vinegar, pickles and sour meat (Fig. 9.12). After the everyday introduction, the usual experiments on solubility, electrical and thermal conductivity, states of aggregation and densities, melting and boiling temperatures are presented.

The same applies to the subsequent chapter of the textbook, which was usually called "Pure Substances, Mixtures and Separation Procedures" (taken from

Table 9.3 Core curriculum Chemistry, Gymnasiums, example of content fields and contexts [4]

Content fields	Professional contexts
Substances and substance changes	Food and drinks—all chemistry?
Mixtures and pure substances Material properties Substance separation procedures Simple particle representation Characteristics of chemical reactions	What's inside? We investigate food, drinks and their components We extract substances from food We change food by cooking or baking

9.3 Teaching Processes: Subject Systematics plus Everyday Chemistry

a

Gems and raw materials

1. Chemistry, the study of substances _____ 6
1.1. Pure substances can be recognized by their properties _____ 7
1.2. What happens when substances are heated? _____ 8
' 1.3 We examine some metals _____ 9

2. **Pure substances, mixtures and separation processes** 10
2.1. Pure substances and mixtures _____
2.2. Solutions are homogeneous mixtures _____
2.3. Mixtures of substances can be separated _____ 12

3. Substances consist of tiny particles _____
3.1. Oa's spherical particle model _____
3.2. The particles move _____
3.3. The states of aggregation in the particle model _____
"3.4 The size of the smallest particles _____

b

2   **We examine food** 24
2.1 Properties of food 2S
2.2 Substances and their properties 26
     Practical course: Examination of monkeys 2
2.3 Investigation van Staffen 2
     Excursus: Detection of substance properties 29
2.4 Aggregate condition 30
     Method: Determination of the melting temperature... 30
     Density 3
     *Method: Determination of density 31
2.5 Substances wanted by warrant. 32
     Practical course: Classification of substances 34
     Method: Working with models 35
     Theory, aggregate states in the particle model\ 36
     Excursus: Aggregate states and
     Temperature change 37
2.6   * Diffusion ....
     Method: Lemen mil Maps 38
     Basic knowledge 39
     Chemistry in our **world** 39
     Pri.1/e your knowledge L,0
     Networked knowledge 41

3   **Food - everything well mixed** 42
3.1 Foodstuffs are mixtures 43
3.2 Classification of mixtures 44
     Overview: Heterogeneous mixtures and
     Homogeneous mixture 45
3.3 *Ice cream - a delicious mixture 46
3.4 L'rlce cream - from a scientific point of view 47
     Internship: Ice cream production 48
3.5 Separating aTlent soup 49
     Overview: Separation process 50
3.6 Wine distillation - a separation process 5
     Pruktikum: Distillation 5

Fig. 9.11 Excerpts from contents of textbooks "Chemistry today" 1984 [38] and 2012 [39]

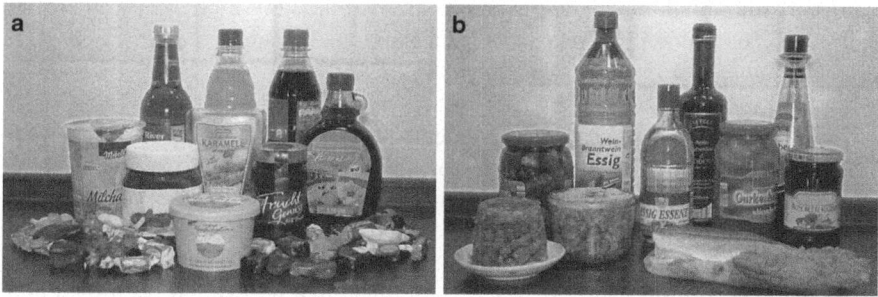

Fig. 9.12 Introduction to the properties of substances in "Chemistry today" 2012 [39]

Fig. 9.11a). In the 2012 textbook, the topic "Food—all well mixed" was chosen in relation to everyday life (Fig. 9.11b) and effervescent powder and household vinegar are considered mixtures, "Ice cream—a delicious mixture" is pointed out, the "Separation of a packet soup" is tackled, even "Wine distillation—a separation procedure" and "Doping control—an application of chromatography" are covered.

The same applies to the following topics: As soon as there is an interesting everyday reference, it is chosen as an introduction to motivate the young people. However, such motivation is always linked to the fact that the chosen everyday mixtures are more complex than the pure substances of the laboratory—so one often starts with the mixtures, which are difficult to describe, before one comes to the pure substances from which the mixtures are built.

In the past, chemistry teachers have always incorporated everyday life into the teaching of a topic—but often not at the beginning, but to round off the topic or as a chemical-technical deepening at the end of a topic. It remains to be seen in long-term empirical studies whether the motivation through everyday references at the beginning of a topic actually contributes to success or whether the deepening through life-world references at the end of each unit is also (or even more) successful.

9.4 Social Reference Fields: Role-Playing and Environmental Education

The societal reference is naturally already addressed in all texts of this chapter. Therefore, only hints remain that societal references to chemistry and environmental education can be effectively conveyed through role-playing. For Hellweger [22], "role-playing games are simulations of discussion rounds in which everyone in the group can or must become active. Depending on whether some players are highlighted—be it as a moderator or discussion leader, be it as an expert who has to answer certain questions in more detail—or whether all participants are equally weighted, the game has more the character of expert questioning or the form of a free discussion round, the game runs more strictly guided or more freely and spontaneously". The following topics have been elaborated in such *role-playing games, performing games* or *decision-making games* [22]:

- **Chemistry lessons—why?** The students discuss from different roles what and how a meaningful chemistry lesson should be, they look for justifications whether everyone should still be obliged to learn chemistry, or whether the subject chemistry should be abolished in favor of other disciplines that are not represented in the subject canon.
- **The Elbe is tipping over!** A river is so polluted that no fish has been seen in it for a long time. Are efforts still worthwhile, for example by relocating industrial companies to other locations, to prevent it from finally collapsing ecologically—even if this endangers many jobs, for example?
- **Meat or grains?** Are there serious arguments that even we in the highly industrialized countries should reduce meat consumption in favor of more vegetarian products? Do we continue to buy products from developing countries, for example to feed our livestock in Europe cheaply?

9.4 Societal Reference Fields: Role-Playing and Environmental Education

- **... and he did drill!** More and more toothpaste is being consumed, yet teeth are getting worse. Is perhaps healthy nutrition more important than dental hygiene? Could one even do without brushing teeth if sugar were treated as a dangerous drug?
- **Everything's fine with butter?** Do you really do something for your health if you limit butter consumption in favor of more margarine? To this end, the audience is once shown in the form of an expert survey that there are two contradictory theories for the development of a heart attack, and on the other hand, two doctor visits demonstrate how these theories can affect everyday life.
- **Energy Forum 2000.** Can the Federal Republic do without the controversial nuclear energy without endangering living standards and jobs? The students simulate a special party conference at which three motions are to be passed on this complex of questions. In the roles of the delegates, they present arguments, facts and pleas for or against the called motion, over which they then vote in a secret ballot according to their personal opinion [22].

Otto [40] comments on the role-playing games in such a way that they "are open to questions about the relationship between scientific knowledge and its consequences for humanity, between scientific research and societal development, between economic interests, environmental pollution and human health". Elsewhere he emphasizes that "in the last 20 years, the natural sciences have become those subjects of instruction for which it should be easiest to prove that this is about content that is not 'for school', but 'for life'. Of course, this is only true if there is a content-related, recognizable connection for the student between what happens in class and what is discussed on television, in the newspaper, in the citizens' initiative. ... Role-playing and decision-making games start from the *existing* problem awareness and expand it, question positions, confront with counter-opinions, differentiate standpoints. Chemistry teaching not only requires student motivation, but must also motivate to continue learning—especially motivated learning *outside* of school".

Finally, the societal reference of everyday chemistry is also to be established for *environmental education*. Demuth [41] formulates:

> For an environmental education in chemistry lessons, which is based on the criteria for effective environmental education determined so far in scientific studies and on the premises formulated at the beginning, it follows: It is not the aim to cover all "environmentally relevant" topics as completely as possible, much more important is to deal intensively with the typical questions in (a few) "environmental projects".

Some project ideas from Demuth are outlined (Fig. 9.13). In a *"Nitrogen Analysis" project*, Demuth [42] offers, in collaboration with some schools, to examine compost and garden soil for the content of ammonium salts and nitrates in a small garden association. Targeted analyses of farmland are also planned, as well as studies on the nitrate uptake of spinach, lettuce, and carrots at different nitrate levels, conducted in the school garden [42].

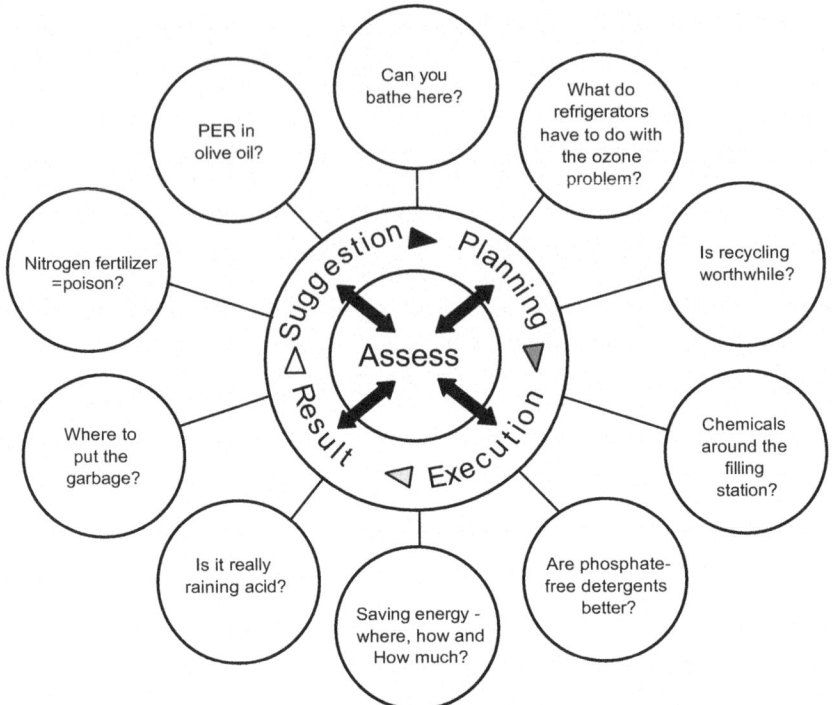

Fig. 9.13 Project ideas for environmental education in chemistry lessons [41]

In these and similar projects, chemistry and everyday life can be optimally combined—it is to be hoped that there will be sufficient time windows available in everyday school life to realize this project-based teaching!

9.5 Exercise Tasks

A9.1
Which areas from everyday life and the world of young people would you base a daily-oriented chemistry lesson on? Give five examples and outline your intentions for this teaching.

A9.2
Household chemicals or products from the bathroom and garden can replace the usual laboratory chemicals at suitable points in the lesson. Name five possibilities and describe the respective teaching context.

A9.3
Many everyday phenomena can be "translated" into chemical processes. Choose five examples of everyday chemistry, formulate reaction symbols and explain the corresponding reaction types that underlie each.

A9.4
Everyday or natural phenomena can serve as motivation and introduction to a topic. Name five such topics and outline a corresponding everyday-oriented and motivationally suitable introduction.

A9.5
Young people's attitudes towards chemistry are often very negative. What do you think are the reasons for this? What would you personally do in chemistry lessons to improve the image of chemistry?

9.6 Experiments

The text to Sect. 9.2 explains phenomena and interpretations of reactions of many everyday chemicals and refers to experiments that are presented and described in more detail here. Since the problem is the same for all experiments and refers to the presentation of everyday chemicals with simple hand experiments, the "Problem" category should be omitted. The "Material" category can also be omitted, as in most cases only test tubes and few specific devices or chemicals are used. Execution and observations are directly given. Finally, it should apply to all everyday chemicals that ingredients are checked on the label and their effects are discussed in connection with the experiments.

V9.1 Drain Cleaner "NaOH/Al-Type"

A spoonful of the cleaner is placed on a watch glass and white salt crystals and silver-colored metal splinters are observed. A small amount of substance is mixed with some water in a test tube: A strongly exothermic reaction begins, gas development and the smell of ammonia are observed. Pieces of a woolen textile are added: They slowly decompose.

In a second experiment, pure sodium hydroxide is reacted with aluminum. Protect your eyes! The resulting gas is collected in a second test tube: When ignited, a bang indicates hydrogen as a reaction product.

V9.2 Toilet Cleaner "HSO_4^--Type"

A small amount of substance is placed in a test tube, a white salt is observed. It is dissolved in water, the solution is tested with universal indicator paper: strongly acidic reaction. A snall amount of calcium carbonate is added to the solution: The sample is dissolved with the development of a gas. A burning splint is extinguished in the gas space above the solution: carbon dioxide.

V9.3 Sanitary Cleaner "ClO^--Type"

A snall amount of the cleaning liquid is added to the test tube, a strip of indicator paper is held inside, methylene blue solution or other organic dyes are added: dyes of both the indicator paper and the methylene blue solution decompose.

A second sample is mixed in the test tube with toilet cleaner (see V9.2): the color and smell of the resulting gas indicate chlorine gas (dilute the solution under the fume hood).

V9.4 Oxi-Cleaner "Sodium Percarbonate-Type" [13]

In a 100 mL beaker, add half a measuring spoon of "Hoffmann's Vanish Oxi" cleaner to 50 mL tap water, insert a thermometer and gently heat the solution with the burner, but not higher than 50 °C. When gas development begins, add a drop of dish soap to the solution and wait until a foam crown about 3 cm high forms. With a glowing splinter of wood, pierce individual bubbles and "stir" the foam. Especially when "stirring", the glowing splinter lights up brightly and begins to burn.

V9.5 Deodorant "Al^{3+}-Type"

One brand of the deodorant type is Hydrofugal Spray. Indicator paper is well wetted with Hydrofugal Spray: acidic reaction. A small amount of aluminum chloride hexahydrate is dissolved in a test tube in water and also tested with indicator paper: acidic reaction.

V9.6 Mineral Tablets "Ca- and Mg-Type"

A tablet is added to water: vigorous effervescence, formation of carbon dioxide gas.

In the gas generator, water is dripped onto a tablet, the resulting gas is collected in a piston sampler (as soon as the 100 mL mark is reached, it is emptied via the three-way valve and the collection continues). The total gas volume is determined, the gas is tested with the burning splinter of wood: the flame goes out.

In the pneumatic tub, the reaction of a tablet is first carried out so that all the gas collects in a 250 mL graduated cylinder filled with water. The volume is marked: it is only about 20 mL. A second tablet is dissolved in the same way: The resulting volume of the gas portion is much larger than before with the same tablet, about 150 mL. The carbon dioxide from the first tablet has led to a saturated solution, so that the gas from the second tablet can rise completely and almost fills the cylinder.

V9.7 Table Salt "Iodine-Type"

Ingredients are checked on the label of the container. A spatula tip of "iodized table salt" is dissolved, combined with potassium iodide solution and acidified with a few drops of dilute sulfuric acid: a brown-colored iodine solution is formed. If starch solution is added, the specific blue coloration also indicates free iodine. The experiment is repeated with pure, diluted sodium iodate solution.

V9.8 Baking Powder "Soda-Type"

The baking powder is suspended in a little amount of water and heated: gas development. It is strongly heated in the dry test tube, the resulting gas is collected in the attached piston sampler and tested with the burning splinter of wood: carbon dioxide.

V9.9 Baking Powder "Hartshorn Salt-Type"

The experiments from V9.8 are repeated with this baking powder and with ammonium carbonate, resulting gas mixtures are tested with moist indicator paper: alkaline reaction. The pungent smell indicates ammonia.

V9.10 Fabric Decolorizer "Dithionite-Type"

Ingredients are checked on the label of the container. A little amount of methylene blue solution is mixed with a spoonful of fabric dye remover powder and shaken. The reaction with pure sodium dithionite is to be repeated in the second test tube: The blue solution is decolorized both times.

V9.11 Developer "Hydroquinone-Type"

In two test tubes, silver chloride is freshly precipitated, one test tube is kept in the dark for 10 minutes, the other in the light: The latter shows a significantly darker color. Alkaline hydroquinone solution is added to both test tubes: The contents turn black.

In the photo lab, a bunch of keys or similar is placed on a photo paper under red light and briefly exposed to the light of the enlarger. The photo paper is placed in the prepared developer solution: The black-and-white image appears within a minute. The photo is to be rinsed with diluted acetic acid as a stop bath, the development is stopped. The photo is to be kept in the red light of the lab—otherwise it would darken further (see V9.12).

V9.12 Fixer "Thiosulfate-Type"

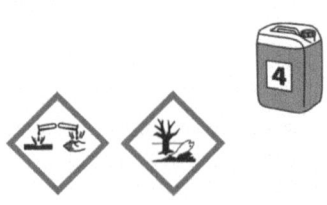

In a test tube, a little amount of silver chloride is freshly precipitated, the suspension is diluted. Concentrated sodium thiosulfate solution is added to the diluted suspension and shaken: The white silver chloride dissolves into a clear solution.

In the red light of the photo lab, the developed photo is dipped in fixation solution and left there for a few minutes. Afterwards, another freshly developed photo paper and the fixed photo are brought into the bright light: The fixed photo remains, the unfixed one turns completely black.

V9.13 Etching Chemical "Fe^{3+}-Type"

In a large test tube, a solution of iron(III) chloride is prepared. A copper-coated plastic strip is marked with a random drawing or a name using a wax pencil and dipped into the solution: After a few minutes, only the drawing or name can be seen, the remaining copper has been dissolved, the solution is colored blue.

V9.14 Accumulator "Type Pb/PbO_2"

A car battery is demonstrated, the voltage of the individual cell and all six cells is measured: Voltage values of 2 V and 12 V are measured.

A beaker is filled three quarters with 20% sulfuric acid solution, two lead plates are placed and fixed in it so that they do not touch. The lead plates are to be connected to the transformer with cables, a direct voltage is to be adjusted so that gas development can be observed: A layer of reddish-brown substance forms on one of the plates. After a few minutes, the transformer is removed, the voltage between both plates is measured: about 2 V. The electric motor is connected: It runs for some time and then stops.

V9.15 Lithium-Ion Accumulator [15]

Two graphite leads are annealed in the non-luminous burner flame to remove the binder. An electrolyte solution is prepared by adding about 10 g of lithium perchlorate to 100 mL of propylene carbonate and stirring with the magnetic stirrer for about 20 minutes until completely dissolved. Both graphite leads are switched as positive and negative poles and dipped 1 cm apart into the solution. It is charged at a direct voltage of about 4 V for three minutes. Then the voltage is measured: It is about 3.5 V.

An electric motor or an LED lamp is connected in series: The motor moves, the lamp lights up for a few minutes. The discharge process can be recorded with suitable computer software: The voltage drops to up to 2.7 V.

Many more experiments are described on the function of the lithium-ion accumulator [15].

V9.16 Flash Cement "Ca(OH)$_2$-Type"

Ingredients are checked on the label of the container. The slurry of the building material is tested with a strip of indicator paper: alkaline reaction.

A small amount of water is added to fresh calcium oxide (safety goggles!): Increase in volume with hissing sounds, strongly exothermic reaction. The white, cooled product is tested with a strip of moist indicator paper: strongly alkaline reaction.

In an Erlenmeyer flask, the product (or calcium hydroxide from the storage bottle) is slurried with water, carbon dioxide from the steel bottle is added, and a flask sampler filled with the same gas is connected in a gas-tight manner. The Erlenmeyer flask is moved so that the slurry is distributed on the glass wall: The piston of the flask sampler moves quickly into the sleeve, the mixture heats up: Calcium carbonate forms.

References

1. Waddington D (2002) (University of York): The Salters Chemistry Projects: 15 years on. Vortrag auf dem 15. Dortmunder Sommersymposium der Chemiedidaktik, 15.6.2000
2. Barke H-D (1987) „Chemieunterricht erscheint nicht so sinnlos, wenn man den Stoff auch im Alltag anwenden kann". In: Lindemann, H.: Alltagschemie. NiU P/C 35(25)
3. Gesellschaft Deutscher Chemiker (1992) Denkschrift zur Lehrerausbildung für den Chemieunterricht auf der Sekundarstufe II. Gesellschaft Deutscher Chemiker, Frankfurt
4. Ministerium für Schule und Weiterbildung des Landes NRW (2011) Kernlehrplan Chemie (G8). Ministerium für Schule und Weiterbildung des Landes NRW, Düsseldorf
5. Pfeifer P, Häusler K, Lutz B (1992) Konkrete Fachdidaktik Chemie. Oldenbourg, München
6. Barke H-D (1996) Lebenswelt und Alltag im Chemieunterricht. In: Behrendt H (Hrsg) Zur Didaktik der Physik und Chemie. Leuchtturm, Alsbach
7. Wanjek J, Barke H-D (1998) Einfluss eines alltagsorientierten Chemieunterrichts auf die Entwicklung von Interessen und Einstellungen. In: Behrendt H (Hrsg) Zur Didaktik der Physik und Chemie. Leuchtturm, Alsbach
8. Müller-Harbich G, Wenck H, Bader HJ (1990) Die Einstellung von Realschülern zum Chemieunterricht, zu Umweltproblemen und zur Chemie. Chimdid 16:151, 233
9. Heilbronner E, Wyss E (1983) Bild einer Wissenschaft. Chemie Ciuz 17:69
10. Hilbing C, Barke H-D (2000) Male dein Bild von der Chemie. Zum Image von Chemie und Chemieunterricht bei Jugendlichen. ChiuZ 34:17
11. Pietsch S (2014) „Male dein Bild zur Chemie" – eine empirische Erhebung zu Einstellungen von Jugendlichen zur Chemie. ChiuZ 48:312
12. Friedrich J (2012) Chemie – Allgemeinbildung und Alltagsbezug. Pdn Chem 8(61):5
13. Zucht U et al (2004) Chemie fürs Leben – Sauerstoff aus Oxireinigern. Chemkon 11:131
14. Asselborn W et al (2010) Chemie heute Sekundarstufe II. Schroedel, Braunschweig
15. Hasselmann M, Oetken M (2011) Elektrische Energie aus dem Kohlenstoffsandwich – Lithium-Ionen-Akkumulatoren auf der Basis redoxamphoterer Graphitintercalations-Elektroden. Chemkon 18:160
16. Oetken M, Hasselmann M (2012) Lithium+. Experimentierset zum Themenfeld „Lithium-Ionen-Akkumulator". Hedinger-Lehrmittel, Stuttgart
17. Daoutsali E, Barke H-D (2011) Der Abgaskatalysator im Chemieunterricht. PdN-Chemie 60(1):33
18. Kompetenzzentrum Nanochem (2009) Experimentierkasten zur Chemischen Nanotechnologie. Hedinger-Lehrmittel, Saarbrücken
19. Harsch N, Estay C, Barke H-D (2011) Treibhauseffekt, Ozon und Saurer Regen. PdN-Chemie Sch 3(66):20
20. Harsch N, Barke H-D (2014) Treibhauseffekt, Ozon und Saurer Regen – eine Soll-Ist-Zustandserhebung und eine darauf aufbauende Unterrichtsreihe für Lernende ab Jahrgangsstufe 10. MNU 67:408
21. Lutz B, Pfeifer P (1989) Chemie in Alltag und Chemieunterricht – Gegensatz oder Chance für ein besseres Chemieverständnis? MNU 42:281
22. Hellweger S (1981) Chemieunterricht 5–10. Skriptor, München
23. Frey K (1982) Die Projektmethode. Beltz, Weinheim
24. Barke H-D (1999) Wasser und Umwelt. In: Münzinger W, Frey K (Hrsg) Chemie in Projekten. Aulis, Köln
25. American Chemical Society (1985) Chemcom, chemistry in the community. American Chemical Society, Washington
26. Salters Advanced Chemistry (1994) Chemical storylines. Heinemann, York
27. Salters Advanced Chemistry (1994) Chemical ideas. Heinemann, York
28. Salters Advanced Chemistry (1994) Activities and assessment. Heinemann, York
29. Salters Chemie (2012) Chemical storylines, Kontexte. Schroedel, Braunschweig
30. Salters Chemie (2012) Chemical ideas, Theoretische Grundlagen. Schroedel, Braunschweig

31. Stanitzki CL et al (1997) Chemistry in context. Applying chemistry to society. McGraw-Hill, Boston
32. Demuth R, Parchmann I, Ralle B (2006) Chemie im Kontext. Sekundarstufe II. Cornelsen, Berlin
33. Huntemann H et al (1999) Chemie im Kontext – ein neues Konzept für den Chemieunterricht? Chemkon 6:191
34. Freienberg J, Krüger W, Lange G, Flint A (2001) "Chemie fürs Leben" auch schon in der Sekundarstufe I – geht das? Chemkon 8:2/67 (und Chemkon 9 (2002), 1/19)
35. Wolf G, Flint A (2000) Rennie räumt nicht nur den Magen auf – Antazida im Chemieunterricht der Sek. I. NiU-Chemie 55:16
36. Anscheit K, Flint A (2012) Chemie fürs Leben – Einführung in die Organische Chemie. PdN-CidS 61(8):27
37. Rossow M, Flint A (2012) Redoxreaktionen ohne Sauerstoff? – Die „Erweiterung" des Redoxbegriffs nach dem Konzept „Chemie fürs Leben". PdN-CidS 61(4):5
38. Frühauf D, Tegen H (1984) Grothe Chemie. Schroedel, Hannover
39. Asselborn W et al (2012) Chemie heute. SI. Schroedel, Braunschweig
40. Otto G (1981) Nachwort: Zur Problemlage in den naturwissenschaftlichen Didaktiken. In: Hellweger S (Hrsg) Chemieunterricht 5–10. Skriptor, München
41. Demuth R (1992) Umwelterziehung im Chemieunterricht – Ziele, Inhalte, Methoden. NiU-Chemie 3:47
42. Demuth R (1992) Stickstoffanalytik im Chemieunterricht der Sekundarstufe I. NiU-Chemie 3:67

History of Chemistry for Student Teachers

10

The history of chemistry is so interesting and exciting for lecturers and insiders of chemistry that it seems sensful to also give students of chemistry teaching an initial overview of important historical milestones in the knowledge of chemistry. Given the particular difficulties in successfully teaching and learning chemistry, it should also be instructive to understand processes of knowledge in the history of chemistry in order to possibly take them into account in current teaching and instruction. For example, if one studies how the understanding of acids, bases, and salts increased dramatically after 1887 with the insights of Arrhenius into the concept of ions, one suspects that the concept of ions also has central importance in our chemistry teaching and that ions should be introduced early on — and not be described with "formula units" or "assembly groups".

The following pages provide an initial overview using a beam of historical dates that depict historical figures and describe their insights with a few keywords. Political events are also mentioned to sketch connections to world history. However, the actual intention is to motivate readers to study the elaborated chapters located in the electronic appendix of the German book (http://www.springer.com/de/book/978-3-662-56491-2):

Chapter	Author(s)	Title
11	H.-D. Barke	Robert Boyle — Father of Experimental Chemistry
12	H.-D. Barke, J.-B. Haas	Cavendish, Scheele, Priestley and Lavoisier — the Discovery of Important Gases
13	G. Harsch, Ch. Bünte	Lavoisier — Oxidation Theory and Conservation of Mass

Chapter	Author(s)	Title
14	G. Harsch, J. Jönssen	Richter and Dalton — Elements, Atoms and Atomic Masses
15	H.-D. Barke	Gay-Lussac and Avogadro — Gases, their Molecules and the Laws of Volume
16	H.-D. Barke	Galvani, Volta, Davy and Faraday — the Introduction to Electrochemistry
17	E. Daoutsali, M. Setter	Berzelius — the Chemical Symbol Language
18	G. Harsch, N. Harsch	Liebig — Elemental Analysis, Formula Language, Agricultural Chemistry
19	G. Harsch, N. Harsch	Wöhler — Aluminium, Urea and More
20	H.-D. Barke	Kekulé and van't Hoff — Structure of Molecules
21	E. Daoutsali, H. Schliephake	Meyer and Mendeleev — the Periodic Table of Elements
22	H.-D. Barke, Ch. Pieper	Arrhenius and Brönsted — Ion Concept and New Acid-Base Ideas
23	H.-D. Barke, F. Bäuerle, M. Krasenbrink	Werner — Creator of Complex Chemistry
24	H.-D. Barke, R. Rölleke	Röntgen, Laue and Bragg — the Structure of Crystals
25	H.-D. Barke	Watson, Crick and the DNA Double Helix — Nobel Laureates Play with Models

The individual chapters only contain such information about the scientists that are of central importance for the teaching of chemistry — all further details about the life of the researchers and many additional insights can be found in the "Book of Great Chemists" by G. Bugge (Verlag Chemie, Weinheim, Germany, Reprint 1965).

For this reason, only the central experiments or structural models of the scientists have been didactically explained in the subchapter "Chemistry Didactic Relevance" and have been reduced so that they can be demonstrated with the simple equipment of a school. The instructions for the experiments have been written only for trained students or active chemistry teachers — there are no detailed instructions for beginners!

German Chapters 11-25 are essentially based on hints from the following literature:

Strube, W.: . Der historische Weg der Chemie.Volumes I and II. Leipzig 1976 (VEB German Publishing House)
Bugge, G.: Das Buch der grossen Chemiker. - Verlag Chemie, Weinheim, Germany, Reprint 1965
Hoffmann, D., et al.: Lexikon der bedeutenden Naturwissenschaftler. . Munich 2003 (Elsevier)
Brock, W. H.: Vieweg's Geschichte der Chemie. Braunschweig 1997 (Vieweg)
Ostwald, W.: Klassiker der exakten Naturwissenschaften. Frankfurt 1996 (German)
Internet: Wikipedia, information and pictures of the quoted natural scientists
Barke, H.-D., Harsch, G.: Chemiedidaktik kompakt. Berlin, Heidelberg 2012 (Springer)
Barke, H.-D.: Chemiedidaktik— Diagnose und Korrektur von Schülervorstellungen. Heidelberg 2006 (Springer)
Asselborn, W., et al.: Chemie heute. Braunschweig 2013 (Schroedel)
Eisner, W., et al.: . Elemente Chemie.Stuttgart 2009 (Klett)
Tausch, M., von Wachtendonk, M.: Chemie 2000+. Bamberg 2004 (Buchner)

The last five titles each refer to statements on "Chemistry Didactic Relevance": Examples from schoolbook literature that have proven to be positive in the teaching of chemistry are added to experiments and models of the scientists from history. Chapter 10 in this book refers to chapters 11-25 of the German electronic appendix — there you will find the detailed text, there the pictures and photos used in this chapter 10 are precisely quoted.

Teaching of the chapters of the German electronic appendix can basically be based on the historical-problem-oriented teaching method according to Jansen []. There is no claim to present the history of science in chemistry lessons, but rather the typical path of knowledge in chemistry should be in the foreground using important examples.

| Chapter 11: ROBERT BOYLE - the father of experimental chemistry | Chapter 12: CAVENDISH, SCHEELE, PRIESTLEY and LAVOISIER - The discovery of important gases |

| BOYLE (1627-1691) | CAVENDISH (1731-1810) | SCHEELE (1742-1786) | PRIESTLEY (1733-1804) |

1654 1661	1766	1772	1774
Theory-based experimentation Pressure-volume law Air and combustion Element concept, corpuscular theory	Discovery of the hydrogen, water synthesis of from the elements reactions	Discovery of the oxygen through Nitrate-acid	Discovery of Oxygen through Decomposition Mercury oxide

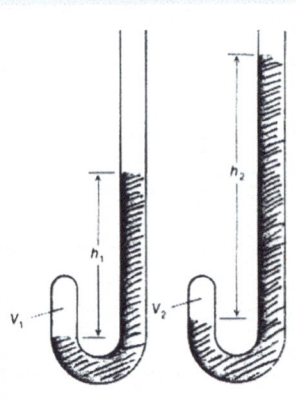
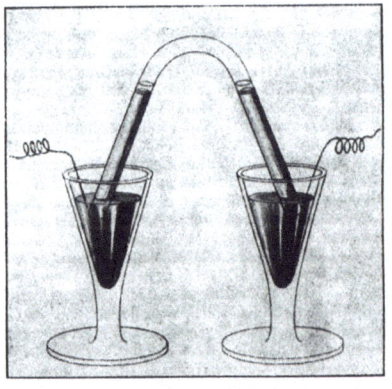

| England became a republic through Cromwell (1648) Peace of Westphalia ends 30-year war (1648) Louis XIV reigns in France (1661-1715) | Start of the Industrial Revolution in England Frederick the Great wages the Seven Years' War (1756-63) Catherine the Great (1762-96) reigns in Russia |

Ch. 13: LAVOISIER - Oxidation theory and conservation of mass	Ch. 14: RICHTER and DALTON - Atoms and atomic masses

LAVOISIER (1743-1793)	JUDGE (1762-1807) DALTON (1766-1844)

1774	1789	1792	1808
Detection of oxygen and Oxidation theory		Law of constant mass ratios	Element and atom concept circular atomic symbols first atomic mass table

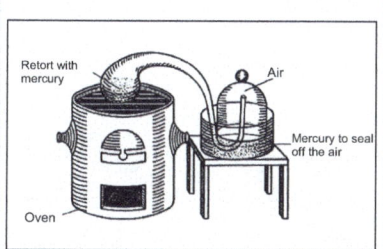

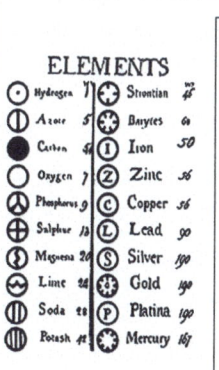

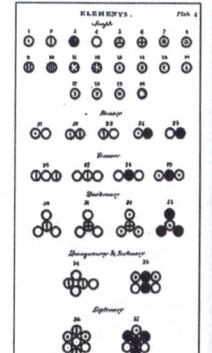

Declaration of Independence of the United States (1776) French Revolution (1789), Napoleon's rise	Continental Blockade against England (1806) Napoleonic Wars in Europe (until 1815)

Chap. 15: GAY-LUSSAC and AVOGADRO – Gases, their molecules and the laws of volume		Ch. 16: GALVANI – Getting started in electrochemistry

GAY-LUSSAC (1778-1850)	AVOGADRO (1776-1856)	GALVANI (1737-1798)
1802 1808	1811	1780
Composition of the air, Temperature- Volume law of volume-law simple numbers	Equal volumes of any gases contain the same number of particles under standard conditions	Frogs' legs twitch when they come into contact with various metals, investigation of thunderbolts

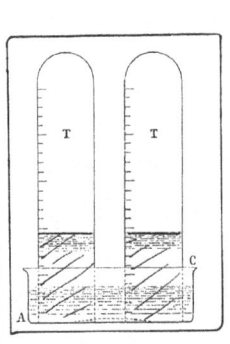 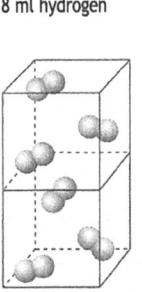 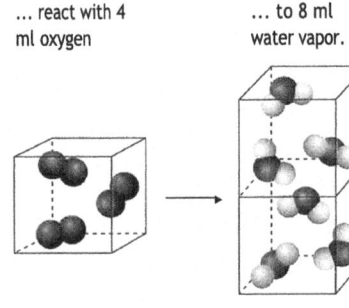

8 ml hydrogen ... react with 4 ml oxygen ... to 8 ml water vapor.

Napoleon rules Europe, battles of Austerlitz (1805), Jena and Auerstedt (1806), end of the Holy Roman Empire of the German Nation, last Roman-German Emperor (1806)

Ch. 16: GALVANI, VOLTA, DAVY AND FARADAY -
Getting started in electrochemistry

VOLTA	DAVY	FARADAY
(1745-1827)	(1778-1828)	(1791-1867)
1800	1808	1834
First voltage source with higher voltages for electrolysis	Melt-flow electrolysis and discovery of the first alkali and alkaline earth metals	Faraday's laws for electrolysis, preliminary ion concept

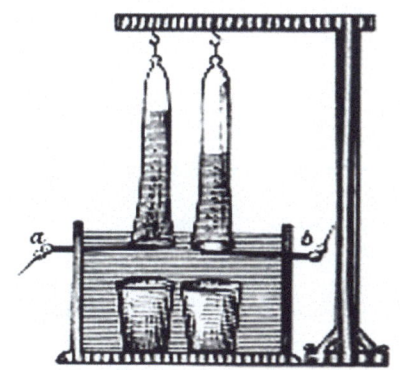

Napoleon's defeats in Russia (1812), Leipzig (1813) and Waterloo (1815)

Congress of Vienna after Napoleon's defeats and the reorganization of borders in Europe (1814-15)

Ch. 17: BERZELIUS - The chemical symbolic language

BERZELIUS
(1779-1848)

1828

Letter symbols
for elements and compounds, dualistic
theory, dualistic symbols

1856

Analysis of mass ratios of elements
in compounds,
Elemental analysis, exact atomic masses

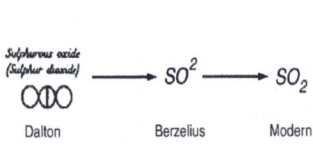

Dalton	Berzelius	Modern

Das Kupferoxydul besteht aus:
                                  Procente.   Atome.
        Kupfer . . . . 88,78 . . 2
        Sauerstoff . . . 11,22 . . 1
Atomgewicht: 891,390 = €uO oder €u.

Das Kupferoxyd besteht aus:
                                  Procente.   Atome.
        Kupfer . . . . 79,83 . . 1
        Sauerstoff . . . 20,17 . . 1
Atomgewicht: 495,695 = CuO oder Ċu.

Peace of Kiel (1814), annexation of Norway to Sweden, first King
of Sweden from the House of Bernadotte (1818)

10 History of Chemistry for Student Teachers

Ch. 18: LIEBIG -
Elementary analysis, formula language, agricultural chemistry

Ch. 19: WÖHLER -
Aluminum, urea and more

LIEBIG (1803-1873)

WÖHLER (1800-1882)

1828	1838	1849	1827	1837
Combustion analysis with the "five-ball apparatus" for the absorption of CO_2	Acid definition (hydrogen theory)	Agriculture chemistry	Discovery of the aluminum, Urea synthesis	SO_3-Catalysis at the lead chamber process

German Revolution and Frankfurt National Assembly in the Paulskirche (1848/49): unsuccessful attempt to unite the individual states in a federal constitution

**Ch. 20: KEKULÉ and VAN'T HOFF –
Structure of the molecules**

KEKULÉ
(1829-1896)

VAN'T HOFF
(1852-1911)

1858	1865	1874
Differentiation of the terms atom and molecule, four-bond nature of the C atom, model kits	Structure of the benzene molecule, oscillation theory, Model concepts	Tetrahedral model of the methane molecule, asymmetric C atom, stereochemistry

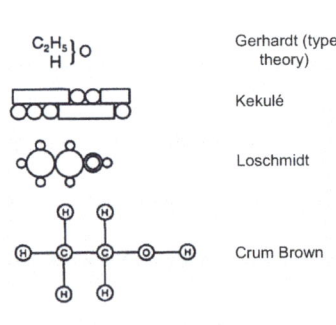

Gerhardt (type theory)

Kekulé

Loschmidt

Crum Brown

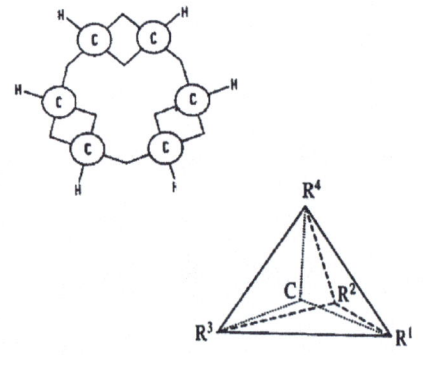

War of North versus South in the USA (1861-65)
Slavery formally ended by law (1868)

USA buys Alaska from Russia (1867)
Wave of immigration from Europe ti America

Ch. 21: MEYER and MENDELEJEW - The periodic table of the elements

MEYER
(1830-1895)

MENDELEEV
(1834-1907)

1864

First draft of a periodic table of the elements, Postulation of gaps in undiscovered elements

1869

Periodic table of the elements, prediction of the properties of scandium, gallium and germanium, which were confirmed years later

4-valent	3-value	2-valent
—	—	—
C = 12,0 16,5	N = 14,04 16,96	O = 16 16,07
Si = 28.5 — = 44,55	P = 31,0 44,0	S = 32,07 46,7
— = 44,55	As = 75.0 45,6	Se = 78.8 49,5
Sn = 117.	6Sb = 120.6	Te = 128.33

Reihen	Gruppe I. — R^2O	Gruppe II. — RO	Gruppe III. — R^2O^3	Gruppe IV. RH^4 RO^2
1	H=1			
2	Li=7	Be=9,4	B=11	C=12
3	Na=23	Mg=24	Al=27,3	Si=28
4	K=39	Ca=40	—=44	Ti=48
5	(Cu=63)	Zn=65	—=68	—=72
6	Rb=85	Sr=87	?Yt=88	Zr=90
7	(Ag=108)	Cd=112	In=113	Sn=118
8	Cs=133	Ba=137	?Di=138	?Ce=140
9	(—)			
10	—		?Er=178	?La=180
11	(Au=199)	Hg=200	Tl=204	Pb=207
12	—			Th=231

Crimean War (1853/54) and Russo-Turkish War (1877/78) against the Ottoman Empire
Abolition of serfdom for peasants in Russia (1861), assassination attempts on Tsar Alexander II.

Ch. 22: ARRHENIUS and BRÖNSTED -
The concept of ions and new acid-base concepts

ARRHENIUS
(1859-1927)

BRÖNSTED
(1879-1947)

1884/87

"In salts" ions, nur " in salt solutions"
grosses H+ statt caustic "alkaline"

1923

"Transfer" (im 1.Fall) statt contain "take"
(2.Fall) kein "absorb"

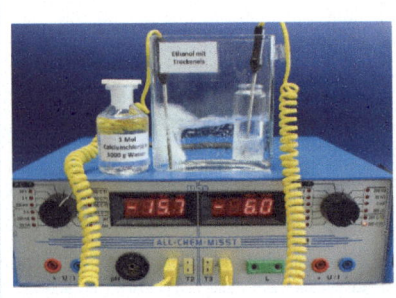

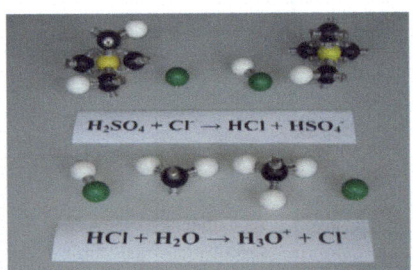

$H_2SO_4 + Cl^- \rightarrow HCl + HSO_4^-$

$HCl + H_2O \rightarrow H_3O^+ + Cl^-$

Prussia's wars against Denmark (1864) and Austria (1866)

Prussian war against France (1870-71) Proclamation of the German Emperor in Versailles (1871)

Ch. 23: ALFRED WERNER -
Creator of complex chemistry

WERNER
(1866-1919)

1892

Historical chain theory: ligands form a chain starting from the metal ion Coordination theory: metal ion is the central particle, ligands surround it spatially and geometrically Coordination number: number of ligands around the central particle, it is 2, 4, 6 or 8

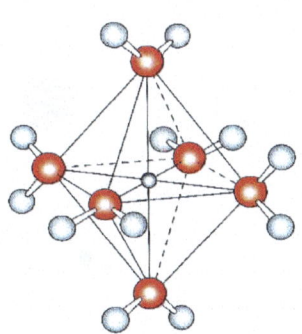
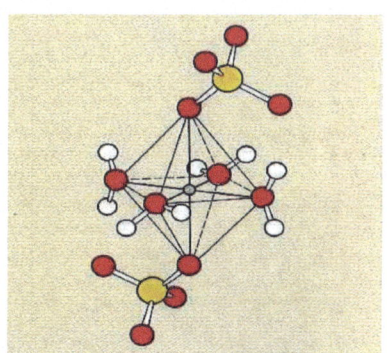

Europe's alliance system through Chancellor Bismarck, Congress of Berlin 1878
Bismarck's dismissal by Kaiser Wilhelm II (1890)

**Ch. 24: RÖNTGEN, VON LAUE and BRAGG -
The structure of crystals**

X-RAY
(1845-1923)

VON LAUE
(1879-1960)

1895

Discovery of penetrating radiation with the cathode ray tube: X-rays,
X-rays, description of the properties, unsuccessful efforts to detect interference of the rays

1912

In the Munich laboratories, there were crystal lattice models for presumed structures, which formed the basis for the idea of creating a crystal lattice.
first diffraction grids for structure analysis of crystals

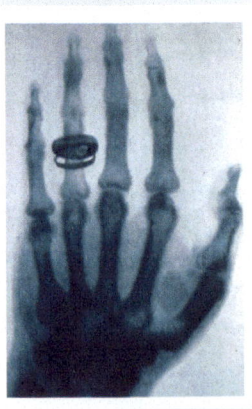

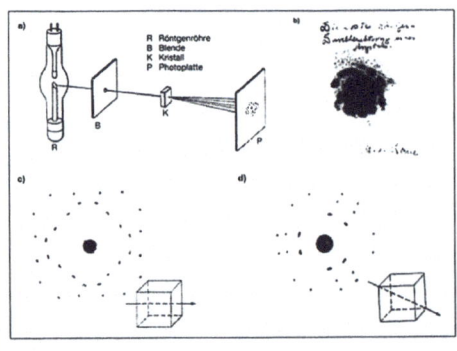

German colonies in Southwest Africa, Cameroon, Togo and East Africa (today Tanzania)

Sarajevo assassination (June 1914), First World War begins (August 1914)

10 History of Chemistry for Student Teachers

Ch. 25: WATSON, CRICK and the DNA double helix - Nobel Prize winners play with models

WATSON
(1928)

CRICK
(1916-2004)

1953

Knowledge of the four bases and Chargaff's rules: (c(adenine) = c(thymine) and c(guanine) = c(cytosine)) X-ray images of DNA crystals (Rosalind Franklin and Maurice Wilkins)
Idea of the double spiral of sugar and phosphoric acid molecules, which are held together by two organic base pairs adenine-thymine and guanine-cytosine according to certain patterns

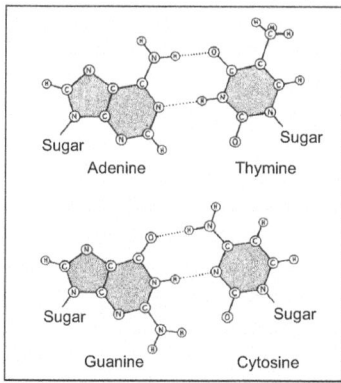
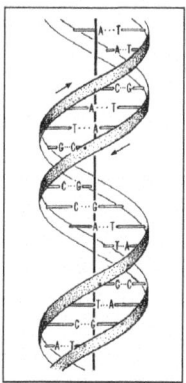

Foundation of the European Economic Community EEC in Rome in 1957 (Treaty of Rome), accession of Great Britain, Ireland and Denmark (1973)

The manufacturer's authorised representative in the EU is Springer Nature Customer Service Centre GmbH, Europaplatz 3, 69115 Heidelberg, Germany. If you have any concerns regarding our products, please contact ProductSafety@springernature.com

Printed and bound by CPI Group (UK) Ltd, Croydon, CR0 4YY

26/03/2026

02078953-0010